Molecular Fluorescent Sensors for Cellular Studies

Edited by

Elizabeth J. New
University of Sydney
New South Wales
Australia

Registered Office
John Wiley & Sons Ltd, The Atrium, Southern Gate, Chichester, West Sussex, PO19 8SQ, UK

Editorial Office
9600 Garsington Road, Oxford, OX4 2DQ, UK

For details of our global editorial offices, customer services, and more information about Wiley products visit us at www.wiley.com.

Wiley also publishes its books in a variety of electronic formats and by print-on-demand. Some content that appears in standard print versions of this book may not be available in other formats.

Library of Congress Cataloging-in-Publication Data Applied for
Paperback ISBN: 9781119749813

Cover Design: Wiley
Cover Image: © Caleb Foster/Shutterstock.com

Set in 9.5/12.5pt STIXTwoText by Straive, Pondicherry, India
Printed and bound by CPI Group (UK) Ltd, Croydon, CR0 4YY

C9781119749813_030822

Contents

List of Contributors

Liam D. Adair
School of Chemistry
The University of Sydney
NSW, Australia

and

Australian Research Council Centre of
Excellence for Innovations in Peptide and
Protein Science
The University of Sydney
NSW, Australia

and

The University of Sydney Nano Institute
(Sydney Nano)
The University of Sydney
NSW, Australia

Amy A. Bowyer
School of Chemistry
The University of Sydney
NSW, Australia

Joy Ghrayche
School of Chemistry
The University of Sydney
NSW, Australia

and

Australian Research Council Centre of
Excellence for Innovations in Peptide and
Protein Science
The University of Sydney
NSW, Australia

Marcus E. Graziotto
School of Chemistry
The University of Sydney
NSW, Australia

and

Australian Research Council Centre of
Excellence for Innovations in Peptide and
Protein Science
The University of Sydney
NSW, Australia

Paris I. Jeffcoat
School of Chemistry
The University of Sydney
NSW, Australia

and

The University of Sydney Nano Institute
(Sydney Nano)
The University of Sydney
NSW, Australia

Jiarun Lin
School of Chemistry
The University of Sydney
NSW, Australia

and

The University of Sydney Nano Institute
(Sydney Nano)
The University of Sydney
NSW, Australia

Elizabeth J. New
School of Chemistry
The University of Sydney
NSW, Australia

and

Australian Research Council Centre of
Excellence for Innovations in Peptide and
Protein Science
The University of Sydney
NSW, Australia

and

The University of Sydney Nano Institute
(Sydney Nano)
The University of Sydney
NSW, Australia

Nian Kee Tan
School of Chemistry
The University of Sydney
NSW, Australia

and

Australian Research Council Centre of
Excellence for Innovations in Peptide and
Protein Science
The University of Sydney
NSW, Australia

and

The University of Sydney Nano Institute
(Sydney Nano)
The University of Sydney
NSW, Australia

Natalie Trinh
School of Chemistry
The University of Sydney
NSW, Australia

and

Australian Research Council Centre of
Excellence for Innovations in Peptide and
Protein Science
The University of Sydney
NSW, Australia

Kylie Yang
School of Chemistry
The University of Sydney
NSW, Australia

and

The University of Sydney Nano Institute
(Sydney Nano)
The University of Sydney
NSW, Australia

Jia Hao Yeo
School of Chemistry
The University of Sydney
NSW, Australia

Jianping Zhu
School of Chemistry
The University of Sydney
NSW, Australia

and

Australian Research Council Centre of
Excellence for Innovations in Peptide and
Protein Science
The University of Sydney
NSW, Australia

1

An Introduction to Small Molecule Fluorescent Sensors

Liam D. Adair[1,2,3], Kylie Yang[1,3], and Elizabeth J. New[1,2,3]

[1] *School of Chemistry, The University of Sydney, NSW, Australia*
[2] *Australian Research Council Centre of Excellence for Innovations in Peptide and Protein Science, The University of Sydney, NSW, Australia*
[3] *The University of Sydney Nano Institute (Sydney Nano), The University of Sydney, NSW, Australia*

Vision has underpinned biological discovery throughout human history, from astronomical observations since prehistoric times, to Mendel's discovery of genetics by observing the colours of sweet pea plants. However, as our understanding of biological systems develops, so does our need to see smaller and smaller structures, and to look deeper and deeper into cells. Today, biology is increasingly investigated at the molecular scale, and there are numerous chemical tools and assays that can be used to observe relevant biomolecules.

Fluorescent sensors are an important tool for biological research as they are able to sense analytes or chemical reactions of interest, and report on their presence through a fluorescence output. Recent advances in technology – from confocal microscopes to flow cytometers to imaging plate-readers – have been accompanied by the development of more selective and sensitive fluorescent sensors, which have enabled the discovery of new biological processes involved in health and disease. While many fluorescent sensors are commercially available, there are many more that have been reported in the literature and are yet to be used widely.

While many imaging technologies have been designed for the simple application of common commercial fluorophores and fluorescent sensors, it is nonetheless important for the end-user to have an understanding of the principles of fluorescence and the mechanisms by which fluorescent sensors operate in order to ensure that sensors are being used appropriately and optimally. This is particularly important for bespoke fluorescent sensors, which may not fit the standard profile of more common commercially available systems. In this chapter, we outline the basic principles of fluorescence and fluorescent sensing. We then explain common classes of fluorophores and the mechanisms by which fluorescent sensors operate.

1.1 What is Fluorescence?

Fluorescence is the emission of light from a substance in an electronically excited state. Specifically, it describes the rapid emission of a photon from a singlet excited state as it returns to the ground state. An electron is promoted from an orbital in the ground state to a higher energy

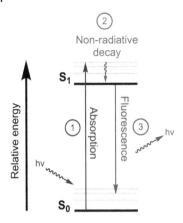

Figure 1.1 Jablonski diagram showing: 1. Absorption of a photon to excite electrons from the S_0 state to the S_1 excited state; 2. non-radiative decay or relaxation from higher vibrational excited states to S_1; and 3. emission of a photon as fluorescence.

empty orbital by absorption of a photon; the ground state S_0 is excited to an excited state S_n. This excited state can then relax to the ground state in several ways, but fluorescence is the process when the molecule returns to the ground state by emission of a photon. A substance that undergoes fluorescence decay is known as a fluorophore.

The orbitals involved in fluorescence are the highest occupied molecular orbital (HOMO) and the lowest unoccupied molecular orbital (LUMO) in the ground state of the molecule. A photon with energy equal to the difference between the HOMO π to LUMO π^* energies can promote an electron, generating an excited state. A Jablonski diagram (Figure 1.1) is a simple schematic often used for illustration of the electronic states of a molecule and the transitions between them.

Vibrational relaxation, or non-radiative decay, is the process in which the excited state relaxes to its lowest vibrational level. This is rapid and takes place on 10^{-12}–10^{-10} second time scale [1]. This rapid relaxation means that emission occurs from the lowest vibrational level.

Internal conversion (IC) is a non-radiative transition between two electronic states with the same spin state, or multiplicity [2]. Kasha's rule states that appreciable luminescence will only be observed from the lowest energy excited state, as IC is orders of magnitude faster than fluorescence (10^{-12} second compared to ~10^{-8} second) [3].

Excited singlet states (denoted S_n) have their excited electron spin paired with the ground state electron. In excited triplet states (T_n), the excited electrons have a parallel spin to the ground state electron and are no longer spin-paired (Figure 1.2).

Another complex excited state process where energy can be dissipated, that competes with fluorescence, is intersystem crossing (ISC). ISC is a non-radiative transition between two electronic states with different spin states, or multiplicities. ISC therefore describes the transition from a singlet state (S_n) to a triplet excited state (T_1). Relaxation from T_1 to S_0 leads to phosphorescence, which is a form of luminescence characterised by having longer-lived emission than fluorescence [4]. This is because emission from the first excited triplet state (*i.e.* phosphorescence) is spin-forbidden, whereas emission from the first excited singlet state

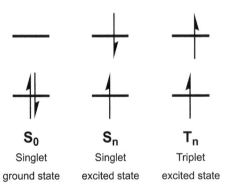

S_0
Singlet
ground state

S_n
Singlet
excited state

T_n
Triplet
excited state

Figure 1.2 Illustration of electron spin states in a singlet ground state S_0, singlet excited state S_n, and triplet excited state T_n.

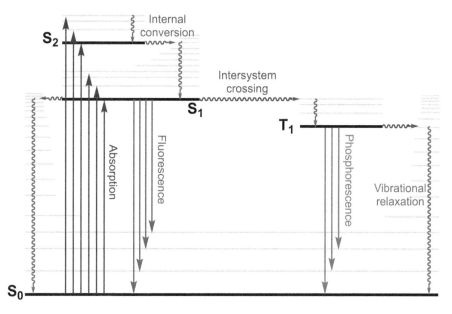

Figure 1.3 Jablonski diagram illustrating absorption, vibrational relaxation, internal conversion, intersystem crossing, fluorescence, and phosphorescence.

(*i.e.* fluorescence) is spin-allowed. Phosphorescence is several orders of magnitude slower than fluorescence emission ($\sim10^{-4}$ second) [4] (Figure 1.3).

Fluorescence spectra are generally presented as a plot of the fluorescence intensity (arbitrary unit) against the wavelength, given in nm (or wavenumber, cm^{-1}). The emission spectra of fluorophores are dependent on the structure of the molecule, and therefore the mechanism by which it fluoresces, and the chemical environment. Fluorescent molecules tend to be aromatic, with extended conjugation and a large system of delocalised π-electrons. These π-electrons are involved in excitation and in general, the more extended the conjugation, the longer the wavelength of photon absorbed, the longer the wavelength emitted.

1.2 Why Is Fluorescence Useful?

There are many characteristics of fluorescence that are advantageous for application in imaging and spectroscopy. Fluorescence is highly sensitive, making it suitable for detection of biological analytes that can be present at extremely low concentrations. This is because fluorophores can be detected in solution at nanomolar to micromolar concentrations, and fluorophores can be identified with high specificity among non-fluorescent material. Furthermore, fluorescence is a fast process, with excitation and emission occurring on the nanosecond time scale, and this allows fluorophores to be imaged with high temporal resolution, enabling the study of detailed dynamic processes in real time. In addition, fluorescence is a convenient method for studying the interactions between different molecular components in a chemically complex system as multiple fluorophores can be used concurrently, which allows for several analytes or molecules and their interactions to be simultaneously tracked.

Fluorescence imaging is amongst the most popular and widely used techniques for molecular imaging. Fluorescence imaging techniques have become indispensable tools allowing the study of the production, localisation, movement, and biological activity of biomolecules *in vitro* and *in vivo*. Fluorescence techniques have found use in biochemistry, biophysics, flow cytometry, diagnostics, DNA sequencing, *etc.* [5–10]. Fluorescence microscopy has become more powerful due to developments such as fluorescence lifetime imaging microscopy (FLIM) and super resolution microscopy [11–15]. The resolution for a standard optical microscope in the visible light spectrum is about 200 nm laterally and 500 nm axially [12]. This has been improved with the use of super resolution microscopy, where the molecular blinking of fluorophores is detected and used to build images that can achieve resolutions of around 20 nm, much below the diffraction limit of visible light [16–21].

As a technique, fluorescence imaging offers many advantages, but its power depends on the availability of fluorescent labels and sensors. Because there are few naturally occurring fluorophores within cells, fluorescence microscopy frequently requires the addition of a stain or probe, which functions to label a structure of interest, certain cell types, organelles, proteins, or chemical species. Microscopy can be performed on fixed specimens, as well as live cells or *in vivo*, as it is minimally invasive, sensitive, and selective. Furthermore, fluorescence imaging provides good spatial and temporal resolution, which allows dynamic biological processes to be monitored in real time.

Recent decades have seen great advances in the field of fluorescence sensing. Brighter fluorophores with improved biological compatibility, robust targeting strategies, more selective sensing groups, and improved strategies to harness various fluorescence mechanisms have resulted in huge improvements in fluorescent sensors. Advances in synthetic chemistry allowing new or more facile chemical transformations underpin these improvements. Better methods for labeling such as biorthogonal labeling, incorporation of unnatural amino acids, and new fluorescent proteins have become available [22–27]. In combination, these breakthroughs have further cemented the importance of fluorescence in scientific discovery.

1.3 What Is a Fluorescent Sensor?

There are numerous fluorescent tools for microscopy and bioimaging. These include fluorescent proteins, fluorescent antibodies, fluorescent nucleic acids, quantum dots, and nanoparticles. Arguably, the most diverse and widely used fluorescent tools are small-molecule fluorescent sensors [28]. Throughout this book, we will use the term 'fluorescent sensor' to signify small-molecule systems, although other fluorescent scaffolds can also be applied for sensing. The fluorescence of a molecule can be drastically modulated by interactions with its surrounding environment. Fluorescent sensors take advantage of this, and interaction with an analyte of choice, either by chemical reaction, binding, or an environmental change, incurs a change in the fluorescent properties which can be observed and measured. Generally, this is a change in emission intensity or wavelength, but can also involve a change in excitation intensity or wavelength, fluorescence lifetime, or a combination of multiple responses.

By way of generalisation, a small-molecule fluorescent sensor can be considered to consist of two parts: the fluorescence-generating group, or fluorophore; and the sensing moiety that interacts with the analyte of interest. Design of a fluorescent sensor therefore requires consideration of both groups, and how they can be tethered in such a way that interaction of the

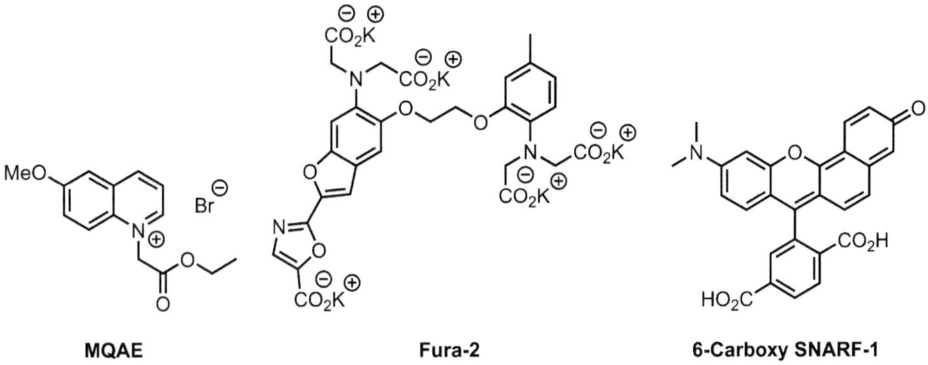

Figure 1.4 Commercial small-molecule fluorescent sensors: **MQAE** chloride sensor; **Fura-2** excitation ratiometric Ca^{2+} sensor; **6-carboxy SNARF-1** pH sensor.

sensing group with the analyte generates a change in fluorescence. This chapter deals with the properties of the fluorophore itself. Subsequent chapters have greater focus on the various design principles for sensing groups.

Responsive small-molecule fluorescent sensors can be designed to sense a wide variety of chemical species and microenvironments. These include reactive oxygen and nitrogen species, ions, unfolded proteins, enzymatic reactions, and factors like viscosity, pH, temperature, and membrane potential. There are many examples of commercially available fluorescent sensors (Figure 1.4).

Small-molecule fluorophores have become more powerful as we have learnt to control photophysical parameters such as brightness, Stokes shift, wavelength of emission, and photostability. Small-molecule fluorophores are relatively simple structurally and established synthetic organic chemistry methods can be exploited to modify these photophysical properties.

To develop sensors, interactions with analytes can be optimised through an understanding of electrostatic interactions, chemical reactivity, supramolecular chemistry, and the photophysical properties of fluorescent molecules. Biological interactions can be fine-tuned through synthetic modifications: the incorporation of targeting groups, or the modification of lipophilicity, solubility, or the pharmacokinetics. Small synthetic fluorophores can also display high biocompatibility, and compared to fluorescent proteins, their diminutive size means they minimally perturb the intracellular environment and therefore have a negligible effect on function. This small size also means they can be bioavailable to cells and tissues.

The term 'fluorescent probe' is often used to describe small-molecule fluorescent sensors: for most intents and purposes, 'fluorescent probe' and 'fluorescent sensor' are synonymous. The terminology is therefore used interchangeably throughout this text.

1.4 General Types of Fluorescent Sensors

Numerous fluorescent tools have been reported in the literature and are available commercially. In general, these can be classified as non-specific dyes and responsive sensors for a particular analyte or environment, depending on whether their fluorescence changes in the

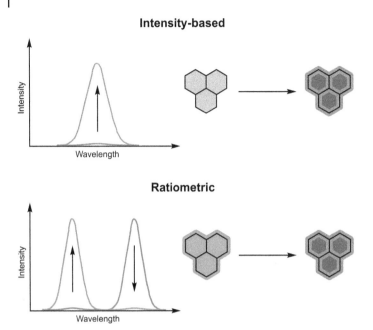

Figure 1.5 Intensity-based vs. ratiometric fluorescence response. An intensity-based sensor undergoes a change in fluorescence intensity, while a ratiometric response changes the fluorescence emission of the sensor.

presence of an analyte. Here, we focus on responsive systems, but many of the principles described herein apply also to fluorescent dyes.

The fluorescence output of a responsive sensor can be characterised as either being intensity-based or ratiometric (Figure 1.5). Intensity-based sensors can have either a turn-on in emission or a turn-off in emission. Turn-off sensors, where fluorescence emission is quenched, *i.e.* switched from on-to-off, are generally more difficult to use in imaging applications, as a loss of emission due to interaction with the desired analyte can be difficult to distinguish from any environmental quenching or loss of sensor. Therefore, turn-on sensors, where fluorescence emission increases, are more desirable.

There are multiple disadvantages to monitoring one emission wavelength. Instrumental parameters will affect the emission, such as fluctuations in excitation intensity. The local concentration of the probe will alter the emission. Aspects of the microenvironment of the probe, such as the pH, temperature, and polarity, will also influence the emission. The propensity for photobleaching will also alter the intensity of the observed emission. Background fluorescence and autofluorescence will interfere and must be taken into consideration when monitoring one emission. These parameters can be accounted for with careful experimental controls. However, quantitative sensing is still challenging.

As the quantitative determination of analytes is important, there has been extensive research into ratiometric fluorescent sensors. These are sensors that undergo a change in emission upon interaction with the analyte of interest (Figure 1.5). A shift in wavelength to a longer wavelength of light, lower in energy, is known as a red-shift or a bathochromic shift. A shift in wavelength to a shorter wavelength of light, higher in energy, is known as a blue-shift or a hypsochromic shift.

Measuring the ratio between two emissions, rather than the intensity, means that the fluorescence output is unaffected by the instrumental parameters or local environment. As a

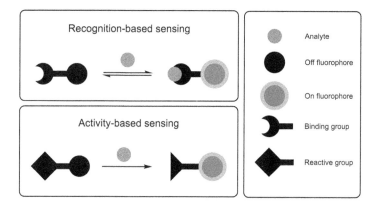

Figure 1.6 Recognition-based vs. activity-based sensing. Recognition-based sensors interact reversibly with the analyte of interest, while activity-based (or reaction-based) sensors undergo an irreversible bond formation or breaking with analyte interaction.

result, ratiometric sensors are self-calibrating [29]. Ratiometric sensing is a technique with the potential to provide precise quantitative analysis. The development of sensors with a ratiometric output is more challenging than an intensity-based change, but there are many notable examples in the literature.

Responsive fluorescent sensors can be generally categorised by their method of sensing as either recognition-based or activity-based sensing (also called reaction-based, with the terms used interchangeably) (Figure 1.6).

In recognition-based sensors, the change in fluorescence is due to an electronic interaction between the sensor and analyte. This process is reversible and does not alter the structure of the sensor, which returns to its initial state after removal of the analyte. Recognition-based sensors can be used for imaging dynamic changes.

Reaction-based, or activity-based, sensors interact with an analyte, and the fluorescence response is due to a covalent bond forming or breaking, inducing a structural modification of the sensor. This reaction is irreversible, in most cases, and analyte concentrations cannot be readily determined. However, very low concentrations of analyte or transient species can best be sensed using activity-based sensors.

1.5 Important Parameters

Fluorescent sensors differ greatly in their photophysical behaviour, which determines the biological questions to which they can be applied and the instrumental methods with which they can be used. The following sections explain the key parameters that are used to describe these photophysical properties.

1.5.1 Excitation Maxima

The excitation spectrum is gathered by monitoring the fluorescence emission at the wavelength of maximum intensity (emission maximum) as the fluorophore is excited over a range of wavelengths lower than the emission (Figure 1.7). The wavelength where the

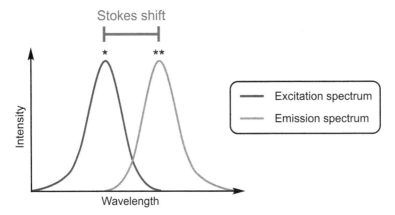

Figure 1.7 Representative excitation and emission spectra, showing the Stokes shift and the excitation (*) and emission (**) maxima.

emission is most intense, or where the excitation is most efficient, is the excitation maximum. Light with wavelengths near the excitation maximum will also cause excitation, but less efficiently.

An absorption spectrum (often termed a UV-visible absorption spectrum) will show every wavelength of light that is absorbed, while an excitation spectrum shows the wavelengths of light that will cause fluorescence emission at a specific wavelength. Typically, the absorption and excitation maxima will be the same wavelength for a given fluorophore, and so absorption spectra are often used to determine excitation maxima.

1.5.2 Emission Maxima

The emission spectrum is gathered by exciting at the wavelength of maximum absorption (excitation maximum) and monitoring emission over a range of wavelengths higher than the excitation (Figure 1.7). The emission maximum is the wavelength at which the emission is most intense. The fluorophore will also emit at wavelengths close to the emission maximum, but less intensely.

Emission spectra generally appear the same regardless of the excitation wavelength used, but the absolute intensities will differ. As described in Section 1.1, according to Kasha's rule, luminescence (whether fluorescence or phosphorescence) takes place from the vibrational ground state of the lowest excited level S_1 or T_1 [4] (Figure 1.3). The energy gap between higher excited states and S_1 is much smaller compared to the energy gap between S_1 and S_0. Rapid IC and vibrational relaxation from S_n to S_1 mean that only fluorescence from this transition will occur at appreciable levels. There are exceptions, for example for some fluorophores there are two distinct emission peaks, corresponding to dual emission from both $S_2 \rightarrow S_0$ and $S_1 \rightarrow S_0$ transitions [30].

1.5.3 Stokes Shift

In fluorescence processes, the energy of emission is lower than that of the absorbed light, meaning the emission maximum wavelength is longer than the excitation maximum, a phenomenon known as the Stokes shift (Figure 1.7). This is due to energy losses, normally from rapid decay to the lowest vibrational energy level of S_1 (Figure 1.3). Excited-state reactions and interactions, excited state complex formation, and energy transfer, amongst others, can all contribute to this effect.

The Stokes shift allows the use of precision bandwidth optical filters that block excitation light from reaching the detector, so light from fluorescence emission is solely observed. However, there is usually an overlap between the excitation and emission spectra, complicating this stratagem. A large Stokes shift, corresponding to less overlap between spectra, is useful for bioimaging sensors as this can minimise the signal bleed through and self-quenching from non-radiative energy transfer. There has been research into anti-Stokes' shift luminescent materials, where low-energy long-wavelength excitation leads to high-energy short-wavelength emission requiring additional energy [31, 32].

1.5.4 Quantum Yield

The quantum yield, Φ, is a measure of the efficiency of a radiation-induced process. For fluorescence, this means the efficiency with which absorbed photons are converted into emitted photons (Figure 1.8). This value is given as the ratio of emitted photons over absorbed photons: the closer to one, the more efficient the fluorophore. If non-radiative decay is much lower than the rate of radiative decay, this will result in a high quantum yield.

Quantum yields are measured experimentally either relative to a standard of known quantum yield or as an absolute quantum yield [33]. Measurement of the absolute quantum yield requires specialised instrumentation; therefore, comparison to a known standard tends to be the more widespread method of determination [34].

1.5.5 Molar Extinction Coefficient

Molar extinction coefficient, ε, or molar attenuation coefficient, is a measure of how strongly light is absorbed at a given wavelength, usually the wavelength of maximum absorbance. The SI unit is $m^2\ mol^{-1}$ but ε is normally given as $M^{-1}\ cm^{-1}$ and since these values are typically large for fluorophores, $\log_{10} \varepsilon$ is often reported. If ε is known, it can be used in the Bouguer-Beer-Lambert Law to obtain the concentration of a sample [35, 36].

Figure 1.8 Visual representation of fluorescence quantum yield.

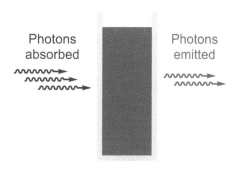

$$\text{Quantum yield } (\Phi) = \frac{\text{Number of photons emitted}}{\text{Number of photons absorbed}}$$

Table 1.1 Comparison of brightness of common fluorophores in PBS buffer pH 7.4.

Fluorophore	Quantum yield Φ	Molar extinction coefficient ε (M^{-1} cm^{-1})	Brightness (M^{-1} cm^{-1})
Fluorescein [38]	0.93	76900	71517
Cy5 [39]	0.27	250000	67500
Coumarin 6 [40]	0.08	15000	1200
Lucifer Yellow [41]	0.22	10471	2303

1.5.6 Brightness

The brightness of a fluorophore is conventionally defined as the product of the quantum yield and the molar extinction coefficient [37]. This parameter can be used to compare different fluorophores as it considers both how strongly the light is absorbed and how efficient the emission is. Fluorophores with high quantum yields and extinction coefficients are therefore usually very bright. The quantum yields, molar extinction coefficients, and brightness values of some widely used fluorophores are given in Table 1.1.

1.5.7 Lifetime

The fluorescence lifetime is the average time the fluorophore spends in the excited state, or the average time between the excitation and emission. This is defined by the time, t, at which the fluorescence intensity has decayed to 1/e of the initial intensity after excitation (Figure 1.9).

This is an intrinsic property of a fluorophore; it is unaffected by excitation wavelength or the length of exposure. However, as it is associated with an excited state, lifetime can be affected by local environmental factors such as temperature, polarity, viscosity, and fluorescence quenchers, as well as internal structural features of the fluorophore [13]. For small-molecule fluorophores, the fluorescence lifetime is typically between 0.5 and 20 ns [13]. Fluorescent sensors can therefore be developed where the lifetime is measured as the output.

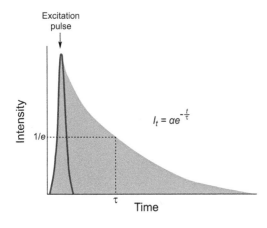

Figure 1.9 Graphical and visual representation of fluorescence lifetime. Not all molecules will decay at exactly the same rate; instead, fluorescence lifetime is defined as the time at which the fluorescence is 1/e of the original value, *i.e.* ~63% of excited molecules have undergone fluorescence decay to the ground state. I_t = fluorescence intensity at time t, τ = fluorescence lifetime, α = normalisation term.

$$I_t = \alpha e^{-\frac{t}{\tau}}$$

Lifetime-based sensing can be advantageous as the measurements are independent of fluorescence intensity, and local concentration of the sensor [42]. This means that lifetime measurements are inherently ratiometric. Fluorescence lifetime imaging allows quantitative sensing with probes that are not wavelength-ratiometric [15].

The measured fluorescence lifetime is normally shorter than the radiative lifetime due to other intramolecular processes or intermolecular interactions that act as quenchers [15].

1.5.8 Photobleaching

A fluorophore can theoretically undergo infinite cycles of excitation and emission of fluorescence. However, there are several pathways for the loss of energy from the excited fluorophore some of which are irreversible chemical processes that lead to the molecule losing its ability to fluoresce (Figure 1.10). This phenomenon is known as photobleaching.

As previously mentioned, an excited fluorophore can undergo inter-system crossing to an excited triplet state. This triplet excited state can be highly reactive, as it is high energy and long-lived compared to the singlet state, increasing the likelihood that side-reactions take place. For example, the long-lived triplet state can react with redox active thiols, amino acids, solvent, and oxygen [43, 44]. Quenching of the triplet state by oxygen gives highly reactive singlet oxygen ($^1O_2$), which has a microsecond lifetime, and can generate further reactive oxygen species (ROS) that can cause photobleaching [45]. There are other, less well-studied, pathways that lead to photobleaching, including reactions with the singlet excited state and fluorophore intermolecular reactions. Deliberate photobleaching of fluorophores can be used experimentally [46, 47], but for small-molecule fluorescent sensors, photobleaching is generally undesirable.

Photostability is therefore an important property for any fluorophore used in fluorescent sensors and much research has gone into the development of fluorophores that are less prone to photobleaching. Modifications that improve photostability ideally must not deleteriously affect other desirable properties like quantum yield and brightness or membrane permeability. The most common strategies to improve photostability are reducing the reactivity of fluorophores toward ROS and shortening of the triplet state T_1 lifetime, as

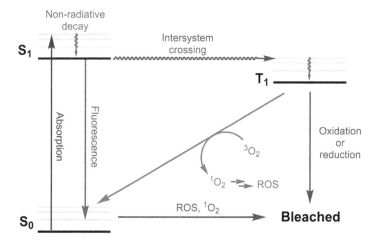

Figure 1.10 Jablonski diagram illustrating pathways to photobleaching.

alterations to this could lessen the effect of the majority of known mechanisms of photobleaching [43]. Commercially available fluorophores with less propensity for bleaching include JaneliaFluors™ [48] and AlexaFluors™ [49].

Experimentally, to minimise the extent of photobleaching, the time of exposure and excitation light intensity are kept to the minimum that still allows for optimal signal detection. Fluorescent probes with high resistance to photobleaching can be continually irradiated for longer, and this can allow for use in specialist techniques (*e.g.* single molecule, super resolution, structured illumination microscopy [SIM], stimulated emission depletion microscopy [STED]) [6, 50].

1.5.9 Anisotropy

The emission intensity of a fluorophore excited with polarised light will not be equal along all axes of polarisation: this is known as fluorescence anisotropy, or fluorescence polarisation [51]. Fluorescence anisotropy is a function of the fluorescence intensities acquired at two orthogonal polarisations, vertical and horizontal, or parallel and perpendicular to the plane of linearly polarised excitation light [52]. Measurement of fluorescence anisotropy allows the estimation of the rates of molecular rotations, or rotational diffusion, during the fluorescent lifetime. This can be used to measure binding constants or reaction kinetics, and fluorescence anisotropy has many applications in clinical chemistry, bioassays, and drug discovery [53].

1.5.10 Quenching

Quenching can be defined as any process in which an external environmental influence results in a lowering of the emission of a fluorophore. These include excited state reactions, energy transfer, ground state complex formation, excited state complex formation, and aggregation-induced quenching. In contrast to photobleaching, which is irreversible, quenching is frequently a reversible process. Quenching can broadly be categorised as either dynamic or static.

Dynamic quenching takes place when the excited fluorophore collides with a species, an atom or molecule, which can facilitate non-radiative decay. This results in the excited fluorophore returning to the ground state without emission of a photon, so no fluorescence is observed [54]. As these are collisional processes, dynamic quenching processes are affected by diffusion.

Static quenching involves complex formation. This can take place if the ground-state fluorophore forms a stable complex with a quencher. When the complex absorbs light, there is no emission of light. Only fluorophore molecules that form a complex are quenched; any free fluorophores would still fluoresce [54].

Static quenching does not change fluorescence lifetime, whereas dynamic quenching does. Fluorescence lifetime measurements can therefore be used as a method to distinguish between static and dynamic quenching [55]. Manipulation of quenching processes form the basis for most small-molecule fluorescent sensors.

Fluorophores can also be self-quenching at high concentrations. Common mechanisms of self-quenching are non-radiative energy transfer, principally in fluorophores with small Stokes shifts, and formation of aggregates.

1.6 Fluorescence Mechanisms Used in Fluorescent Sensors

The preparation of a fluorescent sensor requires a fluorescent molecule, known as a fluorophore, but also requires that interaction with the analyte results in a change in fluorescence output. It is therefore important to understand mechanisms of fluorescence, and fluorescence modulation. This section will detail different fluorescence mechanisms and how they can be harnessed to produce responsive fluorescent sensors.

1.6.1 Photoinduced Electron Transfer

Photoinduced electron transfer (PET) is an electron transfer process, which results in non-radiative dissipation of energy from the excited state [56]. PET may be either oxidative or reductive [57]. In reductive PET, electron transfer takes place from a donor moiety to the excited fluorophore, which is therefore reduced, and the fluorescence quenched. Oxidative PET is when the electron transfer is from the excited fluorophore, which is therefore oxidised, to an electron-deficient acceptor moiety (Figure 1.11). This electron's energy must be between the energies of the π and π^* orbitals of the fluorophore, lowering

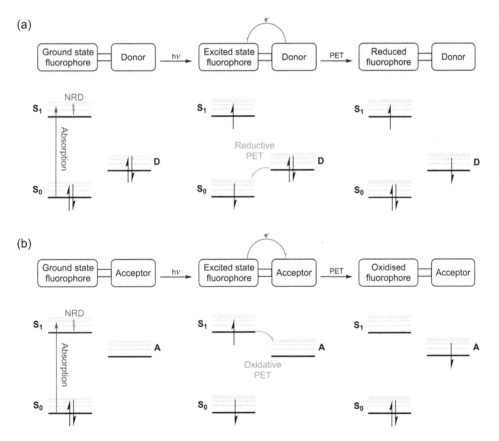

Figure 1.11 Representative photo-induced electron transfer (PET) systems, and energy-level diagrams for molecules undergoing (top) reductive PET and (bottom) oxidative PET.

the energy of the excited state, and blocking the π* to π relaxation and hence the fluorescence emission [57, 58].

Inhibition of PET by interaction of an analyte to generate a fluorescence turn on is a commonly employed sensing strategy. Conversely, PET quenching may be promoted to generate a turn-off sensor. PET-based sensors usually consist of a fluorophore conjugated *via* a linker to an electron-rich, or electron-deficient, sensing group. PET quenching causes only weak fluorescence from the sensor alone, while interaction with a desired analyte causes a recovery in fluorescence and hence a large turn-on. The more efficient the PET quenching, the larger the turn-on response and the more sensitive the sensor [59].

PET sensing is a particularly good strategy for the design of reversible metal and pH sensors [60]. Coordination of a metal cation or protonation of a lone pair of electrons decreases the energy of the lone pair, now involved in a bonding interaction, suppressing PET.

1.6.2 Internal Charge Transfer

Internal (intramolecular) charge transfer (ICT) fluorophores contain an electron-donating or 'donor' moiety in conjugation with an electron-withdrawing or 'acceptor' moiety, forming a 'push-pull' π-system, D-π-A [57]. Following photoexcitation, this ICT state rapidly forms. Electron density from the donor moiety moves to the acceptor moiety, and the charge distribution in the excited state is therefore significantly different from the ground state [61]. As a result, the fluorophore in the excited state is very polarised with a much larger dipole moment than in the ground state. The greater the extent of ICT, the larger the dipole moment, and the greater the bathochromic shift (Figure 1.12).

The photophysical properties, lifetimes, and quantum yields of D-π-A systems, are highly dependent on the donor and acceptor groups, and their electronic properties [62, 63]. ICT typically results in a bathochromic shift of fluorescence emission. ICT fluorophores usually display a large Stokes shift, a useful property for bioimaging, but also often have broad excitation and emission profiles, which can hinder their utility in experiments using multiple fluorophores [64]. Solvatochromism is another feature typical of ICT fluorophores arising from interactions between the solvent and the large dipole moment of the emissive state. This is due to sensitivity to the polarity of the solvent [65]. If the dipole moment of the excited state is larger than in the ground state, there is usually a bathochromic shift in excitation and emission maxima with increasing solvent polarity [66].

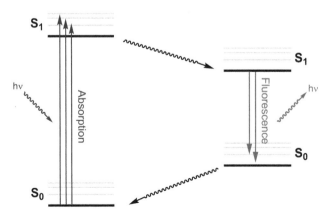

Figure 1.12 Intramolecular charge transfer (ICT) involves a movement of electron density upon excitation, resulting in an increased dipole moment for the molecule. The size of the dipole moment change determines the subsequent emission wavelength.

In ratiometric ICT-based sensors, the analyte can either promote or inhibit the ICT process to give rise to a change in the emission. This could be by interaction with the electron donor, the electron acceptor, or by modulating the conjugation to affect the ICT process. Promotion of ICT leads to a bathochromic shift in emission, whereas suppression of ICT leads to a hypsochromic shift.

ICT fluorophores can often form a twisted intramolecular charge transfer (TICT) state, which is a major mechanism of non-radiative energy loss [67]. This is highly dependent on conformational freedom of the 'donor' auxochrome and so solvent polarity and viscosity and other environmental parameters. Because of this, several sensors have been developed harnessing TICT [68]. However, this is an uncommon strategy and research into inhibiting TICT to improve ICT fluorophores for fluorescent sensing is underway [48, 69, 70].

1.6.3 Förster Resonance Energy Transfer

Förster resonance energy transfer (FRET), often referred to as fluorescence resonance energy transfer, is a non-radiative process. We will restrict the discussion to small-molecule organic fluorophores in this context, but the phenomenon of FRET is widely used in biological imaging with fluorescent proteins [11, 71–73].

In FRET, an excited state donor transfers energy to an acceptor through long-range dipole–dipole interactions, without emission. In fluorescent sensing, the donor fluorophore is excited at a specific wavelength, the energy is then transferred non-radiatively to the acceptor fluorophore. The donor fluorophore returns to the electronic ground (non-excited) state and the fluorescence emission of the acceptor fluorophore is observed (Figure 1.13). This is a good strategy for achieving a large shift between the excitation and emission spectra.

The extent and rate of the energy transfer by FRET is determined by several factors: the degree of spectral overlap between the donor emission and acceptor excitation; the distance between the donor and acceptor fluorophores; the relative orientation of the transition dipoles; and the quantum yield of the donor fluorophore. The brightness of the acceptor fluorophore will determine the brightness of the FRET pair. In the design of small-molecule

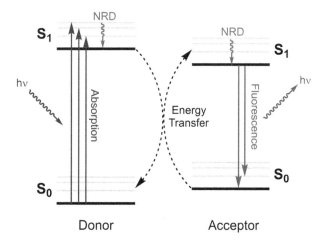

Figure 1.13 A Jablonski diagram for Förster resonance energy transfer (FRET). Dotted arrows indicate the energy transfer between the donor and acceptor molecules. NRD = non-radiative decay.

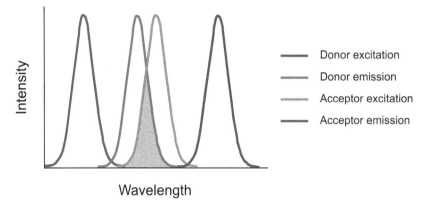

Figure 1.14 'Ideal' FRET donor and acceptor with spectral overlap between donor emission and acceptor excitation shaded.

fluorescent sensors, the spectral overlap between donor and acceptor fluorophores is the most important parameter (Figure 1.14).

Selecting a FRET donor–acceptor pair requires care: the overlap between the donor emission and the acceptor excitation must be maximised to increase the FRET efficiency, whilst the donor emission and acceptor emission overlap must be minimised to allow the measurement of both separately. Since in practice, application of a FRET sensor involves exciting the donor fluorophore and measuring emission of the acceptor fluorophore, the pseudo-Stokes shifts will be much larger than for single fluorophores. This is an attractive property for imaging studies. Furthermore, FRET is an inherently ratiometric technique since there are two emission peaks, corresponding to emission from the donor and acceptor molecules.

FRET is a distance-dependent process, with the efficiency of energy transfer being proportional to $1/d^6$, where d is the distance between donor and acceptor. FRET is therefore highly sensitive to small changes in conformation and can, for example, be used to measure the proximity of chromophores in biomolecules and hence any conformational changes [74]. The distance at which resonance energy transfer is 50% efficient is known as the Förster distance [75].

Several strategies for the design of FRET-based fluorescent sensors can be employed. A cleavable linker can be used to hold the two fluorophores in proximity, and upon interaction with an analyte, the linker is cleaved and the distance between the fluorophores no longer allows for FRET to take place (Figure 1.15a). In many cases, a non-fluorescent or a 'dark' quencher is used: this can be a highly effective strategy to obtain sensors with very-low background fluorescence [76, 77]. This strategy of using a cleavable linker only allows for irreversible sensing.

A second strategy, which allows for the design of reversible sensors, is inclusion of a responsive linker that changes in conformation up interaction with an analyte, therefore altering the distance between the donor and acceptor and changing the FRET efficiency (Figure 1.15b).

Finally, the most common strategy for the design of small-molecule FRET sensors is the use of a responsive fluorophore as either FRET donor or FRET acceptor (Figure 1.15c). The interaction with the analyte can be reversible or irreversible in this design strategy.

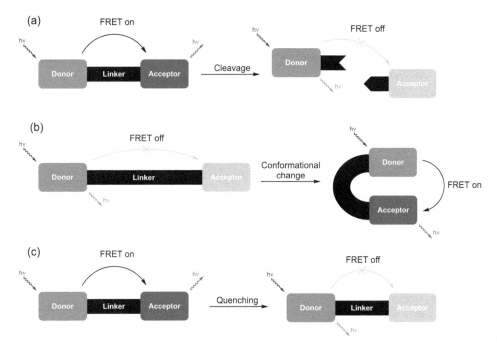

Figure 1.15 Different strategies for the design of FRET-based sensors: (a) A cleavable linker. (b) A conformational change. (c) Quenching of a FRET partner.

1.6.4 Through Bond Energy Transfer

Like FRET, through bond energy transfer (TBET) involves energy transfer between two fluorophores. However, in the case of TBET, the donor and acceptor are electronically conjugated. In such systems, this conjugated linker enables energy transfer but also prevents the donor and acceptor from becoming planar, and acting as a single dye [78]. As energy transfer occurs through the conjugated linker bonds, the spectral overlap requirement vital for FRET is not necessary. This overcomes the limited resolution of the two signals that is a challenge for many FRET sensors; the well-defined fluorescein-rhodamine FRET pair, for example, have only a ~65 nm difference in emission peaks [79]. As for FRET, the TBET donor must have high absorbance at the excitation wavelength and the acceptor component must fluoresce strongly in order to generate a bright sensor.

The absence of a requirement for spectral overlap also means that more fluorophore pairs can be utilised for TBET sensors than for FRET-based sensors, allowing greater flexibility and the possibility of larger pseudo-Stokes shifts [29, 78]. There is also a higher degree of efficiency in TBET than in FRET, and the energy transfer is faster [80]. TBET sensors are also inherently ratiometric, due to the presence of two fluorophores.

One example of a TBET sensor is shown in Scheme 1.1. **CR** contains coumarin and rhodamine fluorophores that exhibit no energy transfer in the absence of Hg^{2+}, so excitation of the coumarin in the ultraviolet leads to the characteristic blue coumarin emission. Hg^{2+} induces formation of an oxadiazole, placing the two fluorophores in conjugation and enabling TBET. As a result, excitation of the coumarin gives rise to red-shifted rhodamine fluorescence.

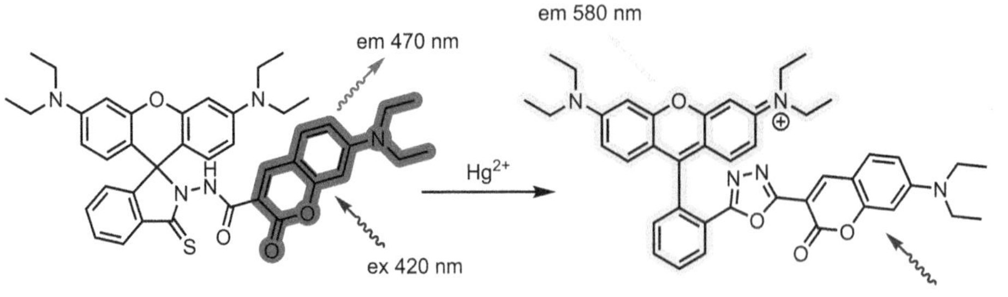

em 580 nm

em 470 nm

Hg²⁺

ex 420 nm

CR

CR-P

ex 420 nm

Scheme 1.1 Ratiometric TBET sensor for Hg²⁺ **CR** employed in live cell imaging [79]. **CR** demonstrates the use of a fluorophore pair with no spectral overlap giving a significant pseudo-Stokes shifts *via* TBET.

1.6.5 Excited-State Intramolecular Proton Transfer

Excited-state intramolecular proton transfer (ESIPT) is an extremely fast ($> 10^{12}\,\text{s}^{-1}$) proton transfer in the excited state of a fluorophore [81]. Fluorophores can exhibit this phenomenon if they contain a hydrogen bond donor and hydrogen bond acceptor in proximity. In this barrierless process, a covalently bonded proton, usually of a hydroxyl or amino group, in the excited-state migrates to a neighbouring hydrogen-bonded atom less than 2 Å away in a 5- or 6-membered ring configuration [82, 83] (Figure 1.16). This is predominantly through keto-enol tautomerism, or amine-imine tautomerism. The enol tautomer is more stable in the ground state, but in the excited state the enol rapidly converts to the keto tautomer. The photo-tautomer emits light, and then there is a reverse proton transfer (RPT) as it thermally equilibrates back to the ground state, where the proton is bonded to its original atom [84]. The initial tautomer shows a higher energy, shorter wavelength emission and a lower energy, longer wavelength emission is seen from the proton-transferred tautomer [85].

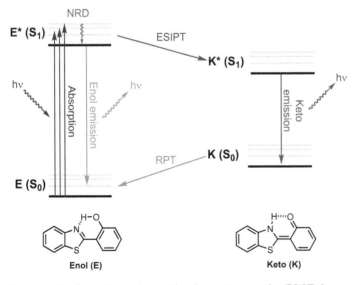

Figure 1.16 Excited-state intramolecular proton transfer (ESIPT) for a representative benzothiazole compound. The enol tautomer is more stable in the ground state, but upon absorption of light, it undergoes ESIPT to the keto excited state, with corresponding emission from this state. After fluorescence emission, the keto ground state undergoes reverse proton transfer (RPT) to the enol S_0.

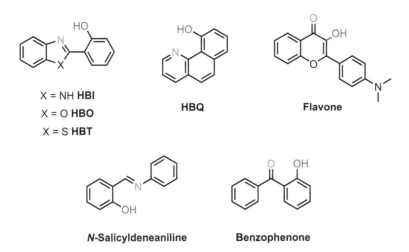

Figure 1.17 Examples of commonly used ESIPT fluorophore scaffolds: 2-(2′-hydroxyphenyl)
benzimidazole (**HBI**), 2-(2′-hydroxyphenyl)benzoxazole (**HBO**), 2-(2′-hydroxyphenyl)benzothiazole
(**HBT**), 10-hydroxybenzo[*h*]quinoline (**HBQ**), flavone, *N*-salicylideneaniline, and benzophenone.

As ESIPT leads to a significant structural reorganisation, the photophysical properties of
the two tautomers can be drastically different [86]. ESIPT often leads to very large Stokes
shifts, often around 200 nm. This means there is an almost complete lack of spectral overlap
between absorption and emission, a very advantageous property for imaging [87]. Along
with the two emission bands, enabling ratiometric sensing, this makes using ESIPT an
appealing strategy in responsive probe design [88, 89].

Responsive fluorescent sensors based on ESIPT typically use a reactive sensing group that
blocks the hydrogen bond donor, thus preventing tautomerisation to the keto form and halt-
ing ESIPT from taking place. This results in fluorescence only being observed from the enol
form. Upon reaction with the analyte, the hydrogen bond donor is unmasked, the keto
tautomer can form, and ESIPT can take place. Several different systems have been reported
for use in ESIPT sensors (Figure 1.17).

ESIPT is highly influenced by the environment, in polar hydrogen-bond donating solvents,
ESIPT is often inhibited. This potential disadvantage can be used to image specific biological
environments or biomolecules that exhibit differences in polarity [89, 90]. Fluorescent sensors based on ESIPT often make use of other fluorescence mechanisms, such as aggregation-
induced emission (AIE).

1.6.6 Aggregation-Induced Emission

Aggregation-induced emission (AIE) is a phenomenon where emission of a fluorophore is
activated by aggregation. This contrasts with the property that most fluorophores exhibit in
being quenched upon aggregation. AIE fluorophores are non-emissive when in dilute solutions and emit intensely upon formation of aggregates. The potential applications of this
class of fluorophores were not realised until Tang and co-workers reported the luminescence
of 1-methyl-1,2,3,4,5-pentaphenylsilole and coined the term AIE [91]. Their subsequent
work explained how restriction of intramolecular motion caused enhanced emission in
these systems, and developed several applications including fluorescent sensing, and now

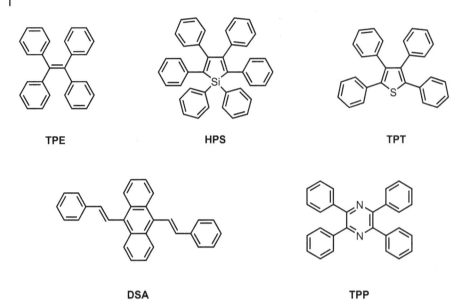

Figure 1.18 Examples of commonly used AIEgens: tetraphenylethene (**TPE**), hexaphenylsilole (**HPS**), tetraphenylthiophene (**TPT**), 9,10-di((*E*)-styryl)anthracene (**DSA**), tetraphenylpyrazine (**TPP**).

AIE has been employed in many fields [92]. Several different AIE fluorogens have now been identified (Figure 1.18).

AIE-based fluorescent sensors typically have large Stokes shifts, high photostability, and negligible background fluorescence as they are non-fluorescent in solution before activation of aggregation [93]. This means AIE sensors have a large turn-on with higher sensitivity than some conventional fluorescent sensors and can often be used in bioimaging with no requirement for washes [94].

Multiple strategies for AIE based sensors have been developed. Interaction with an analyte to produce aggregates through non-covalent interactions is a common sensing platform. For examples, targeting groups can be incorporated resulting in an emission increase due to restriction of intramolecular motion, upon accumulation or interaction with an analyte. Another method is cleavage of a solubilising ligand to decrease solubility and generate aggregates. Reaction-based probes where bond cleavage promotes or inhibits another photophysical process of quenching or fluorescence enhancement used in combination with AIE are also common (Figure 1.19).

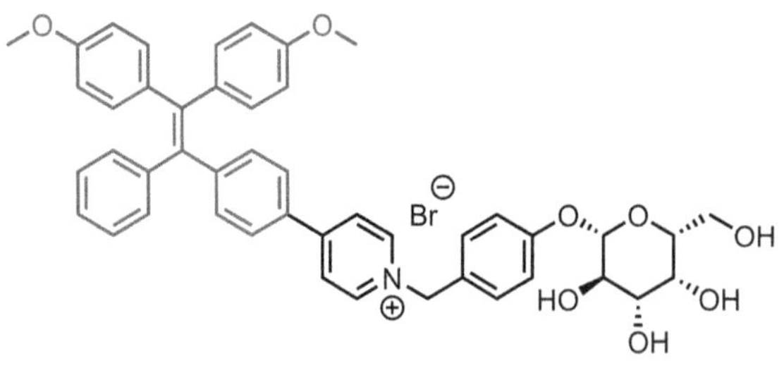

Figure 1.19 AIE-based β-galactosidase sensor **TPE-Gal** [95].

TPE-Gal

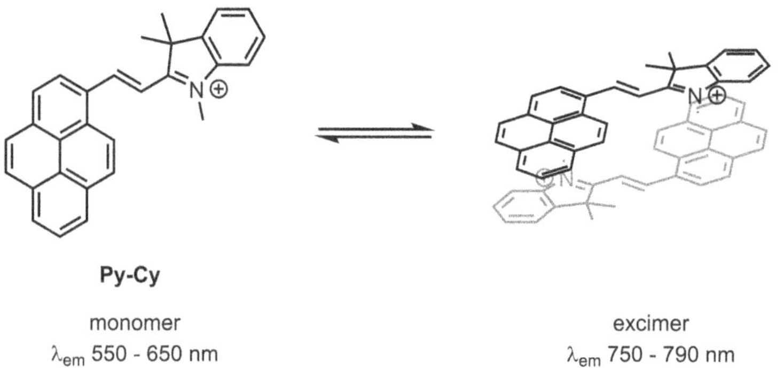

Py-Cy

monomer	excimer
λ_{em} 550 - 650 nm	λ_{em} 750 - 790 nm

Scheme 1.2   Demonstration of monomer and excimer emission shift in **Py-Cy** [100].

1.6.7   Excimer Formation

An excimer, or excited dimer, is a type of complex between two molecules of a fluorophore that exists only in excited electronic states [96]. This complex is formed between a molecule of a fluorophore in the excited state and a molecule of the same fluorophore in the ground state and involves π-π* orbital interactions. Excited-states have half-filled orbitals that can associatively interact with ground-state orbitals [97].

Excimer formation is a diffusion controlled collision process, which is therefore increased when local concentrations are high [98]. This phenomenon is observed for several aromatic molecules. Typically, the emission from the excimer is red-shifted and very broad compared to that of the monomer. This is a convenient strategy for development of a sensor with a ratiometric output.

The most widely studied fluorophore that exhibits excimer emission, and the most commonly used fluorophore for monomer/excimer-based sensors, is pyrene [99]. The monomer emits at 370–410 nm, while the excimer exhibits broad emission around 460–500 nm. The high-energy excitation can limit the use of pyrene-based systems for bioimaging and *in vivo*, and there has been research into more applicable longer wavelength analogues (Scheme 1.2) [100, 101].

In excimer-based sensing, a ratiometric output can be achieved using a single fluorophore and measuring two wavelengths. A common strategy is to link two fluorophore monomers such that a change in separation or orientation will promote or inhibit excimer formation (Figure 1.20).

1.7   Commonly Used Fluorophores

Due to the many useful properties of fluorescent small molecules, there has been a great deal of research into the design, synthesis, characterisation, and application of small-molecule fluorescent sensors. The field is ever-expanding and improvements to existing fluorophores, widening their applicability, and discovery of new fluorophores is continuing.

As described in Section 1.5, there are several properties required for an ideal small-molecule fluorophore. Ideal photophysical properties include bright emission, high molar absorption coefficient, and quantum yield approaching unity. Furthermore, excitation wavelengths should be sufficiently low in energy to not cause damage to cells, and excitation and emission wavelengths should match standard instrumentation. Other desirable properties

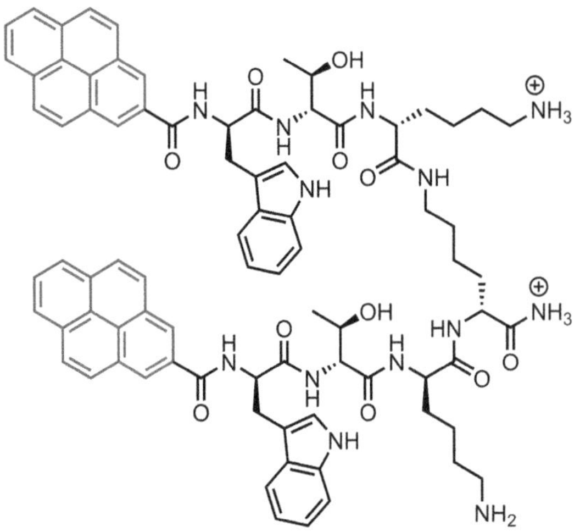

Figure 1.20 Sensor with two pyrene units where disruption of intramolecular excimer formation is used for ratiometric sensing of nucleic acids [102].

include good solubility and stability under physiological conditions, membrane permeability, and biological compatibility. Finally, a fluorophore should be easily synthesised in sufficient quantities and enable the inclusion of sensing groups and targeting groups with synthetic chemistry.

This section covers some of the most common fluorophores employed in sensor design and details their advantages and disadvantages for use in fluorescent sensors.

1.7.1 Fluorescein

Fluoresceins, along with the structurally related rhodamines, make up a large proportion of the fluorophores used in biological imaging to date [103]. This is not only because fluoresceins were amongst the first fluorophores synthesised [104] but also because they remain one of the brightest small-molecule fluorophores available. Fluorescein contains a xanthene core structure substituted with oxygen (Figure 1.21). Derivatisation of the core can be used

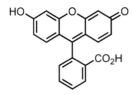

Fluorescein

Eosin B

Rose bengal

Figure 1.21 Structure of fluorescein, and commercially available fluorescein derivatives Eosin B and Rose Bengal.

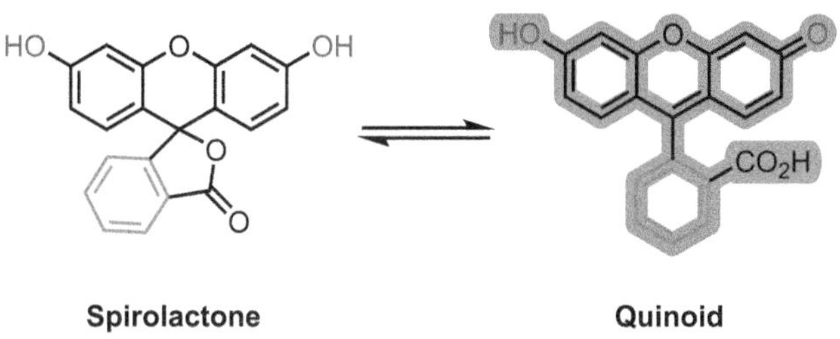

Spirolactone **Quinoid**

Scheme 1.3 Structures of the closed non-fluorescent spirolactone and open fluorescent quinoid forms of fluorescein. Fluorescent sensors are typically prepared by substitution on the hydroxyl groups (red) or the phenyl ring (blue).

to tune the fluorescent properties and append additional functionality such as targeting groups or sensing groups.

Fluorescein is highly water soluble, has a large extinction coefficient, and high quantum yields in aqueous solvents. The absorption and emission of fluorescein and its derivatives are highly pH dependent; it can exist in seven different prototropic states [105, 106].

Fluorescein exists in equilibrium between a closed spirolactone form, where the conjugation is broken and is non-emissive, and an open quinoid form, where the conjugation is restored and is strongly emissive (Scheme 1.3) [107]. Fluorogenic sensors can be designed by substitution on the hydroxyl groups of the xanthene core. If these groups are alkylated or acetylated, the molecule exists in the closed spirolactone form, and therefore does not emit. A sensing group can be attached at these points and either quench fluorescence or lock the sensor in the non-fluorescent closed spirolactone conformation. Upon reaction with the analyte of choice, the fluorescence emission is turned on giving a fluorogenic sensor [108].

Tuning the photophysical properties of fluorescein derivatives with synthetic chemistry is difficult. Functionalisation of the phenolic oxygen does not alter the emission but stops the spirolactone form from opening. The emission of the fluorescein scaffold can be red-shifted by the inclusion of heavy atoms, as seen in Eosin B and Rose Bengal (Figure 1.21), but this increases the incidence of ISC and decreases the fluorescence quantum yield [109]. Rose Bengal is used in synthetic chemistry to generate singlet oxygen from triplet oxygen [110]. Replacing the xanthene bridging oxygen can achieve bathochromic shifts in emission [111, 112]. Fluoresceins are not particularly photostable and suffer from photobleaching [113].

Removal of the carboxylic acid that forms the lactone results in the structurally related fluorone derivatives [103]. If the acid is removed, this means the energy barrier for rotation of the aryl group is much lower, and the quantum yield is lowered and fluorones are less emissive. Substitution at this position can increase the fluorescence, by restricting rotation of the phenyl ring which keeps it orthogonal to the xanthene, as seen in the fluorone derivative Tokyo Green [114] (Figure 1.22).

Figure 1.22 Fluorone and substituted analogue Tokyo Green [114].

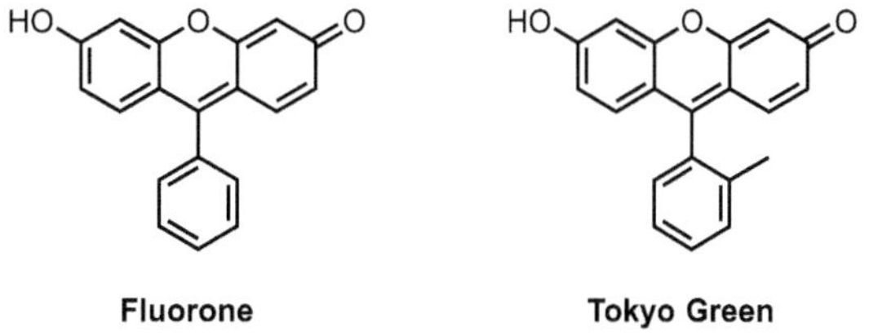

Fluorone **Tokyo Green**

Other functionalised sensors or biological labels with a fluorescein scaffold usually have substitution on the phenyl ring as this is the most straightforward method synthetically. Modifications of the phenyl ring can also be used to modulate the reduction potential, for development of responsive sensors, and fluorescent properties [115].

1.7.2 Rhodamine

Rhodamines share the xanthene core with the structurally similar fluoresceins (Figure 1.23), and therefore they also exhibit high absorption and very high quantum yields, although slightly lower than those of fluoresceins. Rhodamines contain amino groups in place of the phenolic oxygens. They do differ in that they are less pH sensitive and have lower water solubility than fluorescein [116]. They are also more photostable [117]. Introduction of rigidified systems at the amino group auxochromes lead to improved quantum yields, approaching unity [48, 118, 119].

In an analogous mechanism to that seen in fluorescein, the spirolactone form of rhodamines is non-fluorescent, as the conjugation of the molecule is disrupted. Upon ring opening, the conjugation is restored, and the ring-open form is strongly emissive (Scheme 1.4). An amine can be introduced to the rhodamine scaffold giving a spirolactam ring-closed form. A sensing group can be introduced at this position using an amine to form the spirolactam. Interaction with the desired analyte leads to opening of the ring and hence a turn-on in emission. This is a widely used strategy for the design of fluorogenic rhodamine-based sensors [108, 120].

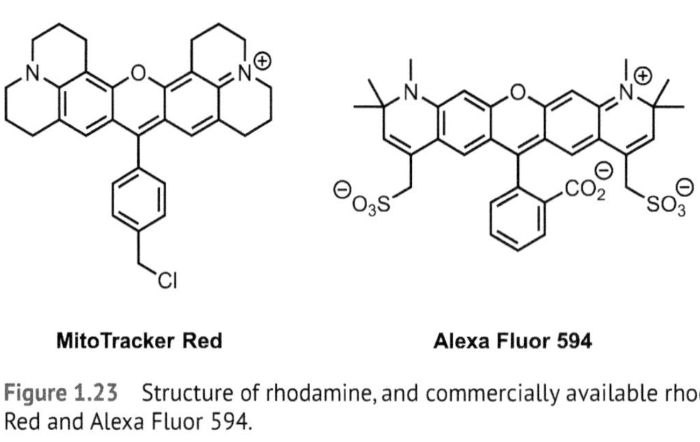

Rhodamine

MitoTracker Red

Alexa Fluor 594

Figure 1.23 Structure of rhodamine, and commercially available rhodamine derivatives MitoTracker Red and Alexa Fluor 594.

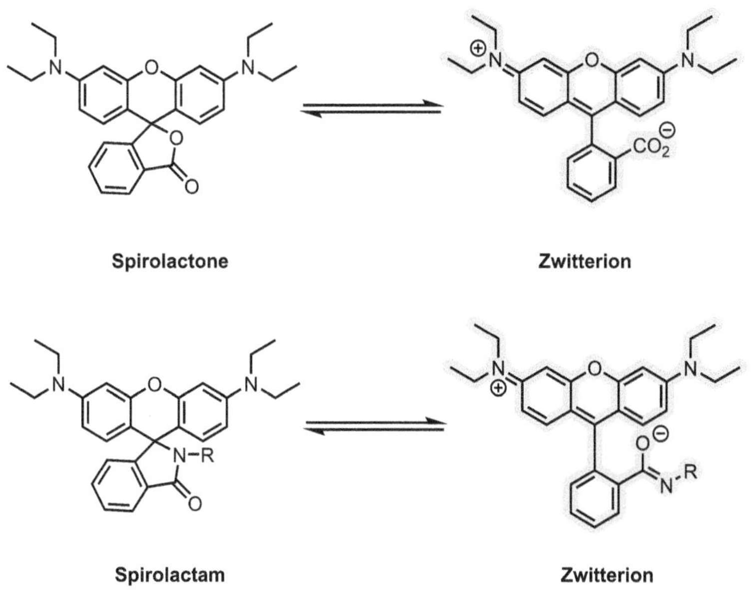

Scheme 1.4 Closed spirolactone and open zwitterion of rhodamine B; Rhodamine with lactam sensing group inhibiting ring opening until interaction with analyte.

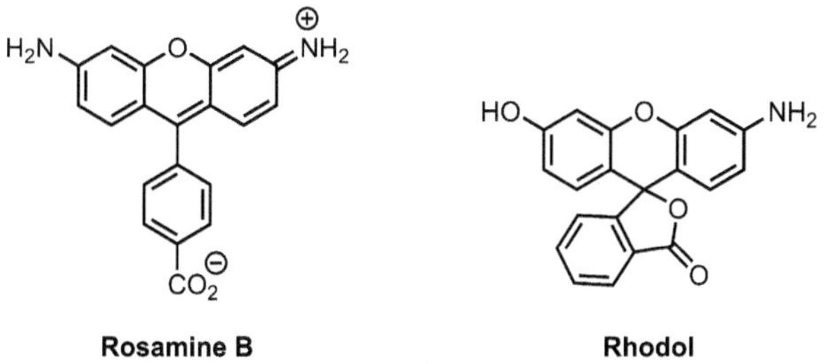

Figure 1.24 Structures of rosamine B and rhodol.

As with fluorescein, removal of the carboxylic acid group that forms the spirolactone results in a structurally related series of analogues called rosamines, which are less fluorescent unless substituted at the same position on the aryl group [103]. A rhodol is the derivative when one of the aniline nitrogens of a rhodamine is replaced with a phenolic oxygen [121] (Figure 1.24).

Rhodamines can be functionalised easily in other positions to yield responsive sensors [122]. This is predominantly through the inclusion of an amine bonded to the carboxylate group, as mentioned previously to generate fluorogenic dyes, or *via* substitution on the phenyl ring [123].

The bridging oxygen atom in the xanthene unit can be replaced, and this can dramatically alter the emission wavelength, for example the silicon [124, 125], and the carbon [112], and tin or germanium-bridged rhodamines [126]. Multiple hybrid rhodamine derivatives incorporating other fluorophore scaffolds have also been developed allowing access to emissions ranging from visible to NIR [127].

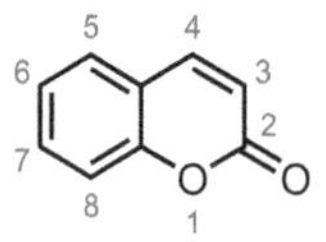

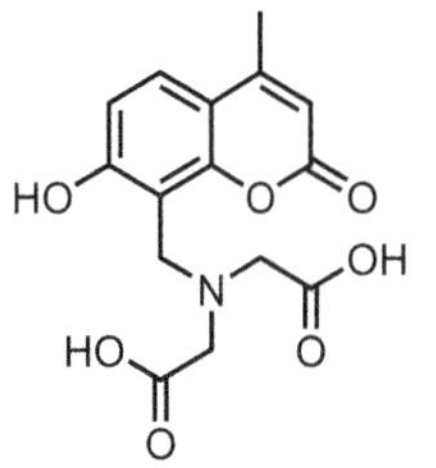

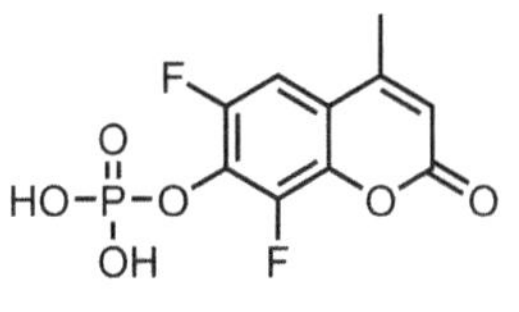

Figure 1.25 Coumarin core structure and commercially available coumarin derivatives **Calcein Blue** and **DiFMUP**. Most fluorescent coumarins have a hydroxyl, alkoxy, or amino substituent at the 7-position.

Calcein Blue   DiFMUP

1.7.3 Coumarin

Coumarins, 2*H*-chromen-2-ones, are amongst the most studied fluorophores (Figure 1.25). This scaffold is ubiquitous and found in natural products and pharmaceutical drugs [128]. The parent compound is non-fluorescent but substituted coumarins can be very emissive and many fluorescent sensors incorporate coumarins. The most emissive analogues are substituted at the 7-position with an amino or hydroxy substituent [129].

Coumarins possess good photostability and large Stokes shifts, a property characteristic of ICT fluorophores. They also have good cell membrane permeability, water solubility, and are sensitive to the polarity of their environment or solvent [130]. Many have high quantum yields, but coumarin brightness can be low, due to the weak absorbance of the small conjugated system [131]. They normally possess high-energy UV excitation, and emission from ~400 to 480 nm, which can limit their application in cellular imaging. However, this makes them useful as the donor in FRET pairs, and they are commonly used in enzymatic assays [131]. The emission of coumarin derivatives can be red-shifted by increasing the extent of conjugation. Coumarins containing difluoro-methylene groups in the 4-position of the core have demonstrated dramatically red-shifted emission, being the lowest molecular weight dyes to emit beyond 700–900 nm [132].

Coumarins are easily synthesised, and there are several routes to synthesise the core [133]. Functionalisation of the coumarin core is also practical, with the 3- and 4-positions the most commonly modified to incorporate sensing groups and reaction-based sensing groups at the 7 position to mask the ICT-donor and form fluorogenic probes, also a common strategy for sensor design [129].

1.7.4 Naphthalimide

1,8-Naphthalimides (benz[*de*]isoquinolin-1,3-diones) have been used as DNA binders, anti-cancer compounds, organic LEDs, imaging agents, and chemosensors [134–136]. Structurally, they comprise of a naphthalene ring and a dicarboxyl imide (Figure 1.26).

Naphthalimides are versatile fluorophores with good photochemical properties for bioimaging [137]. Substituted analogues, especially those with a 4-amino substituent, have high quantum yields, large Stokes shifts and good photostability [136, 138]. The typical excitation

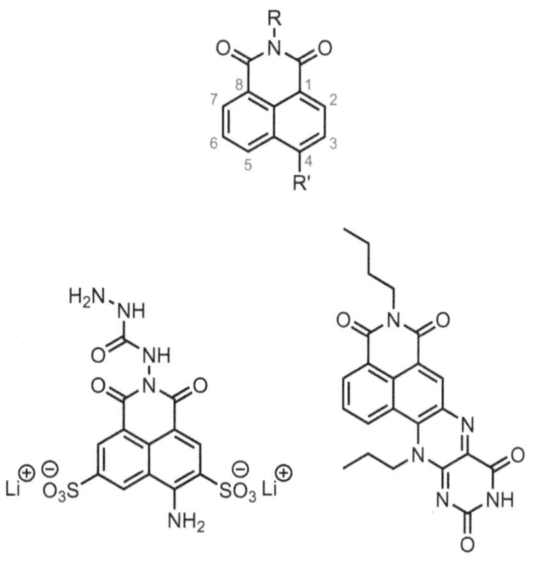

Figure 1.26 1,8-Naphthalimide core structure and commercially available derivatives **Lucifer Yellow** and **NpFR1**. Fluorescent 1,8-naphthalimides typically contain a hydroxy, alkoxy, or amino group normally at the 4-position.

Lucifer Yellow   NpFR1

and emission wavelengths are around 440 and 520 nm, respectively, but the range of emission accessible can be expanded by substitution on the naphthalene ring [139, 140].

Naphthalimides generally have poor aqueous solubility and low quantum yields in aqueous solvent systems, but structural modification and functionalisation can moderate this [135]. They exhibit strong solvatochromism, being red shifted in polar protic solvents, as is typical for ICT fluorophores. Naphthalimides tend to have very broad excitation and emission spectra, which can make imaging with multiple fluorophores difficult.

1.7.5 BODIPY (4,4-Difluoro4-bora-3*a*,4*a*-diaza-*s*-indacene)

One of the most extensively used classes of fluorophores in bioimaging is BODIPY: the common name for 4,4-difluoro-4-bora-3*a*,4*a*-diaza-*s*-indacene. The structure consists of a boron difluoride and dipyrromethene (Figure 1.27). BODIPYs have a high extinction coefficient and a narrow absorption and emission, giving rise to a characteristically small Stokes shift. They are also insensitive to their environment, possessing negligible solvatochromism and pH sensitivity, and quantum yields that often approach unity even in aqueous solvent [141, 142]. Additionally, BODIPY fluorophores generally possess high photostability and are stable to physiological conditions. BODIPY derivatives have been extensively used in fluorescent dyes and sensors [143]. This is not only due to their favourable fluorescent properties, but also their relative ease of functionalisation allowing for extensive derivatisation [144]. This allows for incorporation of functional groups for sensing, and for tuning of their photophysical properties. Therefore, a wide variety of BODIPY-based fluorescent sensors and organelle dyes have been reported [143, 145]. For example, the commercially available ER-Tracker Red, LysoTracker Red, and LysoTracker Green are all based on BODIPY scaffolds (Figure 1.27).

Despite their wide application and many beneficial characteristics, BODIPYs also possess some undesirable properties. They usually have very poor aqueous solubility, and emission

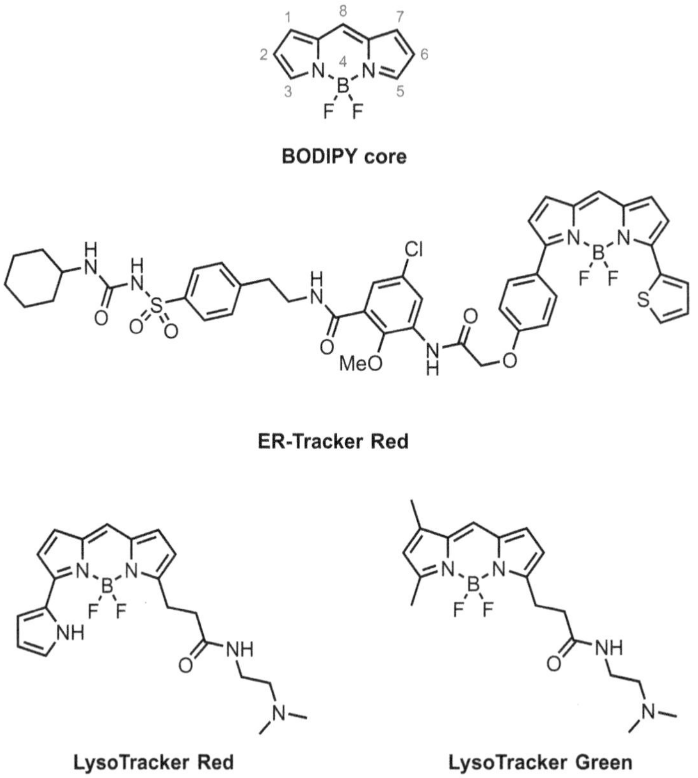

Figure 1.27 BODIPY core structure, and commercially available BODIPY derivatives, ER-Tracker Red, LysoTracker Red, and LysoTracker Green.

maxima that are typically below 600 nm, limiting their application for *in vivo* imaging. Due to the ease of post-synthetic functionalisation of the BODIPY core, several analogues have been reported where the emission is significantly red-shifted [146, 147]. Their small Stokes shift and narrow excitation and emission spectra can lead to high levels of self-quenching [148].

1.7.6 Cyanine

The term cyanine is commonly used to describe the family of tetramethylindo(di)-carbocyanines that consist of a polymethine chain containing an odd number of carbon atoms between two tertiary amines (Figure 1.28). Cyanines have been studied widely and are one of the oldest synthetic fluorophores [149].

Cyanines tend to have complex excited-state chemistry leading to quantum yields that are relatively low, but they possess extremely high molar extinction coefficients that are amongst the largest known for small molecules [150, 151]. This makes them amongst the brightest fluorophores [152]. They have a typically small Stokes shift, which can cause self-quenching and signal bleed-through. More recently, analogues with large Stokes' shift have been developed [153, 154].

They have emission maxima ranging from orange to far-red and near-infrared (NIR), making them popular for *in vivo* imaging as they are one of the few molecular scaffolds that can emit at these longer wavelengths. Longer polymethine chains lead to longer emission

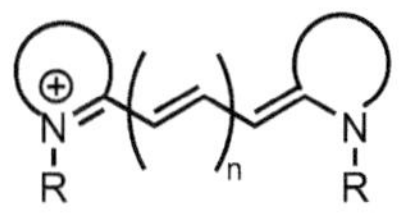

Cyanine

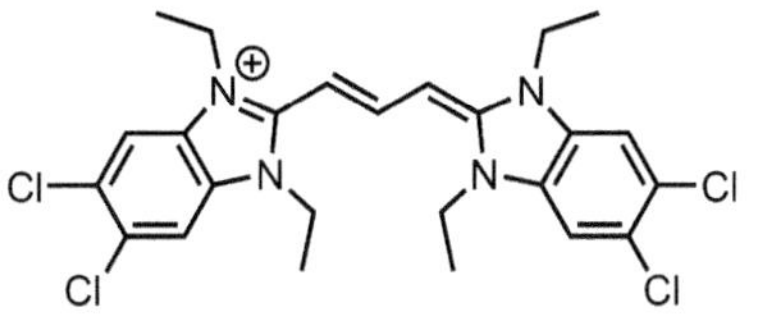

JC-1

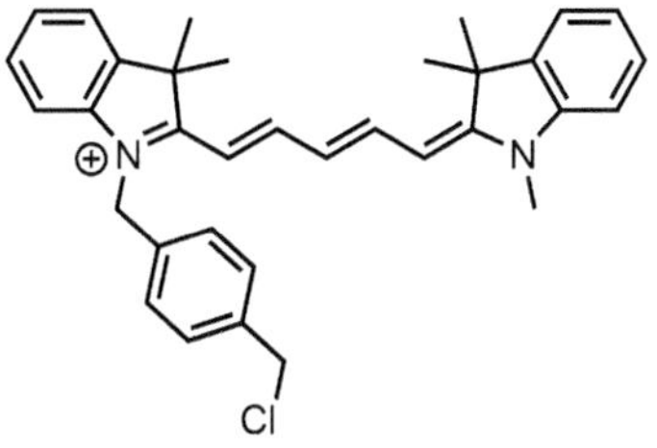

MitoTracker Deep Red

Figure 1.28 Cyanine general structure, and commercially available cyanine derivatives **JC-1** and **MitoTracker Deep Red**.

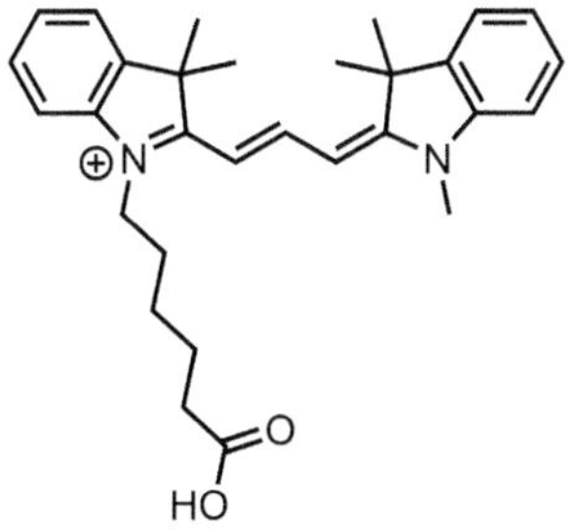

Cy3

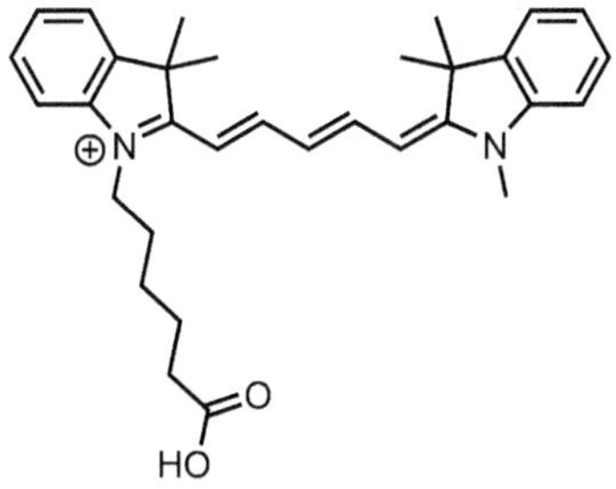

Cy5

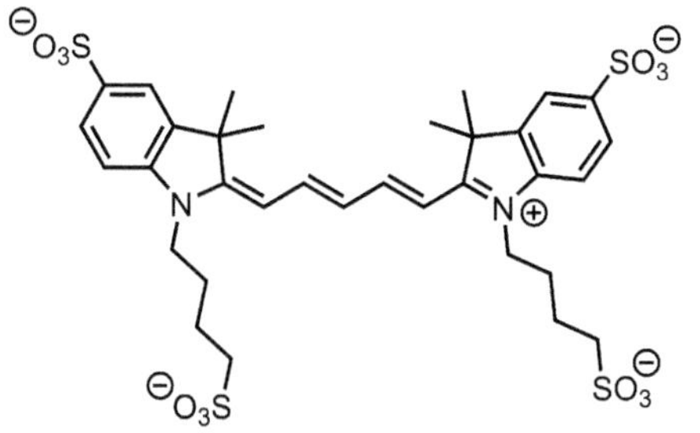

Alexa Fluor 647

Figure 1.29 Structures of **Cy3** and **Cy5** derivatives, and the sulfonated cyanine derivative **Alexafluor 647**.

wavelengths, with each additional methine unit increasing the emission by approximately 100 nm [155]. For example, Cy3 dyes have maximum emission at 570 nm, while the next cyanine in the series, Cy5 dyes, emit at 670 nm (Figure 1.29).

Photostability and chemical stability decreases with increasing polymethine chain length and cyanines with extended chains ($n > 3$; Figure 1.28) require structural modifications for stability. Furthermore, increasing the chain length increases the propensity for photoisomerisation, a non-emissive process where energy is lost from the excited state, lowering the quantum yield in solution [44, 156]. The long-conjugated systems can be prone to oxidation and in rare cases susceptible to nucleophilic attack, but overall cyanines possess high chemical stability both in aqueous buffer and in biological systems [150].

Cyanines are also prone to aggregation, and the most common remedy is the introduction of sulfonate groups, which also improves their typically low aqueous solubility [149]. As sulfonates are charged, this can result in non-specific binding or other undesired electrostatic interactions with analytes or fluorophores. The commercially available Alexa Fluor 647 is an example of a sulfonated cyanine (Figure 1.29).

1.8 Summary

Fluorescent compounds are found throughout nature and have fascinated researchers for centuries. Even today, armed with a UV torch, nature enthusiasts are discovering more and more plants and animals that are equipped with fluorescent molecules, although for many their role is not yet understood. Synthetic chemistry has armed us with an even greater supply of fluorophores, across the full range of the visible spectrum, and ongoing research in the field is developing even brighter systems.

This growing arsenal of fluorophores finds use in many applications: from fluorescent brighteners in papers and washing powders, to markers and stains for forensic application, to the highlighters ubiquitous in school pencil cases. Perhaps no application has been more impactful, however, than the use of fluorescence to better understand biology. To successfully apply fluorescent sensors, it is essential to understand how they work, and we have sought here to provide an overview of fluorescence and fluorescence mechanisms, as well as common classes of fluorophores, in order to provide such understanding. Throughout subsequent chapters of this book, tried, and tested fluorescent sensors will be presented alongside the exciting new systems that push the limits of sensitivity and resolution.

References

1 Owrutsky, J.C., Raftery, D., and Hochstrasser, R.M. (1994). *Annual Review of Physical Chemistry* 45 (1): 519–555.

2 Bixon, M. and Jortner, J. (1968). *The Journal of Chemical Physics* 48 (2): 715–726.

3 Kasha, M. (1950). *Discussions of the Faraday Society* 9: 14.

4 Lewis, G.N. and Kasha, M. (1944). *Journal of the American Chemical Society* 66 (12): 2100–2116.

5 Wouters, F.S., Verveer, P.J., and Bastiaens, P.I.H. (2001). *Trends in Cell Biology* 11 (5): 203–211.

6 Ha, T. and Tinnefeld, P. (2012). *Annual Review of Physical Chemistry* 63 (1): 595–617.

7 Vermes, I., Haanen, C., Steffens-Nakken, H., and Reutellingsperger, C. (1995). *Journal of Immunological Methods* 184 (1): 39–51.

8 Ramakers, C., Ruijter, J.M., Deprez, R.H.L., and Moorman, A.F.M. (2003). *Neuroscience Letters* 339 (1): 62–66.

9 Hong, G., Antaris, A.L., and Dai, H. (2017). *Nature Biomedical Engineering* 1 (1): 0010.

10 Smith, L.M., Sanders, J.Z., Kaiser, R.J. et al. (1986). *Nature* 321 (6071): 674–679.

11 Sekar, R.B. and Periasamy, A. (2003). *Journal of Cell Biology* 160 (5): 629–633.

12 Huang, B., Bates, M., and Zhuang, X. (2009). *Annual Review of Biochemistry* 78 (1): 993–1016.

13 Berezin, M.Y. and Achilefu, S. (2010). *Chemical Reviews* 110 (5): 2641–2684.

14 Sarder, P., Maji, D., and Achilefu, S. (2015). *Bioconjugate Chemistry* 26 (6): 963–974.

15 Szmacinski, H. and Lakowicz, J.R. (1995). *Sensors and Actuators B: Chemical* 29 (1): 16–24.

16 Dempsey, G.T., Vaughan, J.C., Chen, K.H. et al. (2011). *Nature Methods* 8 (12): 1027–1036.

17 Bates, M., Huang, B., Dempsey, G.T., and Zhuang, X. (2007). *Science* 317 (5845): 1749–1753.

18 Huang, B., Babcock, H., and Zhuang, X. (2010). *Cell* 143 (7): 1047–1058.

19 Huang, B., Wang, W., Bates, M., and Zhuang, X. (2008). *Science* 319 (5864): 810–813.

20 Gustafsson, M.G.L. (2005). *Proceedings of the National Academy of Sciences* 102 (37): 13081–13086.

21 Gustafsson, M.G.L. (2000). *Journal of Microscopy* 198 (2): 82–87.

22 Lang, K. and Chin, J.W. (2014). *Chemical Reviews* 114 (9): 4764–4806.

23 Sletten, E.M. and Bertozzi, C.R. (2009). *Angewandte Chemie International Edition* 48 (38): 6974–6998.

24 Shieh, P. and Bertozzi, C.R. (2014). *Organic & Biomolecular Chemistry* 12 (46): 9307–9320.

25 Schneider, A.F.L. and Hackenberger, C.P.R. (2017). *Current Opinion in Biotechnology* 48: 61–68.

26 Giepmans, B.N.G. (2006). *Science* 312 (5771): 217–224.

27 Liu, J. and Cui, Z. (2020). *Bioconjugate Chemistry* 31 (6): 1587–1595.

28 New, E.J. (2016). *ACS Sensors* 1 (4): 328–333.

29 Lee, M.H., Kim, J.S., and Sessler, J.L. (2015). *Chemical Society Reviews* 44 (13): 4185–4191.

30 Shi, L., Yan, C., Guo, Z. et al. (2020). De novo strategy with engineering anti-Kasha/Kasha fluorophores enables reliable ratiometric quantification of biomolecules. *Nature Communications* 11: 793. https://doi.org/10.1038/s41467-020-14615-3.

31 Zhu, X., Su, Q., Feng, W., and Li, F. (2017). *Chemical Society Reviews* 46 (4): 1025–1039.

32 Zhou, J., Liu, Q., Feng, W. et al. (2015). *Chemical Reviews* 115 (1): 395–465.

33 Würth, C., Grabolle, M., Pauli, J. et al. (2013). *Nature Protocols* 8 (8): 1535–1550.

34 Rurack, K. and Spieles, M. (2011). *Analytical Chemistry* 83 (4): 1232–1242.

35 Mayerhöfer, T.G., Pahlow, S., and Popp, J. (2020). *ChemPhysChem* 21 (18): 2029–2046.

36 Mayerhöfer, T.G. and Popp, J. (2019). *ChemPhysChem* 20 (4): 511–515.

37 Lavis, L.D. and Raines, R.T. (2008). *ACS Chemical Biology* 3 (3): 142–155.

38 Bazylevich, A., Patsenker, L.D., and Gellerman, G. (2017). *Dyes and Pigments* 139: 460–472.

39 Mujumdar, R.B., Ernst, L.A., Mujumdar, S.R. et al. (1993). *Bioconjugate Chemistry* 4 (2): 105–111.

40 Komatsu, K., Urano, Y., Kojima, H., and Nagano, T. (2007). *Journal of the American Chemical Society* 129 (44): 13447–13454.

41 Middleton, R.W., Parrick, J., Clarke, E.D., and Wardman, P. (1986). *Journal of Heterocyclic Chemistry* 23 (3): 849–855.

42 Becker, W. (2012). *Journal of Microscopy* 247 (2): 119–136.

43 Zheng, Q. and Lavis, L.D. (2017). *Current Opinion in Chemical Biology* 39: 32–38.

44 Stennett, E.M.S., Ciuba, M.A., and Levitus, M. (2014). *Chemical Society Reviews* 43 (4): 1057–1075.

45 Schweitzer, C. and Schmidt, R. (2003). *Chemical Reviews* 103 (5): 1685–1758.

46 Braeckmans, K., Peeters, L., Sanders, N.N. et al. (2003). *Biophysical Journal* 85 (4): 2240–2252.

47 Ishikawa-Ankerhold, H.C., Ankerhold, R., and Drummen, G.P.C. (2012). *Molecules* 17 (4): 4047–4132.

48 Grimm, J.B., English, B.P., Chen, J. et al. (2015). *Nature Methods* 12 (3): 244–250.

49 Panchuk-Voloshina, N., Haugland, R.P., Bishop-Stewart, J. et al. (1999). *Journal of Histochemistry and Cytochemistry* 47 (9): 1179–1188.

50 Zheng, Q., Juette, M.F., Jockusch, S. et al. (2014). *Chemical Society Reviews* 43 (4): 1044–1056.

51 Mocz, G. (2006). *Journal of Fluorescence* 16 (4): 511–524.

52 Weber, G. (1952). *Biochemical Journal* 51 (2): 145–155.

53 Jameson, D.M. and Ross, J.A. (2010). *Chemical Reviews* 110 (5): 2685–2708.

54 Lakowicz, J.R. (2006). Mechanisms and dynamics of fluorescence quenching. In: *Principles of Fluorescence Spectroscopy*, 331–351. Boston, MA: Springer.

55 Albani, J.R. (2004). Fluorescence quenching. In: *Structure and Dynamics of Macromolecules: Absorption and Fluorescence Studies*, 141–192. Amsterdam: Elsevier Science.

56 Daly, B., Ling, J., and De Silva, A.P. (2015). *Chemical Society Reviews* 44 (13): 4203–4211.

57 De Silva, A.P., Gunaratne, H.Q.N., Gunnlaugsson, T. et al. (1997). *Chemical Reviews* 97 (5): 1515–1566.

58 De Silva, A.P. (2011). *The Journal of Physical Chemistry Letters* 2 (22): 2865–2871.

59 Chi, W., Chen, J., Liu, W. et al. (2020). *Journal of the American Chemical Society* 142 (14): 6777–6785.

60 De Silva, A.P., Moody, T.S., and Wright, G.D. (2009). *The Analyst* 134 (12): 2385.

61 Zachariasse, K.A., Grobys, M., von der Haar, T. et al. (1996). *Journal of Photochemistry and Photobiology A: Chemistry* 102 (1, Supplement 1): 59–70.

62 Zhu, H., Li, M., Hu, J. et al. (2016). *Scientific Reports* 6 (1): 24313.

63 Gong, Y., Guo, X., Wang, S. et al. (2007). *The Journal of Physical Chemistry A* 111 (26): 5806–5812.

64 Mudalige, G., Fedor, S., Travis, G. et al. (2011). *Biophysical Journal* 101 (6): 1522–1528.

65 Reichardt, C. (1994). *Chemical Reviews* 94 (8): 2319–2358.

66 Kucheryavy, P., Li, G., Vyas, S. et al. (2009). *The Journal of Physical Chemistry A* 113 (23): 6453–6461.

67 Grabowski, Z.R., Rotkiewicz, K., and Rettig, W. (2003). *Chemical Reviews* 103 (10): 3899–4032.

68 Rettig, W. and Lapouyade, R. *Fluorescence Probes Based on Twisted Intramolecular Charge Transfer (TICT) States and Other Adiabatic Photoreactions*, 109–149. Kluwer Academic Publishers.

69 Liu, X., Qiao, Q., Tian, W. et al. (2016). *Journal of the American Chemical Society* 138 (22): 6960–6963.

70 Song, X., Johnson, A., and Foley, J. (2008). *Journal of the American Chemical Society* 130 (52): 17652–17653.

71 Jares-Erijman, E.A. and Jovin, T.M. (2003). *Nature Biotechnology* 21 (11): 1387–1395.

72 Piston, D.W. and Kremers, G.-J. (2007). *Trends in Biochemical Sciences* 32 (9): 407–414.

73 Roy, R., Hohng, S., and Ha, T. (2008). *Nature Methods* 5 (6): 507–516.

74 Sahoo, H. (2011). *Journal of Photochemistry and Photobiology C: Photochemistry Reviews* 12 (1): 20–30.

75 Latt, S.A., Cheung, H.T., and Blout, E.R. (1965). *Journal of the American Chemical Society* 87 (5): 995–1003.

76 Myochin, T., Hanaoka, K., Iwaki, S. et al. (2015). *Journal of the American Chemical Society* 137 (14): 4759–4765.

77 Le Reste, L., Hohlbein, J., Gryte, K., and Kapanidis, A.N. (2012). *Biophysical Journal* 102 (11): 2658–2668.

78 Fan, J., Hu, M., Zhan, P., and Peng, X. (2013). *Chemical Society Reviews* 42 (1): 29–43.

79 Gong, Y.-J., Zhang, X.-B., Zhang, C.-C. et al. (2012). *Analytical Chemistry* 84 (24): 10777–10784.

80 Ueno, Y., Jose, J., Loudet, A. et al. (2011). *Journal of the American Chemical Society* 133 (1): 51–55.

81 Barbara, P.F., Walsh, P.K., and Brus, L.E. (1989). *The Journal of Physical Chemistry* 93 (1): 29–34.

82 Henary, M.M. and Fahrni, C.J. (2002). *The Journal of Physical Chemistry A* 106 (21): 5210–5220.

83 Das, K., Sarkar, N., Ghosh, A.K. et al. (1994). *The Journal of Physical Chemistry* 98 (37): 9126–9132.

84 Wu, C.-H., Karas, L.J., Ottosson, H., and Wu, J.I.C. (2019). *Proceedings of the National Academy of Sciences* 116 (41): 20303–20308.

85 Iijima, T., Momotake, A., Shinohara, Y. et al. (2010). *The Journal of Physical Chemistry A* 114 (4): 1603–1609.

86 Tang, K.-C., Chang, M.-J., Lin, T.-Y. et al. (2011). *Journal of the American Chemical Society* 133 (44): 17738–17745.

87 Wu, J., Liu, W., Ge, J. et al. (2011). *Chemical Society Reviews* 40 (7): 3483.

88 Zhao, J., Ji, S., Chen, Y. et al. (2012). *Physical Chemistry Chemical Physics* 14 (25): 8803–8817.

89 Sedgwick, A.C., Wu, L., Han, H.-H. et al. (2018). *Chemical Society Reviews* 47 (23): 8842–8880.

90 Bertman, K.A., Abeywickrama, C.S., Baumann, H.J. et al. (2018). *Journal of Materials Chemistry B* 6 (31): 5050–5058.

91 Luo, J., Xie, Z., Lam, J.W.Y. et al. (2001). *Chemical Communications* 18: 1740–1741.

92 Mei, J., Leung, N.L.C., Kwok, R.T.K. et al. (2015). *Chemical Reviews* 115 (21): 11718–11940.

93 Gao, M. and Tang, B.Z. (2017). *ACS Sensors* 2 (10): 1382–1399.

94 Ding, D., Li, K., Liu, B., and Tang, B.Z. (2013). *Accounts of Chemical Research* 46 (11): 2441–2453.

95 Jiang, G., Zeng, G., Zhu, W. et al. (2017). *Chemical Communications* 53 (32): 4505–4508.

96 Förster, T. (1969). *Angewandte Chemie International Edition in English* 8 (5): 333–343.

97 Callan, J.F., De Silva, A.P., and Magri, D.C. (2005). *Tetrahedron* 61 (36): 8551–8588.

98 Birks, J.B., Dyson, D.J., and Munro, I.H. (1963). *Proceedings of the Royal Society of London. Series A. Mathematical and Physical Sciences* 275 (1363): 575–588.

99 Winnik, F.M. (1993). *Chemical Reviews* 93 (2): 587–614.

100 Wu, Y., Wang, J., Zeng, F. et al. (2016). *ACS Applied Materials & Interfaces* 8 (2): 1511–1519.

101 Han, G., Kim, D., Park, Y. et al. (2015). *Angewandte Chemie International Edition* 54 (13): 3912–3916.

102 Wu, J., Zou, Y., Li, C. et al. (2012). *Journal of the American Chemical Society* 134 (4): 1958–1961.

103 Lavis, L.D. (2017). *Annual Review of Biochemistry* 86 (1): 825–843.

104 Baeyer, A. (1871). *Berichte der Deutschen Chemischen Gesellschaft* 4 (2): 555–558.

105 Klonis, N. and Sawyer, W.H. (1996). *Journal of Fluorescence* 6 (3): 147–157.

106 Sjöback, R., Nygren, J., and Kubista, M. (1995). *Spectrochimica Acta Part A: Molecular and Biomolecular Spectroscopy* 51 (6): L7–L21.

107 Orndorff, W.R. and Hemmer, A.J. (1927). *Journal of the American Chemical Society* 49 (5): 1272–1280.

108 Zheng, H., Zhan, X.-Q., Bian, Q.-N., and Zhang, X.-J. (2013). *Chemical Communications* 49 (5): 429–447.

109 Fleming, G.R., Knight, A.W.E., Morris, J.M. et al. (1977). *Journal of the American Chemical Society* 99 (13): 4306–4311.

110 Neckers, D.C. (1989). *Journal of Photochemistry and Photobiology A: Chemistry* 47 (1): 1–29.

111 Egawa, T., Koide, Y., Hanaoka, K. et al. (2011). *Chemical Communications* 47 (14): 4162.

112 Grimm, J.B., Sung, A.J., Legant, W.R. et al. (2013). *ACS Chemical Biology* 8 (6): 1303–1310.

113 Song, L., Hennink, E.J., Young, I.T., and Tanke, H.J. (1995). *Biophysical Journal* 68 (6): 2588–2600.

114 Urano, Y., Kamiya, M., Kanda, K. et al. (2005). *Journal of the American Chemical Society* 127 (13): 4888–4894.

115 Ueno, T., Urano, Y., Setsukinai, K.-I. et al. (2004). *Journal of the American Chemical Society* 126 (43): 14079–14085.

116 Lavis, L.D. and Raines, R.T. (2014). *ACS Chemical Biology* 9 (4): 855–866.

117 McKay, I.C., Forman, D., and White, R.G. (1981). *Immunology* 43 (3): 591–602.

118 Karstens, T. and Kobs, K. (1980). *The Journal of Physical Chemistry* 84 (14): 1871–1872.

119 Kubin, R.F. and Fletcher, A.N. (1982). *Journal of Luminescence* 27 (4): 455–462.

120 Chen, X., Pradhan, T., Wang, F. et al. (2012). *Chemical Reviews* 112 (3): 1910–1956.

121 Whitaker, J.E., Haugland, R.P., Ryan, D. et al. (1992). *Analytical Biochemistry* 207 (2): 267–279.

122 Beija, M., Afonso, C.A.M., and Martinho, J.M.G. (2009). *Chemical Society Reviews* 38 (8): 2410.

123 Nguyen, T. and Francis, M.B. (2003). *Organic Letters* 5 (18): 3245–3248.

124 Grimm, J.B., Brown, T.A., Tkachuk, A.N., and Lavis, L.D. (2017). *ACS Central Science* 3 (9): 975–985.

125 Fu, M., Xiao, Y., Qian, X. et al. (2008). *Chemical Communications* (15): 1780–1782. https://pubs.rsc.org/en/content/articlelanding/2008/cc/b718544h.

126 Koide, Y., Urano, Y., Hanaoka, K. et al. (2011). *ACS Chemical Biology* 6 (6): 600–608.

127 Wang, L., Du, W., Hu, Z. et al. (2019). *Angewandte Chemie International Edition* 58 (40): 14026–14043.

128 Borges, F., Roleira, F., Milhazes, N. et al. (2005). *Current Medicinal Chemistry* 12 (8): 887–916.

129 Cao, D., Liu, Z., Verwilst, P. et al. (2019). *Chemical Reviews* 119 (18): 10403–10519.

130 Marini, A., MuñOz-Losa, A., Biancardi, A., and Mennucci, B. (2010). *The Journal of Physical Chemistry B* 114 (51): 17128–17135.

131 Fu, Y. and Finney, N.S. (2018). *RSC Advances* 8 (51): 29051–29061.

132 Matikonda, S.S., Ivanic, J., Gomez, M. et al. (2020). *Chemical Science* 11 (28): 7302–7307.

133 Salem, M.A., Helal, M.H., Gouda, M.A. et al. (2018). *Synthetic Communications* 48 (13): 1534–1550.

134 Banerjee, S., Veale, E.B., Phelan, C.M. et al. (2013). *Chemical Society Reviews* 42 (4): 1601.

135 Duke, R.M., Veale, E.B., Pfeffer, F.M. et al. (2010). *Chemical Society Reviews* 39 (10): 3936.

136 Kolosov, D., Adamovich, V., Djurovich, P. et al. (2002). *Journal of the American Chemical Society* 124 (33): 9945–9954.

137 Aveline, B.M., Matsugo, S., and Redmond, R.W. (1997). *Journal of the American Chemical Society* 119 (49): 11785–11795.

138 Saha, S. and Samanta, A. (2002). *The Journal of Physical Chemistry A* 106 (18): 4763–4771.

139 Leslie, K.G., Jacquemin, D., New, E.J., and Jolliffe, K.A. (2018). *Chemistry – A European Journal* 24 (21): 5569–5573.

140 Rudebeck, E.E., Cox, R.P., Bell, T.D.M. et al. (2020). *Chemical Communications* 56 (50): 6866–6869.

141 Loudet, A. and Burgess, K. (2007). *Chemical Reviews* 107 (11): 4891–4932.

142 Karolin, J., Johansson, L.B.A., Strandberg, L., and Ny, T. (1994). *Journal of the American Chemical Society* 116 (17): 7801–7806.

143 Boens, N., Leen, V., and Dehaen, W. (2012). *Chemical Society Reviews* 41 (3): 1130–1172.

144 Ulrich, G., Ziessel, R., and Harriman, A. (2008). *Angewandte Chemie International Edition* 47 (7): 1184–1201.

145 Kowada, T., Maeda, H., and Kikuchi, K. (2015). *Chemical Society Reviews* 44 (14): 4953–4972.

146 Umezawa, K., Nakamura, Y., Makino, H. et al. (2008). *Journal of the American Chemical Society* 130 (5): 1550–1551.

147 Lu, H., Mack, J., Yang, Y., and Shen, Z. (2014). *Chemical Society Reviews* 43 (13): 4778–4823.

148 Prlj, A., Vannay, L., and Corminboeuf, C. (2017). *Helvetica Chimica Acta* 100 (6): e1700093.

149 Levitus, M. and Ranjit, S. (2011). *Quarterly Reviews of Biophysics* 44 (1): 123–151.

150 Gorka, A.P., Nani, R.R., and Schnermann, M.J. (2018). *Accounts of Chemical Research* 51 (12): 3226–3235.

151 Bricks, J.L., Kachkovskii, A.D., Slominskii, Y.L. et al. (2015). *Dyes and Pigments* 121: 238–255.

152 Wessendorf, M.W. and Brelje, T.C. (1992). *Histochemistry* 98 (2): 81–85.

153 Peng, X., Song, F., Lu, E. et al. (2005). *Journal of the American Chemical Society* 127 (12): 4170–4171.

154 Sissa, C., Painelli, A., Terenziani, F. et al. (2020). *Physical Chemistry Chemical Physics* 22 (1): 129–135.

155 Benson, R.C. and Kues, H.A. (1977). *Journal of Chemical & Engineering Data* 22 (4): 379–383.

156 Sanchez-Galvez, A., Hunt, P., Robb, M.A. et al. (2000). *Journal of the American Chemical Society* 122 (12): 2911–2924.

2

The Applications of Responsive Fluorescent Sensors to Biological Systems

Jia Hao Yeo[1] and Elizabeth J. New[1,2,3]

[1] School of Chemistry, The University of Sydney, NSW, Australia
[2] Australian Research Council Centre of Excellence for Innovations in Peptide and Protein Science, The University of Sydney, NSW, Australia
[3] The University of Sydney Nano Institute (Sydney Nano), The University of Sydney, NSW, Australia

Fluorescent sensors have found widespread use throughout research and clinical laboratories. New innovations in sensors and imaging techniques are enabling the elucidation of new biochemical processes and understanding of how exogenous species interact with cells. In fact, fluorescent sensors are so ubiquitous in biomedical studies that the most common systems are provided to the end-user in the form of kits, with detailed instructions. However, it is important that all researchers – those who make new sensors and those who apply them – have a good understanding of the biological interactions of these sensors. This chapter aims to provide a general overview of biological factors that must be considered when applying fluorescent sensors to ensure that meaningful and robust data is collected.

Fluorophores are compatible with a range of techniques that provide spatial and temporal information about the chemistry of the cell. For example, state-of-the-art imaging systems such as super-resolution microscopes and advanced confocal techniques (*e.g.* Airy-scans and multiphoton excitation) provide unprecedented contrast in studying live cells. Flow cytometry complements these imaging systems by enabling the high-throughput study of cells, while fluorescence plate-reader techniques are rapid and simple assessments for biological experiments. This chapter discusses the criteria required for the biological application of responsive fluorescent sensors, outlining the instrumentation that can be applied to their study and the types of samples that are appropriate for analysis. It also identifies common challenges and misconceptions in the use of such sensors for biological study.

2.1 Criteria for Biologically Relevant Fluorescent Sensors

Despite an abundance of responsive fluorescent sensors reported in the literature, not all sensors are compatible with live cells. To be biologically compatible, sensors must fulfil the following requirements:

Molecular Fluorescent Sensors for Cellular Studies, First Edition. Edited by Elizabeth J. New.
© 2022 John Wiley & Sons Ltd. Published 2022 by John Wiley & Sons Ltd.

1) Adequate response in aqueous competitive media, requiring:

a) *Sufficient water solubility*

Water is the most abundant molecule in cells, and essential for life. It is crucial that sensors are sufficiently soluble and functional in aqueous solution. Dissolving sensors in high volumes of non-aqueous solvents may perturb cellular physiology, resulting in a less-accurate representation of the native cell. However, some water-soluble dyes interact with lipid bilayers (even when dyes are charged), and hence any data should be interpreted with caution [1].

b) *Sufficient fluorescence emission in biologically relevant buffers*

Biological buffers like phosphate-buffered saline (PBS) have been optimised to keep cells osmotically intact while mimicking homeostatic physiology. To be able to observe environmental changes, 'turn-on' and 'always on' sensors must be sufficiently fluorescent in such buffers. Ideally, sensors should also be also bright enough in low concentrations to avoid toxicity issues and/or induce unwarranted cellular effects in these buffers while still enabling distinction between signal and background (see Section 2.2.1.2). Many organic small-molecule sensors have reduced fluorescence aqueous environments, and it is therefore essential to carry out *in vitro* studies in aqueous media.

c) *Response maintained in the presence of potential interferents*

The cell contains many small molecules and macromolecules that can act as potential interferents to sensor response. These interfering biomolecules are often in higher concentrations than both the analyte of interest and the fluorescent sensor. Hence, one should test the fluorescence response of the sensor to these biomolecules to confirm that there are no fluorescent artefacts in the collected images. Furthermore, it is important to test that the analyte of interest can still be sensed in the presence of these interferents, and biologically relevant concentrations.

In addition to native cellular biomolecules, some protocols require additional treatment such as fixation, additional blocking steps with serum or application of substrate for better cell-surface adherence. These steps can be very damaging to cells and could affect the response of the sensor. Therefore, these chemicals are potential interferents for fluorescence experiments. See Section 2.5 for more details.

2) Good cellular uptake

While passive diffusion is thought to be the main mechanism of cell permeation, active transport and endocytotic pathways may also be responsible for sensor uptake in cells [2]. Although the uptake of intracellular sensors might vary between cell types, good intracellular sensors should ideally be able to rapidly penetrate cells. Passive diffusion of sensors is often preferable to active as there is less need for washing, and uptake will be independent of temperature. Larger molecular sensors may require permeabilisation such as the use of detergent (*e.g.* Triton-X or Tween): such protocols are not ideal for intracellular sensing applications.

3) Minimal perturbation of biological systems

It is crucial that sensors have minimal effect on cells, such that the biological effect observed is not an artefact of the sensor itself. The sensor and the solvent in which it is administered should not cause any cell death. Nuclear-labelling live-cell dyes are infamous for being cytotoxic and influences on cellular behaviour, due to the intercalating interaction of these dyes to the DNA. Hence, prolonged imaging experiments with nuclear-labelling dyes are not feasible [3]. Sensors that require UV wavelengths for excitation can also cause cellular death and damage.

4) Suitable cellular/organelle localisation

Eukaryotic cells contain compartmentalised organelles with distinct chemical environments. The ability to target specific cell types, organelles, or sub-organelle structures is important to understand cellular behaviour and its molecular processes such as cell–cell interaction, intracellular trafficking, or mitochondria biology. A good sensor should be highly selective for a specific sub-cellular location. Strategies for sensor localisation to common organelles are discussed in Chapter 3.

5) Suitable photophysical behaviour to match available instrumentation

The selection of a suitable fluorescent sensor requires matching of the excitation and emission wavelengths with the available instrumentation. Most conventional instruments cater for commonly available commercial sensors, providing optimal excitation wavelengths, and detectors with appropriate filters for sensors, such as fluorescein (green) and rhodamine (red). Although there are many promising sensors on the market and in the literature, the challenge in applying these sensors often lies in finding suitable instrumentation for their application.

For non-commercialised or specialised sensors, users would need to source appropriate instruments to match excitation/emission wavelengths and ensure that spectral crosstalk is minimised. The following sections summarise the main types of instruments that are used for the study of fluorescent sensors.

2.2 Microscopy for Visualising Fluorescent Sensors

Advances in biology and medicine throughout history have been underpinned by advances in microscopy, from the development of the first microscope to the latest cutting-edge techniques. This section outlines the key microscopy techniques that enable the study of fluorescent sensors, beginning with general factors and considerations, which affect microscopy results.

2.2.1 Important Considerations in Microscopy

2.2.1.1 Resolution in Microscopy

One critical limitation of imaging techniques is resolution. Simply, resolution refers to the ability to differentiate two objects. This (lateral) resolution, also known as the Abbe diffraction limit, was also mathematically defined by Abbe as

$$d = \frac{\lambda}{2NA}$$

Where d is Abbe's diffraction limit, λ is the wavelength of light, and NA is the numerical aperture.

The numerical aperture of a microscope is the angular measure for incoming light. The numerical aperture is about 1.4–1.6 using objective lenses, so the approximate Abbe limit is $d = \lambda/2.8$. For green light (approximately 500 nm), this corresponds to approximately 180 nm, less than the size of most biological cells (approximately 10–100 μm). However, viruses (100 nm), proteins (10 nm), and nanoparticles (1 nm) require shorter wavelengths to obtain sufficient resolution, such as those achieved by UV and X-ray microscopes. As tissues and cells are three-dimensional, two resolutions must be considered: lateral (xy-resolution) and axial (z-resolution). Lateral resolution is directly proportional to the numerical aperture of the objective, while axial resolution is approximately proportional to the square of the objective's numerical aperture.

Traditional methods of achieving high resolution have included the use of dyes that provide contrast, and application of non-conventional optical components such as annular devices and polarisers to bend light. The latter strategy evolved into modules that we now know as differential interference contrast (DIC), dark field, and phase contrast.

When an image is taken digitally, the image captured is divided into multiple squares, known as pixels. Therefore, a digital image is made up of numerous pixels. The number of pixels taken by a camera is normally shown in the x and y dimensions, with a higher pixel count correlating to a greater resolution (Figure 2.1). The term spatial resolution is used to describe the number of pixels in constructing and rendering a digital image. The number of pixels that represent cellular structures must satisfy the Nyquist Sampling Criterion. This means that the pixel size of an image needs to be at least 2.3 times smaller than the cellular structure being resolved in both xy-lateral and z-axial dimensions.

While magnification from the objective lens can increase the resolution of a cell image (by increasing the numerical aperture), the term 'magnification' from the objective lens should not be confused with 'digital zoom', which can affect μm^3/pixel resolution.

In the context of live cell imaging or videography, temporal resolution refers to the time between each image acquired. This is crucial for visualising rapid biomolecular events and measuring biochemical kinetics, such as calcium fluxes or action potentials in neurons. To demonstrate rapid biological processes, one would often record the entire process via a video. The time-lapse resolution must also meet the Nyquist Sampling Criterion; the time interval in a time-lapse stack should be 2.3 times smaller than the time required for the smallest resolved object to travel in its own diameter.

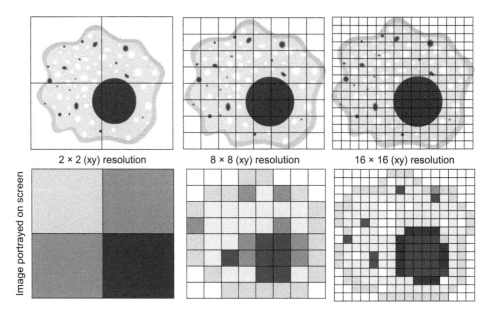

Figure 2.1 Representation of a cell by the number of pixels defines resolution in digital imaging. Illustration of number of pixels representing a cell. The more pixels to represent a cell, the higher the resolution to distinguish between two objects in a cell.

2.2.1.2 Understanding the Competition Between True Signal and Noise

Distinguishing true signal and noise is crucial for achieving optimum imaging performance and quantitative fluorescence measurements [4]. The precision of quantitative intensity and spatial fluorescence measurements is limited mainly by the signal-to-noise ratio (SNR) (Figure 2.2). To distinguish signal from noise, the signal must be significantly higher than the noise level on a digital image. As the signal increases relative to the noise level, measurements of the signal become increasingly more precise. SNR is also be determined by factors such as the brightness of the sensor and instrumentation [5].

In order to make quantitative measurements, true signals should solely be from the fluorophore(s) under study, which will give an intensity value for every representative pixel (Figure 2.1). However, intensity values are also influenced by background and noise. Background adds to the true signal and can arise from a variety of sources, such as endogenous chemicals (autofluorescence), substrates on which the specimen is mounted [6] and out-of-focus illumination. The presence of background reduces SNR and hence should be subtracted from microscopy images during analysis [4].

In contrast to background, noise causes variance in the intensity values and cannot be eliminated. Unlike background, noise is not a mathematical constant and hence cannot be subtracted from a digital image. Digital fluorescence microscopy images are affected by Poisson noise (statistical uncertainty associated with photon measurements), thermal noise (caused by the generation of heat within the detector), and detector read noise (generated by the electronics) [4]. Frame averaging may reduce noise. By increasing laser power, or detector gain on the instrument, the fluorescence intensity is increased, but the SNR can remain low with higher background or noise generated (Figure 2.2, blue labels).

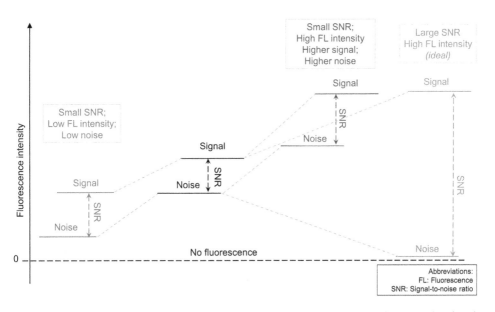

Figure 2.2 Signal-to-noise ratio (SNR) Schematic explaining the relationship between signal and noise. Ideally, the signal and noise are separated and highly distinctive from each other. However, improper use and adjustments of microscopy settings often result in a small SNR.

One way to increase SNR is to increase the sensor incubation concentration or time to ensure that more sensor accumulates within the cell. However, even with a higher overall fluorescence, the SNR can remain unchanged, if the excess concentration of sensors starts binding to the glass/substrate, resulting in higher background noise. Removing excess sensor *via* washing can reduce background but can simultaneously reduce true signal by altering sensor distribution equilibrium between intracellular and extracellular environments. For more in-depth understanding on the mathematical aspects of signal-to-noise ratio, the authors refer the readers to the excellent chapter by Sheppard and colleagues [7].

2.2.1.3 Phototoxicity in Cells

Fluorescent cells in a dark incubator can be maintained in the dark for weeks. However, under light illumination, samples can be subjected to phototoxicity. Furthermore, the fluorescent marker itself might be photobleached by illumination.

While phototoxicity and photobleaching generally go hand-in-hand, these are two distinct phenomena and can occur alone [8]. The process of photobleaching arises during the imaging process when a fraction of excited fluorescent molecules enter the triplet excited state, losing the ability to emit energy as fluorescence. This process can result in the production of singlet oxygen species ($^1O_2$) [9], which in turn will damage proteins, lipids, and other biomolecules, inducing phototoxicity [10]. Phototoxicity in fluorescence microscopy can also originate from natural cellular components. Endogenous cellular biomolecules, such as flavins and porphyrins, can generate damaging ROS upon irradiation.

Phototoxicity (and by extension, photobleaching) depends on several variables [7]:

1) *Photochemical properties of the fluorophore.* The relative lifetimes and energy levels of singlet and triplet excited states will determine the phototoxicity of the fluorophore.
2) *Subcellular location of the fluorescent molecule.* Photodamage is limited to the small diffusional range of $^1O_2$ [11]. With its short lifetime (<0.5 ms) in cells, $^1O_2$ only has a radius of action of <50 nm [7]. Fluorophores in the nucleus and mitochondria also tend to induce greater toxicity than those located elsewhere, because these organelles contain components that are most sensitive to oxidative stress.
3) *Fluorophore concentration.* The higher the local concentration of fluorophore, the higher the level of phototoxicity that can be induced.
4) *Excitation intensity.* Numerous studies associate phototoxicity to the excitation of sensors [8, 9]. Repeated excitation of fluorophores (usually ≥10 repeated scans) leads to photobleaching and phototoxicity.
5) *Cell culture media.* Standard culture media components can also interfere with phototoxicity. Some early studies indicated a phototoxic effect from inadequate bicarbonates in 4-(2-hydroxyethyl)-1-piperazineethanesulfonic acid (HEPES)-containing media. Other chemicals such as riboflavin/vitamin B2, phenol-red, and tryptophan may also contribute to phototoxicity [7].

A reduction in the levels of photobleaching can be achieved by decreasing excitation time, intensity of light sources, and acquisition time. However, these compromises reduce signal and can also lead to greater noise (Figure 2.2, orange labels). One rapid imaging technique that can overcome phototoxicity is spinning disk confocal microscopy [12].

2.2.2 Common Microscopy Techniques

2.2.2.1 Fluorescence Microscopy

At the beginning of the twentieth century, Ellinger and Hirt devised 'intravital microscopes' by treating living organisms with fluorescent substances. The pair used UV-light for illumination and introduced filters between objective and eyepiece, which reflected the exciting rays and transmitted the red-shifted fluorescent light. This approach is still similar in principle to that of a modern fluorescence microscopy.

Traditionally, there are two modes of fluorescence microscopy: epi-fluorescence and trans-fluorescence. The difference lies in the position of the illumination source. In trans-fluorescence, illumination is in the direct path of visualisation. To date, there are limited trans-fluorescence microscopes available. In contrast, the illumination is often not in the direct path of visualisation in epifluorescence microscopy. This is achieved by dichroic mirrors and filters that reflect excitation light and/or emitted fluorescence from the sample (Figure 2.3). Modifications were made to widefield epi-fluorescence (*e.g.* illumination source, number of filters, introduction of pinholes, sensitive detectors) to make the modern fluorescent microscopes we know today.

In a widefield epi-fluorescence microscope, a light source is used to illuminate the sample and excite the fluorophore. The excitation filter limits the wavelengths that are reflected by the dichroic mirror and pass through the objective lens to illuminate the specimen. Emitted light from the specimen is detected, *via* the objective lens, by the detector/camera, after passing through a dichroic mirror and emission filter.

The main components of epi-fluorescence microscope are:

- *Illumination source*: Various illumination sources are used for fluorescence microscopy. Mercury lamps provide bright intensity light across the visible spectrum, but they can cause UV damage, generates enormous heat, and have limited lifespans. Cheap halogen lamps are long-lived, cause no UV damage and require no bulb alignments, but are not as bright as mercury lamps. Metal halide lamps are relative stable light sources with uniform

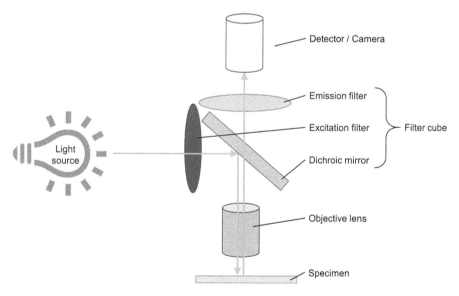

Figure 2.3 Classic set-up of a simple widefield epi-fluorescence microscopy.

emission, but are expensive and have a UV emission component. Expensive LEDs are long-lasting (up to 10 000 hours), generate negligible heat and no UV damage, with discrete excitation bands.

- *Filters and mirrors:* The excitation filter limits the wavelengths of incoming light that will excite the fluorophore-containing sample. The dichroic beam-splitter/mirror bends the excitation light to the sample and transmits only the emitted light from the sample back to the detector. The emission filter enables detection of the emitted fluorescence from the sample. The entire filter system (excitation filter, dichromatic mirror, and barrier filter) is usually contained in one unit called the filter block (Figure 2.3).
- *Detectors and cameras:* Microscope cameras contain silicon arrays of photo-sensitive elements that detect photons.

In selecting fluorescent sensors for epi-fluorescence microscopy, it is important to consider the available filter blocks. These are generally limited to the excitation and emission wavelengths of the most common commercial fluorophores (such as DAPI, FITC, and TRITC) and so may not be appropriate for the bespoke fluorescent sensors described throughout this book. Filter blocks can be readily modified using relatively inexpensive filters available from optics suppliers.

2.2.2.2 Confocal Microscopy

Confocal microscopy is an optical imaging technique that removes out-of-focus light by introducing pinholes at both the illumination source and at the detector. This spatial filtering technique allows 'true' signal to be detected, increasing resolution. Confocal microscopy offers several distinct advantages over traditional epi-fluorescence microscopy, including selective illumination of a specific field depth, elimination of background that is not in focus, and optical sectioning. Optical sectioning refers to imaging at different depths of a sample, allowing the reconstruction of three-dimensional structures within an object.

Modern confocal microscopy relies on a confocal laser scanning method. A confocal laser scanning microscope focuses the illumination beam to a specimen and detects the photons that arise from the immediate region being excited by the focused laser bean. This can be achieved by galvanometric or resonance scanners [13]. This is repeated in a raster scan method, throughout the field of view, forming an image. Because the photon multiplier tube collects the number of photons, the fluorescence is not the true fluorescence emitted from the sample. Other ways to remove 'stray light' to increase resolution include multiphoton excitation, the merging of blinking events such as ground-state depletion super-resolution microscopy, and application of mathematical algorithms.

One disadvantage of confocal microscopy is the harmful nature of high-intensity laser irradiation to living cells and tissues. Confocal microscopy generally uses argon or helium-neon lasers to provide sufficient light with a specific excitation wavelength. In addition to the safety issues inherent with the use of lasers, these lasers can induce phototoxicity. Finally, the high cost of purchasing and operating multi-user confocal microscope systems, which can range up to an order of magnitude higher than comparable widefield microscopes, often limits their implementation in smaller laboratories.

The selection of fluorescent sensors for use with a particular confocal microscope is also limited by the laser excitation sources available. Commercial laser sources tend to be those that match commonly used fluorophores (*e.g.* 488 nm for fluorescein; 561 nm for Texas Red), and these will not be appropriate for all fluorescent sensors. Because of the phototoxicity associated with shorter wavelength lasers, many set-ups are not equipped with a blue laser necessary for the excitation of many fluorescent sensors.

2.2.2.3 Multiphoton Microscopy

Two-photon excitation of a fluorophore describes the simultaneous absorption of two photons: each about twice the single-photon excitation wavelength, corresponding to half the energy. The use of longer wavelengths in biology has several advantages. Photodamage and phototoxicity caused by such wavelengths is negligible compared to short, ultraviolet wavelengths. There are also far fewer endogenous molecules that absorb photons or fluoresce in the NIR wavelengths. The diminished light scattering of NIR wavelengths in biological samples also allows deeper penetration into the specimen. Therefore, multi-photon microscopy is a popular choice for imaging of living, intact biological tissues. While high-order multiphoton microscopy is possible, two-photon systems predominate, so the terms 'multiphoton microscopy' and 'two-photon microscopy' tend to be used interchangeably.

The two photons collide in a femtolitre-magnitude volume, exciting the fluorophores only in that volume of the sample. The fluorophores outside this region are not excited. Therefore, unlike confocal microscopy, there is no need for an aperture pinhole to remove the out-of-focus light. As a result, photobleaching is restricted to the plane-of-focus.

The fluorescence emission from two-photon excitation is proportional to the square of the excitation intensity in two-photon absorption. However, the high-excitation intensities of multiphoton excitation from the laser could heat up the samples. Therefore, short-pulsed femtosecond lasers with high repetition rates (typically ~80 MHz) are commonly used. Because of the pulsing and wavelength-tuneable lasers required, multiphoton microscopy can be used in conjunction with other microscopy techniques such as fluorescence lifetime imaging and fluorescence correlation spectroscopy [14]. However, wavelength-tuneable lasers also make switching excitation wavelengths for fluorophores less straightforward than in conventional confocal microscopy.

A great advantage of multiphoton microscopy is that it enables lower energy excitation of sensors that require ultraviolet excitation. While multiphoton microscopy is a popular choice for imaging live thicker biological specimens, progress is hindered by the lack of efficient two-photon excitable sensors. Two-photon excitable sensors should have good multiphoton action cross-sections – a measure of a molecule to absorb two photons simultaneously with other determinants such as quantum yield and blue-shifting of the sensors [15–17]. However, this parameter is not often reported for sensors on the market and in the literature, and a newly acquired sensor may therefore need to be first screened for suitability for multiphoton microscopy.

2.2.2.4 Fluorescence Lifetime Imaging Microscopy

Intensity-based fluorescence imaging provides information on the spatial distribution of fluorophores and can discriminate between fluorophores with distinct spectral properties [18, 19]. In contrast to intensity-based fluorescence microscopy, fluorescence lifetime imaging microscopy (FLIM) can be used to discriminate spectrally similar fluorophores and detect changes in the molecular environment of fluorophores. The fluorescence lifetime of a molecule describes the decay time from the S_1 excited state to the S_0 ground state. FLIM measures fluorescence decay curves for each pixel by using a pulsed excitation source. The

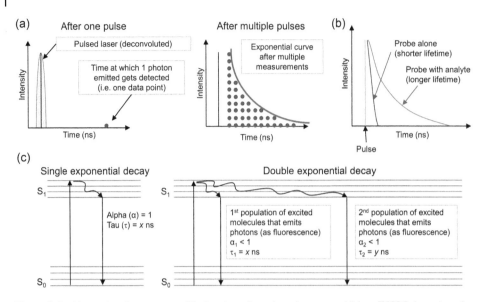

Figure 2.4 Measuring fluorescence lifetime in a time-domain system. (a) In a TCSPC time-domain measurement of fluorescence lifetime/decay, a pulsed laser will excite the fluorophore and the time at which the emitted photon reaches the detector is measured. Repetition of this process gives an exponential curve that can be fitted to determine average fluorescence lifetime. (b) Fluorescent sensors can exhibit changes in fluorescence lifetime in the presence of the analyte of interest. (c) In single exponential decays, all molecules have a specific τ value/fluorescence lifetime. In multiple exponential decays, more than one τ value can be calculated.

collected decay curves can be fitted to the expected number of decay processes (typically two) corresponding to the sensor fluorescence lifetime in the presence and absence of the analyte of interest (Figure 2.4). Similar to lifetime spectrometers, standard confocal FLIM microscopes use time-correlated single-photon counting (TCSPC) method for measuring fluorescence lifetime. Exponential curves are measured at each pixel, and lifetime maps can therefore be built for the area being imaged (Figure 2.5).

The fluorescence lifetime of a molecule will depend on the mobility of the molecule in its excited state, which is influenced by factors such as temperature, aromaticity of the molecule, solubility, and viscosity of solvent. In the biological context, physical factors such as temperature and solvent viscosity play a less significant role. Multiple exponentials are common in FLIM; there are other organic molecules including free/bound NADH, FAD, and riboflavins, which all have fluorescence lifetimes, and if the excitation wavelength of the fluorophore of interest overlaps with these biomolecules, they will contribute to the decay [18, 20]. FLIM images with multiple-exponential decays are often presented on phasor plots to illustrate complexities of pixels with the different lifetimes and ratios [21].

Modern FLIM microscopes measure the time (nanoseconds) after excitation at which the photon reaches the detector. A laser beam is pulsed over the image pixel by pixel, and signal is recorded with a point detector, such as a PMT. This pulsing is performed multiple times, and an average tau value or lifetime is estimated based on fitting onto exponential histograms. This will be performed across all pixels.

FLIM is independent of sample thickness, light source noise, photobleaching, and excitation intensity, making it an attractive imaging technique. However, photobleaching may generate new fluorescing photoproducts, which may introduce new decay components and

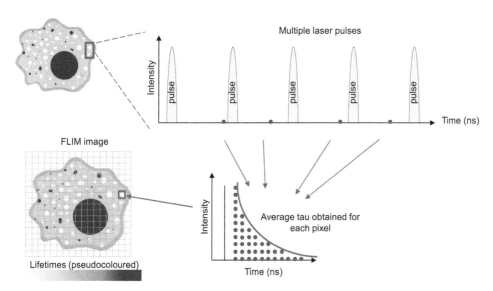

Figure 2.5 FLIM measurement is a pixel-by-pixel measurement of fluorescence lifetime.

affect the decay data quality. The current consensus is that FLIM is not affected by the concentration of fluorophores. However, high concentrations of fluorophores may affect the fluorescence lifetime due to π–π stacking, which is common in molecules, such as pyrene-based sensors [18].

One of the disadvantages of the technique is the interpretation required to analyse FLIM data. A change in τ or α values, or additional components (*i.e.* having higher exponentials), could be a result of other autofluorescent molecules in cells rather than the sensor. Another disadvantage is the time required for acquiring data to obtain sufficient photon counts for curve fitting. This is a considerable handicap in time-limiting experiments with multiple treatment conditions and is exacerbated when multiple fields-of-view (as replicates) are required for each dosing condition.

Despite its many advantages, FLIM is not commonly used for the study of fluorescent sensors. This is largely due to the fact that few sensors are explicitly presented as lifetime-based sensors, although it is likely that many of the sensing systems presented throughout this book do indeed undergo changes in fluorescence lifetime upon analyte interaction. More recent examples of sensors designed specifically for FLIM application are outlined in Chapter 8. At present, FLIM is most commonly used to determine Förster resonance energy transfer (FRET) between two fluorophores, most often fluorescent proteins, in order to determine the interaction of two molecules within a biological system [21].

2.2.2.5 Other Advanced Microscopy Techniques

2.2.2.5.1 Non-super resolution Microscopy Techniques
This subsection aims to provide a brief introduction to popular microscopy techniques, which are trending in current research. Often, these microscopic techniques require special equipment or accessories.

Fluorescence recovery after photobleaching (FRAP) is a method for determining the kinetics of diffusion of a fluorophore through tissue or cells. FRAP is often performed by photobleaching a region of interest. Reverse FRAP is a modification of this technique, whereby regions other than the region of interest are photobleached. These techniques allow tracking of a fluorophore and hence determination of the kinetics of the labelled biomolecule of interest [22].

Total internal reflection fluorescence (TIRF) microscopy allows selective excitation of fluorophores just above the sample surface. Thus, TIRF enables the visualisation of dynamic, molecular events that take place close to the cell surface. These events include biophysical studies of surface molecule on membranes, convergence points of cytoskeletal filaments, and cellular mobility. This technique works through the reflection and transmission of light as it travels through different media. In TIRF microscopy, light is presented to the slide-sample interface through the objective or using prisms such that fluorophores are excited only at the interface [23, 24].

Tomographic phase microscopy is a trending microscopy technique and currently marketed as NanoLive. Using a combination of digital holography, rotational scanning with a prism wedge, and algorithms, tomographic phase microscopy is a no-label system for monitoring live cells [25, 26]. Currently, this modality can be used in conjunction with fluorophores for molecular labelling.

2.2.2.5.2 Super-resolution Microscopy
Super-resolution microscopy is gaining popularity due to its superior resolving capability compared to fluorescence microscopy. Super-resolution microscopy surpasses the diffraction limit of light, providing extra resolving power [27]. Most fluorescence microscopes have a limiting xy-lateral resolution of approximately 250 nm and a z-axial resolution limit of 600 nm [24], so the resolution size of the fluorescently labelled biomolecule is limited to these dimensions. Specific super-resolution techniques are summarised below, and details of sensors for super-resolution microscopy can be found in Chapter 8.

Structured illumination microscopy (SIM) excites fluorophores in samples using a pre-determined illumination pattern (usually in stripes). The stripes' position and orientation will be changed at various times, and emitted fluorescence at these positions and orientation are acquired. The resulting images can be reconstructed into 2D or 3D SIM images using algorithms [24, 28]. The resolution is approximately 100 and 250 nm in the xy-lateral and z-axial planes, respectively.

Stochastic optical reconstruction microscopy (STORM) [29] and *photoactivation localisation microscopy (PALM)* [30] use the switching of individual fluorophores. In each imaging cycle, only a fraction of the fluorophores are excited, allowing their positions to be determined. The fluorophore positions obtained from a series of imaging cycles (also known as "blinking events") can then be reconstructed into images. The reconstructed image has a resolution of approximately 20 and 50 nm in xy-lateral and z-axial planes [28–30]. This resolution is highly dependent on the fluorophore used and the labelling density in the sample. Currently, the 'blinking' properties of fluorophores are poorly understood. Some fluorophores that have been successfully used in photo-switching imaging experiments include rhodamines, caged fluoresceins, and cyanines [27].

Stimulated emission depletion (STED) microscopy is based on the principle of the confocal laser-scanning microscopy. A STED microscope is equipped with a second STED laser, with a doughnut-shaped pulse at longer wavelength. During image acquisition, the normal excitation laser pulse is closely followed by the STED beam. Excited fluorophores exposed to the STED beam are immediately 'bleached' back to the ground state. Therefore, only molecules in the centre of the STED beam (the 'hole in the doughnut') emits fluorescence. While the technique improved resolution in the lateral and axial planes to approximately 50 and 100 nm respectively, fluorophores suffer heavily from photobleaching and phototoxicity [24, 28].

2.3 Other Instrumental Techniques for Studying Cells Treated with Fluorescent Sensors

2.3.1 Flow Cytometry

Flow cytometry is another tool used to measure and quantitate fluorescence signals from biological samples. While microscopy gives information about the localisation of fluorescence sensors, quantitative analyses of fluorescence in a large population of cells is challenging. To obtain statistically robust data, flow cytometry is an indispensable tool in rapid analyses of large number of suspension cells. Flow cytometry enables high-throughput screening of cells (up to 15 000 cells per second in most modern systems) with high statistical robustness. A conventional benchtop flow cytometer can measure up to six parameters in a population of cells. This multi-parametric analysis provides more information such as the phenotypical characteristics of each cell and is usually based on fluorescence labelling. Another advantage of flow cytometry is the ability to sort cells, known as fluorescence-activated cell sorting (FACS). FACS is a specialised type of flow cytometry that enables physical separation and isolation of cell of interest based on a phenotype defined by users.

2.3.1.1 Principles of Flow Cytometry

The simplest flow cytometry set-up measures how particles interact with light, with each interaction termed an 'event'. In addition to measuring the fluorescence of exogenously applied fluorophores, flow cytometers collect two types of light scattering – forward and side scatter (Figure 2.6). Forward scatter indicates the size of an event or cell. Side scatter usually indicates the granularity of the event or cell. It is common, in post-collection analysis, to apply gates to the sample to virtually select a specific population. Gating often uses one or both scattering parameters.

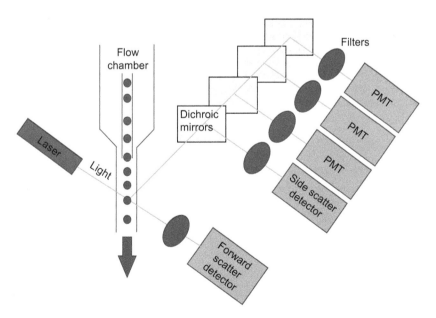

Figure 2.6 Schematic of a five-colour flow cytometer equipped with one laser.

There are three main components of a flow cytometer:

1) *Fluidics:* The fluidics system is responsible for transporting the cellular suspension from the sample tube to the flow cell, such that cells are presented to the laser in single file (Figure 2.6). Once through the flow cell (and past the laser), the sample is either sorted (in FACS) or transported to waste.

2) *Optics:* The components of the optical system include excitation light sources (typically two to four lasers), dichroic mirrors to reflect light for the appropriate detectors, filters to collect specific wavelength of the fluorescence light, and a photomultiplier tube to detect the fluorescence signal (Figure 2.6). Photodiodes are used to detecting forward and side scattered light.

3) *Electronics:* The electronic system digitises acquired data and processes the data for subsequent analysis. During acquisition, the detectors (PMT and photodiodes), signal amplifiers and analogue-to-digital conversion system converts the photon signal to electric current (also known as photocurrent), thereby digitising the data.

Briefly, cells are transported into the flow chamber in single file. As cells passed through the laser beam, light scattering and fluorescence will be detected. Dichroic mirrors and emission filters enable transmission of specific wavelengths and to the respective detectors.

There are a few major disadvantages in the use of flow cytometers. The expertise in interpreting flow cytometry data is challenging for amateurs, especially for non-classical experiments and polychromatic systems. Sufficient cell numbers are required for successful data collection: low cell numbers can cause unreasonably high background and false-positive events. Furthermore, the diameter of the sample injection sensor limits the size of systems that can be studied.

In contrast to microscopy, flow cytometry does not provide information on spatial localisation of the fluorescence, limiting the information that can be determined. The current resolution of imaging flow cytometers is inferior to the spatial resolution of a conventional confocal microscope. Unlike microscopy, specialised fluorescence measurements (such as time-resolved fluorescence) are not commercially available on flow cytometers, and add-on equipment for these specialised measurements are built in-house.

2.3.1.2 Understanding Flow Cytometry Data with Small-molecule Sensors

Flow cytometry studies tend to be predominated by the use of antibody-labelled systems [31]. However, fluorescent sensors are highly appropriate for use in conventional flow cytometers. The selection requirements for appropriate sensors match those for confocal microscopy, with the most common limitation being the available excitation lasers. This section briefly summarises considerations for applying fluorescent sensors in flow cytometry.

Firstly, in a homogenous sample, the fluorescence intensities should be uniform across sampling time. In addition to ensuring a smooth flow run to inspect for bubbles or clots, huge deviation (a coefficient of variance ≥ 25–30% from the mean) and significant fluctuations/shifts of the fluorescence signals should be noted. These are indicators of improper set-up of fluorescence sensors in flow cytometry, and the protocol requires to be reviewed. Secondly, data interpretation is often complicated for novel sensors. It can be difficult to distinguish sensor uptake from fluorescence output, in particular for ratiometric sensors.

One key difference between microscopy and flow cytometry is that in the latter, it is standard practice to perform compensation protocols to ensure that there is no spectral spill-over from one detector to another. Compensation is a mathematical numeric matrix that adjusts the ratios of fluorescence signal sensed by two fluorescence detectors. This compensation matrix does not alter the fluorescence properties of the sensors nor the wavelengths in which the detectors is detecting but is instead a virtual mathematic algorithm that adjusts ratios between two detectors. This is essential in experiments using multiple fluorophores, as spectral overlap is common in fluorophores with broad fluorescence. For proper compensation to be carried out, a set of appropriate control samples must be run.

2.3.1.3 Recent Advances in Flow Cytometry

Modern flow cytometry instruments are equipped with multiple lasers and detectors for multiparametric, polychromatic flow cytometric experiments. However, spectral spill-over of fluorophores and lack of spatial information of fluorescence in the cells are still the limitations with these optical flow cytometers. Over the past few decades, advanced flow cytometer systems have been engineered to overcome these limitations, as summarised briefly here.

Spectral flow cytometers (SFC) captures the entire emission spectrum of each cell analysed [32]. The technique therefore differentiate combinations of fluorophores that conventional systems cannot. With similar fluidics and lasers, SFC uses multichannel detectors (usually CCD) rather than the traditional optics (dichroic mirrors, optical filters, and PMTs) used in conventional flow cytometers.

Imaging flow cytometry (IFC) combines the high-throughput, multiparameter capabilities of conventional flow cytometry with morphological and spatial information, all at single-cell resolution [33]. Multichannel digital images of hundreds of thousands of individual cells can be captured within minutes with CCD cameras. These include several fluorescence channels, as well as bright-field and dark-field information. The throughput of IFC means that it is especially well suited to the analysis of rare cell types such as circulating tumour cells [34] and transition states such as cell cycle phases (mitosis) [35]. The various technologies and configurations of imaging flow cytometers have been helpfully reviewed by Hans and colleagues [36].

2.3.2 Fluorescence Plate-readers

Microscopy and flow cytometric measurements often require specialised expertise and expensive equipment, which is a limitation for many laboratories. Fluorescence plate-readers are a simple, cost-effective, and rapid technique that enable multiple experiments to be run in parallel in a controlled environment. Fluorescence plate-readers can generate quantitative time-course or dose-dependent data by measuring absorbance and/or fluorescence. These numerical data can provide information of cell growth or death in response to drug treatments or represent expression of genetically encoded fluorescent reporters. The most common microplate format used is 96-well (12×8 matrix) with a typical reaction volume between 100 and 300 μl well^{-1}.

In general, fluorescence plate-readers only provide the mean absorbance or fluorescence of the population of cells, and are insensitive to differences in subpopulations of cells. In addition to fluorescence intensity readings, other common detection modes for microplate assays include absorbance, luminescence, and time-resolved fluorescence.

2.3.2.1 Standard Plate-reader Assays

The ability to assess cell health accurately and efficiently is often crucial as readout for drug-dosing experiments. Plate-reader assays are often used for screening libraries of compounds to assess the effects on cellular health such as cytotoxicity or cellular proliferation. These assays give information as to how many viable cells are remaining at the end of the experiment. In interpreting the results of such assays, it is important not to over-interpret viability data as an indication of proliferation.

Common cell viability methods include tetrazolium reduction, resazurin reduction, firefly luciferase ATP detection, and DNA synthesis. Tetrazolium reduction (such as MTT assays [37]) and resazurin reduction (such as Alamar blue assays [38]) rely on general cellular metabolism or an enzymatic activity as a marker of viable cells. These assays require incubating viable cells with reagents or substrates that will be converted to a fluorescent product that can be measured with a plate-reader. The fluorescence intensity from the fluorescent products is proportional to the number of viable cells present. Other common plate-reader assays are the glycylphenylalanyl-aminofluorocoumarin (GF-AFC) assay for protease activity [39] and the firefly luciferase assay for adenosine triphosphate (ATP), which requires a plate-reader capable of luminescence measurements, *i.e.* the collection of emitted light without any excitation. [40].

2.3.2.2 High-content Imaging (HCI) Plate-readers

Although fluorescence plate-readers have clear advantages in terms of speed, simplicity, and suitability for homogeneous assays, conventional plate-readers are incapable of identifying and measuring cellular heterogeneity in a sample well. In recent years, HCI instruments with automated analysis have enabled the large-scale study of cellular physiology. HCI plate-readers are superior to conventional fluorescence plate-readers, as they are capable of image-based assays, such as detecting changes in subcellular organelles and biomolecule translocations [41].

One challenge for the rapid adoption of HCI in experiments is the algorithm development that could analyse images rapidly: automated extraction and processing of numerical data from micrographs. This data processing includes spatial localisation of fluorescence, temporal localisation of fluorescence reflecting cellular behaviour, fluorescence intensity corresponding to target-of-interest, surface area, and shape of cells [42]. Evolving software has improved these image-processing pipeline with increasing precision. However, HCI remains a challenge in terms of acquisition time, image processing complexity, and storage [43].

Although most HCI is performed on fixed cells, live-cell imaging on HCI platforms has been made possible with controlled environmental chambers and onboard pipetting [44]. In general, HCI systems operate on the principles of fluorescence microscopy, with filter-based control of excitation and emission wavelengths. More recently, confocal-based systems have been developed, along with systems with higher magnification objectives for higher resolution images [45]. However, simultaneous, high-resolution imaging of live cells in different conditions still remains a challenge.

2.4 Biological Samples to Which Fluorescent Sensors Can Be Applied

2.4.1 Cultured Mammalian Cells

Biological cultured cells can derive from various sources. These can be immortalised cell lines or primary cells. Immortalised cell lines are cost effective, easy to use, providing a pure population of cells and material, and bypass ethical concerns associated with the use of animal and human tissue. An immortalised or continuous cell line can proliferate indefinitely.

However, maintaining cell lines to represent *in vivo* organisms still presents as a challenge. Cell lines are limited to serial passages, which can cause genotypic and phenotypic variation over time, resulting in heterogeneity in cultures.

Primary cells can be isolated directly from human or animal tissue using enzymatic or mechanical methods. Once isolated, they are placed in an artificial environment in plastic or glass containers supported with specialised medium containing essential nutrients and growth factors to support proliferation. Although primary cells usually have a limited lifespan, greater ethical concerns, and high donor-to-donor variability, they have several advantages over immortalised cell lines. Primary cells contain higher relevance to understand physiology and avoid inter-species differences compared to animal models, such as in anatomy, molecular pathways, and metabolism.

2.4.1.1 Adherent Mammalian Cells

Adherent cells require attachment for growth and are commonly known as anchorage-dependent cells. Due to the anchoring surfaces, adherent cells are best used for microscopy and plate-reader assays, due to the ease in removing excess fluorescent sensors with washing.

General protocol for application of fluorescent sensors to cells for microscopy.
1) Prepare a sterile glass bottom dish.
2) Seed adherent cells (approximately 2×10^4 cells cm^{-2} of surface area) in appropriate media and allow cells to adhere overnight.
3) Preparing sensors:
   a) Dissolve lyophilised sensors in appropriate solvent (*e.g.* aqueous media, DMSO, and methanol).
   b) Dilute the stock solution into suitable concentrations (typically range: 10–50 μM) with media. (Note: if the sensor is sensitive to serum or other alternatives, consider using other alternatives, such as serum-free media.)
4) Wash the cells with warm PBS three times.
5) Stain the cells with sensors for appropriate incubation time at 37 °C.
6) Wash the cells with warm PBS three times.
7) Prepare the cells for imaging with phenol-free Fluorobrite® DMEM media. To maintain viability during imaging, media should be supplemented with 10% FBS and 2 mM glutamine.

During the imaging experiment, it is crucial to look for signs of cell death, including physical signs such as blebbing, loss of attachment to the surface, cell shrinkage, excessive cytoplasmic protrusions, swelling of nuclei and cells, and the presence of multiple, large vacuoles/lysosomes at the perinuclear region. Furthermore, fluorescence staining within the nucleus (unless the sensor targets nuclear material) is often indicative of cytotoxicity.

2.4.1.2 Non-adherent Cells

Suspended, or non-adherent, cells do not require attachment for growth and are said to be anchorage-independent cells. Cell suspensions are highly suitable for flow cytometry and can be used for microscopy with proper imaging tools/aids. Numerous cell lines are non-adherent or can be a mixture of adherent and non-adherent. Some common challenges in imaging non-adherent cells include difficulty focussing on cells and changes to cellular phenotype and behaviour when adherence occurs.

2.4.1.3 Multi-cellular Models

Three-dimensional cellular models are better representation of biology *in vivo*, as compared to monolayer adherent cells. These three-dimensional cell models include blastocysts, trans-wells, and tumouroids. As these models are volumetric, there are several factors to consider, including penetration of sensors, cytotoxicity, and preparation for imaging. Some additional challenges in imaging trans-wells include autofluorescence of the trans-well matrix, identifying two different adherent cells adhered to either side of the trans-well (*i.e.* cell type A on the top, cell type B at the other end of the matrix), and layering matrixes for imaging without destroying the sample. Conventional microscopes could have issues with exciting the fluorophores. Hence, multiphoton microscopy is recommended for better sample penetration.

The key to successful imaging of three-dimensional cellular models lies in not destroying the three-dimensional ultra-architecture of the model. Options include using Vaseline to lift adherent surfaces to glass slides and using glass bottom dishes with inverted microscopes. Good, gentle-handling technique is essential in mounting the cells for imaging.

2.4.2 Bacteria

Fluorescence is emerging as an important tool in clinical bacteriology. Applications range from classic live-dead assays to serving as markers for isolating bacterial persisters, which are responsible for antibiotics resistance and recurrent infections [46]. Unlike eukaryotic mammalian cells, prokaryotic bacteria have different physiochemical properties, such as lack of compartmentalisation, structural differences (circular DNA and cell membrane), and much smaller in size.

The uptake of the fluorescent dyes differs between mammalian cells and bacteria, due to the lipid membrane composition, transmembrane potential, and the presence of a cell wall. Bacterial membranes possess high amounts of negatively charged phospholipids, such as cardiolipin and phosphatidylglycerol [47]. In Gram-positive bacteria, phosphate-containing, negatively charged teichoic acids are covalently linked to either the peptidoglycan or to the underlying membrane. In contrast, Gram-negative bacteria have an outer membrane of phospholipids and lipopolysaccharides, which impart the negative charges to Gram-negative bacterial cells [48]. Furthermore, unlike most mammalian cells, many bacteria such as *Pseudomonas aeruginosa* [49] and *Brocadia fulgida* [50] are also known to autofluoresce.

Most fluorescent sensors for bacteria rely on probing non-specific intracellular targets, such as nucleic acids (such as propidium iodide and commercial SYTO dye series [46]) and cytoplasmic redox environment (such as carboxyfluoresceins [49]). While there are antibiotic-inspired chemical sensors and peptidoglycan-inspired chemical sensors [51], these sensors are pre-dominantly suitable for Gram-positive bacteria. Common fluorescent sensors used in mammalian systems are often poor fluorophores for imaging Gram negative bacteria, as these sensors are not readily taken up without using permeabilising agents such as Triton-X [52]. A good review of fluorescence imaging of bacteria is provided by Cambré and Abram [24].

2.4.3 Plants

While fluorescent proteins are often favoured in botanical studies, sensors have been successful in assessing intracellular environments [53] and DNA ploidy [54] in plant specimens. Botanical specimens pose significant challenges for using fluorescence as a research tool, due to autofluorescence and challenges in penetrating the cell wall [55]. Further complexities arise with imaging different anatomical regions of the botany specimen. Significant endogenous autofluorescence arises from a variety of biomolecules, including anthocyanins and flavonols. Among these biomolecules, chlorophyll is the major contributor to the autofluorescence. Fluorescent labels near the autofluorescence can be excited by autofluorescence [7]. Unless the cell wall is the specimen of interest, the cellulose cell walls posed a significant challenge for protein-based sensors such as antibodies or large molecule sensors. Often, enzymes must be employed to break down the cell wall for fluorophores to enter.

2.4.4 Multi-cellular Organisms

Fluorescence studies on translucent or transparent organoids and multicellular organisms, such as *Caenorhabditis elegans* and zebrafish embryos, comes with challenges such as penetration by both sensors and excitation light source [56] and temperature dosing conditions [57]. Spectral characteristics of the sensor will also be affected by the thickness and the transparency of tissues. As a rule-of-thumb, the use of long wavelength sensors reduces light scattering and increases tissue penetration [58]. Two-photon microscopy and light sheet fluorescence microscopy (LSFM) tend to be platforms of choice for imaging organisms.

The limiting factor for assessing these multicellular organisms or organoids by flow cytometry relies on the SIP diameter for delivering the sample to the flow chamber, suitable optical configuration for the light scatters, and sufficient cell numbers. Successful flow cytometric analyses (imaging and sorting) of these multicellular organisms/organoids using fluorescence have been reported in the literature [59, 60].

2.4.5 Towards *In Vivo* Imaging

Cellular biochemical events may not truly represent the entire physiology of an organism. However, conventional microscopy techniques alone are not enough to penetrate non-transparent whole organisms. Intravital administration such as injection of fluorescent sensors and/or labelled cells into the organ-of-interest is a popular choice for fluorescence labelling. Anaesthesia, invasive irreversible surgery, animal restraints techniques, hair-shaving, endoscopy probes, and optical window models are some common ways to gain access and expose organs/tissues for fluorescence imaging [61–64].

Cytotoxicity of the fluorophore is often a limitation for intravital imaging. Most cytoplasmic and nuclear dyes are rarely used for intravital microscopy due to the high cytotoxicity. While photostable membrane dyes present low cytotoxicity, membrane dyes have shown to be transferred from labelled cells to other cell types [3]. For example, dialkylcarbocyanines and carbocyanines have lower cytotoxicity *a priori* and have been widely used for *in vivo* imaging [3]. Another concern for *in vivo* imaging is dye dilution. Dilution of dyes upon cell division is widely exploited *in vitro* and *in vivo* to study cell proliferation. Cellular fluorescence will be halved following each cell division, which can be easily observed and quantified *ex vivo* [46, 65, 66]. However, for *in vivo* imaging studies, the fluorescent signal obtained from cells within a tissue depends not only on the fluorophore brightness but also on the nature of the surrounding tissue. Fluorescent 'debris' (such as those that were exocytosed or diffused out of labelled cells) could accumulate within the tissue and be taken up tissue macrophages actively *in vivo* causing imaging artefacts [67].

Depending on the tissue or organ, microscopy modalities with high-penetration capabilities such as light sheet and multiphoton microscopy are often used. However, a technical challenge arises from the physiological motion of animals, such as breathing and heart beats. These motions introduce blurring artefacts during imaging. The most common solution is to fixate tissue on rigid surfaces to ensure sufficient motion stability without interfering with normal cellular physiology [64]. A more in-depth discussion on *in vivo* imaging can be found in Chapter 8.

2.5 Common Challenges and Misconceptions in the Applications of Fluorescent Sensors

2.5.1 Important Considerations in Applying Sensors

1) *Dosing conditions such as temperature and duration of labelling.* Mammalian cells would require a temperature close to 37 °C for sensor labelling. In contrast, non-mammalian cellular targets would need different temperature and incubation conditions (such as anaerobic conditions for anaerobic bacteria, or room temperature for *C. elegans*). Mammalian cells behave differently at different temperatures, in particular the immune cells. The release of biomolecules such as metabolites, cytokines, or heat shock proteins at different incubation conditions could potentially interfere with fluorescence behaviour of sensors and/or cellular behaviour to sensors. Some specialised sensors may require fixation (*e.g.* DAPI and phalloidin) or permeabilisation (*e.g.* sensors targeting nuclear transcription factors). In contrast, sensors that target the extracellular space do not generally require temperature control.

2) *Fluorescence intensity and spectral overlap.* Titrating different concentrations of sensors can be used to determine the desired levels of fluorescent intensities. However, novel sensors may exhibit low quantum yields. Unless conjugated to classic fluorophore scaffolds, finding an appropriate instrument with optimal excitation wavelengths for most synthesised sensors is challenging. Synthesised sensors may require excitation source with wavelengths that are <400 or >700 nm for excitation, and hence specialised instruments are required. These excitation wavelengths cannot be achieved by conventional filter-cube-based fluorescence microscopes or benchtop flow cytometers. Combinatorial labelling using synthesised sensors will depend on the excitation and emission wavelength of sensors. Many of these sensors have broad fluorescence spectra, which may have severe spectral overlapping issues.

3) *Inter-experiment variability for reproducibility.* There can be great variabilities in fluorescence response arising from instruments (such as laser output and detector stability), which may make it difficult to compare data collected on different days. Fluorophores can degrade over time or inappropriate storage and should be visibly inspected before use. A simple *in vitro* test with addition of the analyte of interest can verify that the sensor is still operating as expected.

4) *Appropriate controls.* It is difficult to determine appropriate 'one-size-fits-all' controls for every fluorescence experiment, as the questions asked differ vastly. However, it is important to have both biological and technical controls (not to be confused with technical replicates) in every experiment. It is crucial to ensure that the experimental samples are

assessed with similar instrumental settings and dosing conditions as the controls. Basic recommended controls are:

a) No stain and vehicle controls

These controls assess whether there is significant background fluorescence. Vehicle controls are necessary to determine if the solvent in which the sensor is in causes cytotoxicity in cells. This technical control is usually performed when organic solvents are used (such as DMSO, ethanol, or methanol). After initial experiments validating that the vehicle had negligible effect on experiments, vehicle controls can replace no-stain controls.

b) Single stains or fluorescence-minus-one (FMO) controls

These are classic technical controls for flow cytometry for compensation due to spectral spill-overs and still serve as good controls for microscopy and other fluorescence experiments. Single stain controls are highly recommended for co-localisation experiments. FMO controls are usually to help spectral spill-over and aid compensation in flow cytometric analyses. FMO would also be useful in microscopy if the sensor is suspected to act on cells in an unexpected manner, such as altering cellular behaviour.

c) Positive and negative controls

These are usually biological controls that check for selectivity and specificity of the sensor in the presence or absence of biomolecule of interest. The typical positive control is the exogenous addition of the analyte of interest, and ideally, a negative control using an agent that starves the cell of the analyte will also be used. Other forms of controls are modified sensor scaffolds. For example, a sensor molecule with a modified sensor scaffold can act as a good negative control, while for a reaction-based sensor, the product form of the sensor can also be a valuable control.

2.5.2 Common Misconceptions About the Use of Sensors – The Bridge Between Multiple Disciplines

This section aims to clarify common misconceptions regarding the use of fluorescent sensors, as well as common questions when using sensors. Firstly, it is important to have a good understanding of common fluorophore classes, as outlined in Chapter 1. Experimental quality and outcome of fluorescence-based experiments is always highly dependent on the fluorophore used. Unfortunately, there is no standard 'one-size-fits-all' protocol for using fluorescent sensors in either microscopy or flow cytometry. The following comments relate to the use of fluorophores and may serve as an important starting point for troubleshooting, whenever necessary.

On many instruments, many fluorophores can be excited by the same excitation source and have very similar emission profiles. Similar structures may have similar spectral properties (such as brightness and photostability), but may have different electrostatic interactions (hydrophobicity), causing non-specific binding such as the glass surfaces or adhesive coating [68].

Poor solubility properties of sensors can result in their precipitation prior to or during imaging experiments. This can result in erroneous interpretation of quantitative fluorescence experiments. Sensors that are too hydrophobic can cause solubility issues, which can lead to unspecific staining of cellular organelles, thus increasing the background [2]. One must take caution when choosing the solvent in which to dissolve lyophilised sensors. Most polar solvents such as dichloromethane and xylene are highly toxic to cells or causes a phenotypic change in cells. Generally, sensors are dissolved in DMSO or aqueous solvents (such as PBS buffers, water or even complete media). However, this rule-of-thumb can be bent. Sensors that are steroid derivatives or insoluble in water can be dissolved in ethanol or methanol before dosing cells. However, it is important to note that these alcohols based can alter cellular behaviour. Additives such as Pluronic F-127 can be added at 0.1% (w/v) to the solvent to help dissolve the sensors [55].

Re-freezing reconstituted sensors have a high tendency of fluorescent sensors precipitating from solution. On flow cytometric experiments, small aggregation of fluorescent sensors can be observed. Sensor aggregates are even more evident with fluorescence microscopy. Adding more solvent in attempt to dissolving the sensors, sonicating and warming up the sensors may remedy the situation. However, the authors recommend using freshly dissolved aliquots of sensor to minimise these issues.

1) Cell matters – Not all cells behave the same way

   Not all cells will respond similarly to sensors, and it is therefore essential to establish the optimal conditions for each new cell line. Consider a scenario where a sensor is being validated for its applicability to label apoptotic cells. As controls, one will need cells to be dosed with apoptotic inducing drug, such as cisplatin. However, some cell lines will be more susceptible to apoptosis. Hence, the dosing conditions (of both the drug and the sensor) will need to be modified. It is crucial to look for change in cellular morphology other than signs of apoptosis such as blebbing and nuclear swelling. Different cell types also have different phototoxicity thresholds [69, 70]. Fluorescence intensities may be crucial, but subcellular distribution and correlating to brightfield/phase-contrast images will provide insights to the response of the sensor inside the cells.

   It is also common for the constituents of the media (such as DMEM or RPMI) to be overlooked as a variable in cell experiments. Indeed, the contents of the media are rarely questioned beyond the supplements being added. These supplements include glutamine (which hydrolyses rapidly), foetal bovine serum, sodium pyruvate, monothioglycerol (mainly for maintaining pluripotency of differentiable cells), vitamins (which are known to cause autofluorescence), and other growth factors. Small-molecule sensors are very sensitive to the media constituents. One key component of many commonly used physiological buffers (such as PBS and Hank's balanced salt solution) contain cations and anions that will affect studies of such analytes. Other components such as albumin in foetal bovine serum may also potentially affect experimental results. Mueller-Hinton broth for bacterial cultures contain beef extracts, which might interfere with DNA-binding or amino-acid-labelling dyes. Hence, it is advisable to find out the exact content of the media used prior to experimental design. It is also essential to ensure that, in replicating reported studies, identical buffer conditions are used.

2) Autofluorescence – a friend and a foe

   Autofluorescence can be observed in both live and fixed cells. However, the source of the autofluorescence is not the same. In fixed samples, the autofluorescence is likely from aldehyde-based fixation. This can be remedied using 0.1% (w/v) glycine in PBS washes after aldehyde fixations and/or using sodium borohydride. In live mammalian cells, the endogenous autofluorescence could be flavins or bound NAD. This autofluorescence can be used to our advantage in live cells and can be measured and visualised with FLIM (see Section 2.2.2.4). Currently, there are no known quenchers for the autofluorescence of these biomolecules in live cells, without having to add a separate chemical that alters cellular behaviour and introduces additional variables to the experiments. One potential source of autofluorescence can also arise from the adhesives on the glass. Commercialised glass slides such as post-labelled cells on SuperFrost® slides can show non-cellular auto-fluorescence in the red-wavelengths region (approximately 550 to >700 nm).

3) Fixation

Live cells offer a holistic perspective into cellular physiology compared to fixed cells in biological research. A common question regarding the use of fluorescent sensors is their compatibility with fixation protocols: this is certainly true for sensors of cellular conditions that will be perturbed by fixation.

There are multiple ways to preserve tissue and cells. Aldehyde based fixatives remain a popular choice for preserving subcellular structures and molecules. Paraformaldehyde is the most popular choice to preserve tissues and cells rapidly for light microscopy. Glutaraldehyde is often used as an electron microscopy fixative; however, protocols might include glutaraldehyde for CLEM studies. While it is rarely used in other light microscopies, glutaraldehyde does autofluoresce strongly in the 400–500 nm wavelength region. As paraformaldehyde are smaller molecules than to glutaraldehyde, the penetration into cells is more rapid. However, glutaraldehyde has two aldehyde groups and hence serves as a better fixative for structural preservation. Formalin is a paraformaldehyde-containing methanol, which acts as a stabiliser for paraformaldehyde and a dehydrator and fixative. Peptide-based sensors could be affected by these aldehyde fixatives or be affected by the autofluorescence. Commercial phycobiliprotein-tandem dyes (such as PE-Cy7) is affected by aldehyde fixation. Whole non-peptide-based sensors are unlikely to be affected by fixation, post-fixation washing steps might influence fluorescence intensities and localisation since the cell membrane has become permeabilised.

Methanol and ethanol are common alcohol-based fixatives. These alcohols also rapidly dehydrate cells. While the preservation of structural integrity is poor, alcohol-based fixatives allow non-polar sensors/dyes (such as eosin) to enter cells more readily. Tissues and cells appear to be still well preserved in 70% (w/v) alcohol for long-term storage. Cryogenic techniques are more suited for preserving tissue with high lipid content such as the mammary glands and adipose tissue. Cryogenic techniques require glycerol and glycols to prevent cells from bursting upon freezing. It is currently unclear if these cryo-preservatives affect sensor uptake or fluorescence emission.

4) Wash or no wash

Due to the equilibrium established between sensors inside and outside the cell, washing steps might influence the fluorescence intensities of fluorophores. This is highly dependent on factors such as the affinities of the target-and-sensor interaction. High affinities to substrate surfaces (*i.e.* binding of sensors to glass surfaces) are rare for small-molecule sensors. The authors do not advise blocking of surfaces pre-staining in live cell imaging unless necessary. This is because blocking steps can alter cellular behaviour and introduce experimental variables. For flow cytometry experiments, a flow cytometry plot monitoring fluorescence intensity measurement over acquisition time is highly recommended.

5) Microwaving techniques

Microwaving cells (<10 seconds) to enhance uptake of dyes is a common technique, especially for increasing uptake of dyes in cells to increase contrast for electron microscopy and correlative light-electron microscopy studies. However, microwaving is a form of irradiation. Hence, the authors do not recommend microwaving live cells to enhance sensor uptake, no matter how short the duration is. Microwaving may also destroy sensors by causing radicals, contributing to phototoxicity in live cells.

2.6 Conclusions

From the state-of-the-art super resolution microscopies to high-throughput fluorescence plate-reading, fluorescent sensors play a huge role in biology. Modern instruments used in biological experiments focus on high-throughput readouts and rapid data output for single-cell analysis and are trending towards investigating whole organisms or organs. With the expanding number of novel sensors and labelling methods as detailed in other chapters of this book, we anticipate more instruments engineered to accommodate these sensors and labelling methods for biological experiments in the future. The bottleneck of a one-for-all instrument still lies in the biological specimen. These include factors such as intrinsic anatomical/histological differences (*e.g.* different parts of a botanical sample, different organelles of a cell) and physiological/biochemical processes (*e.g.* how fast is the cellular event). As no one single instrument can provide answers to all biological questions, we believe that only the combinatorial use of instruments will continue to push the boundaries in biomedicine, drug discovery, and novel discoveries in cell biology.

References

1 Hughes, L.D., Rawle, R.J., and Boxer, S.G. (2014). *PLoS One* 9 (2): e87649.
2 Wang, L., Frei, M.S., Salim, A., and Johnsson, K. (2019). *J. Am. Chem. Soc.* 141: 2770–2781.
3 Progatzky, F., Dallman, M.J., and Celso, C.L. (2013). *Interface Focus* 3: 20130001.
4 Waters, J.C. (2009). *J. Cell Biol.* 185: 1135–1148.
5 Murray, J.M., Appleton, P.L., Swedlow, J.R., and Waters, J.C. (2007). *J. Microsc.* 228: 390–405.
6 Zanetti-Domingues, L.C., Martin-Fernandez, M.L., Needham, S.R. et al. (2012). *PLoS One* 7: e45655.
7 Pawley, J.B. (2006). *Handbook Of Biological Confocal Microscopy*, 3ee. Boston, MA: Springer US.
8 Icha, J., Weber, M., Waters, J.C., and Norden, C. (2017). *Bioessays* 39: 1–15.
9 Stennett, E.M.S., Ciuba, M.A., and Levitus, M. (2014). *Chem. Soc. Rev.* 43: 1057–1075.
10 Laloi, C. and Havaux, M. (2015). *Front. Plant Sci.* 6: 1–9.
11 Greenbaum, L., Rothmann, C., Lavie, R., and Malik, Z. (2000). *Biol. Chem.* 381: 1251–1258.
12 Stehbens, S., Pemble, H., Murrow, L., and Wittmann, T. (2012). *Methods Enzymol.* 504: 293–313.
13 Jonkman, J. and Brown, C.M. (2015). *J. Biomol. Tech.* 26: 54–65.
14 Stutzmann, G.E. and Parker, I. (2005). *Physiology* 20: 15–21.
15 Kim, H.M., Fang, X.Z., Yang, P.R. et al. (2007). *Tetrahedron Lett.* 48: 2791–2795.
16 Podgorski, K., Terpetschnig, E., Klochko, O.P. et al. (2012). *PLoS One* 7: e51980.
17 Bestvater, F., Spiess, E., Stobrawa, G. et al. (2002). *J. Microsc.* 208: 108–115.
18 Berezin, M.Y. and Achilefu, S. (2010). *Chem. Rev.* 110: 2641–2684.
19 Datta, R., Heaster, T.M., Sharick, J.T. et al. (2020). *J. Biomed. Opt.* 25 (7): 071203.
20 Lakowicz, J.R., Szmacinski, H., Nowaczyk, K., and Johnson, M.L. (1992). *Proc. Natl. Acad. Sci. U. S. A.* 89: 1271–1275.
21 Levitt, J.A., Poland, S.P., Krstajic, N. et al. (2020). *Sci. Rep.* 10: 5146.
22 Reits, E.A.J. and Neefjes, J.J. (2001). *Nat. Cell Biol.* 3: E145–E147.
23 Axelrod, D. (1981). *J. Cell Biol.* 89: 141–145.
24 Cambré, A. and Aertsen, A. (2020). *Microbiol. Mol. Biol. Rev.* 84: 1–76.

25 Jin, D., Zhou, R., Yaqoob, Z., and So, P.T.C. (2017). *J. Opt. Soc. Am. B* 34: B64–B77.

26 Cotte Y, Pavillon N, Depeursinge C. US 8937722B2, 2015.

27 Huang, B., Babcock, H., and Zhuang, X. (2010). *Cell* 143: 1047–1058.

28 Ha, T. and Tinnefeld, P. (2012). *Annu. Rev. Phys. Chem.* 63: 595–617.

29 Rust, M.J., Bates, M., and Zhuang, X. (2006). *Nat. Methods* 3: 793–795.

30 Betzig, E., Patterson, G.H., Sougrat, R. et al. (2006). *Science* 313: 1642–1645.

31 Herzenberg, L.A., Tung, J., Moore, W.A. et al. (2006). *Nat. Immunol.* 7: 681–685.

32 Bonilla, D.L., Reinin, G., and Chua, E. (2021). *Front. Mol. Biosci.* 7: 1–10.

33 Doan, M., Vorobjev, I., Rees, P. et al. (2018). *Trends Biotechnol.* 36: 649–652.

34 Ogle, L.F., Orr, J.G., Willoughby, C.E. et al. (2016). *J. Hepatol.* 65: 305–313.

35 Blasi, T., Hennig, H., Summers, H.D. et al. (2016). *Nat. Commun.* 7.

36 Han, Y., Gu, Y., Zhang, A.C., and Lo, Y.H. (2016). *Lab Chip* 16: 4639–4647.

37 Mosmann, T. (1983). *J. Immunol. Methods* 65: 55–63.

38 Huet, O., Petit, J.M., Ratinaud, M.H., and Julien, R. (1992). *Cytometry* 13: 532–539.

39 Riss, T.L., Moravec, R.A., Niles, A.L. et al. (2004). *Assay Guidance Manual* (ed. S. Markossian, G.S. Sittampalam, A. Grossman, et al.). Eli Lilly & Company and the National Center for Advancing Translational Sciences: Bethesda, MD.

40 McElroy, W.D. (1947). *Proc. Natl. Acad. Sci. U. S. A.* 33: 342–345.

41 Bushway, P.J., Mercola, M., and Price, J.H. (2008). *Assay Drug Dev. Technol.* 6: 557–567.

42 Shariff, A., Kangas, J., Coelho, L.P. et al. (2010). *J. Biomol. Screen.* 15: 726–734.

43 Kozak, K., Rinn, B., Leven, O., and Emmenlauer, M. (2018). *Methods in Molecular Biology*, vol. 1683 (ed. P.A. Johnston and O.J. Trask), 131–148. Springer.

44 Esner, M., Meyenhofer, F., and Bickle, M. (2018). *Methods Mol. Biol.* 1683: 149–164.

45 Mandavilli, B.S., Aggeler, R.J., and Chambers, K.M. (2018). *Methods in Molecular Biology* (ed. P.A. Johnston and O.J. Trask), 33–46. Springer.

46 Wong, F.H.-S., Cai, Y., Leck, H. et al. (2020). *Antimicrob. Agents Chemother.* 64: e01712–e01719.

47 Sohlenkamp, C. and Geiger, O. (2015). *FEMS Microbiol. Rev.* 40: 133–159.

48 Silhavy, T.J., Kahne, D., and Walker, S. (2010). *Cold Spring Harb. Perspect. Biol.* 2: a000414.

49 Hoefel D, Grooby WL, Monis PT, Andrews S, Saint CP. 2003; J *Microbiol Methods* 52:379–388.

50 Kartal, B., Van Niftrik, L., Rattray, J. et al. (2008). *FEMS Microbiol. Ecol.* 63: 46–55.

51 Kocaoglu, O. and Carlson, E.E. (2016). *Nat. Chem. Biol.* 12: 472–478.

52 Mason, D.J., Shanmuganathan, S., Mortimer, F.C., and Gant, V.A. (1998). *Appl. Environ. Microbiol.* 64: 2681–2685.

53 Swanson, S.J., Choi, W.G., Chanoca, A., and Gilroy, S. (2011). *Annu. Rev. Plant Biol.* 62: 273–297.

54 Vrána J, Cápal P, Bednářová M, Doležel J. In: Nick P, Opatrny Z, editors. *Applied Plant Cell Biology* Springer-Verlag, Berlin, Heidelberg; 2014. p. 395–430.

55 Qu, H., Xing, W., Wu, F., and Wang, Y. (2016). *PLoS One* 11: e0152320.

56 Galas, L., Gallavardin, T., Bénard, M. et al. (2018). *Chemosensors* 6 (3): 40.

57 Wang, H., Karadge, U., Humphries, W.H., and Fisher, A.L. (2014). *Methods* 68: 508–517.

58 Pak, Y., Swamy, K., and Yoon, J. (2015). *Sensors* 15: 24374–24396.

59 Le Bihanic, F., Di Bucchianico, S., Karlsson, H.L., and Dreij, K. (2016). *Mutagenesis* 31: 643–653.

60 Rosenbluth, J.M., Schackmann, R.C.J., Gray, G.K. et al. (2020). *Nat. Commun.* 11: 1711.

61 Mazzoni, F., Müller, C., DeAssis, J. et al. (2019). *Sci. Rep.* 9: 1590.

62 Holtmaat, A., de Paola, V., Wilbrecht, L. et al. (2012). *Cold Spring Harb. Protoc.* 7: 694–701.

63 Ju, L., McFadyen, J.D., Al-Daher, S. et al. (2018). *Nat. Commun.* 9: 1087.

64 Choi, M., Kwok, S.J.J., and Yun, S.H. (2015). *Physiology* 30: 40–49.

65 Banks, H.T., Sutton, K.L., Thompson, W.C. et al. (2011). *Bull. Math. Biol.* 73: 116–150.

66 Terrén, I., Orrantia, A., Vitallé, J. et al. (2020). *Methods Enzymol.* 631: 239–255.

67 Pawelczyk, E., Jordan, E.K., Balakumaran, A. et al. (2009). *PLoS One* 4: e6712.

68 Zanetti-Domingues, L.C., Tynan, C.J., Rolfe, D.J. et al. (2013). *PLoS One* 8: e74200.

69 Forman, H.J. (2003). *Signal Transduction by Reactive Oxygen and Nitrogen Species: Pathways and Chemical Principles*. Dordrecht: Springer Netherlands.

70 Manders, E.M.M., Kimura, H., and Cook, P.R. (1999). *J. Cell Biol.* 144: 813–822.

3

Methods to Control the Subcellular Localisation of Fluorescent Sensors

Jiarun Lin[1,2], Kylie Yang[1,2], and Elizabeth J. New[1,2,3]

[1] *School of Chemistry, The University of Sydney, NSW, Australia*
[2] *The University of Sydney Nano Institute (Sydney Nano), The University of Sydney, NSW, Australia*
[3] *Australian Research Council Centre of Excellence for Innovations in Peptide and Protein Science, The University of Sydney, NSW, Australia*

3.1   Introduction

Of the biochemical processes that occur within the cell, many take place within membrane-bound subcellular compartments, known as organelles. Organelles carry out specialised roles within the cell to enable regular cellular function (Figure 3.1). For example, the nucleus carries the genetic material and is responsible for cellular reproduction, while the mitochondria are the site of energy production and cellular respiration. In addition to the mitochondria, several other organelles are involved in metabolism: the lysosomes and other digestive vesicles are responsible for the breakdown of biomolecules; the peroxisomes are involved in oxidation reactions; and the lipid droplets are responsible for energy storage. In terms of protein synthesis, the endoplasmic reticulum is involved in protein folding and glycosylation, and the Golgi apparatus is the site of further post-translational modification and protein sorting. In addition to these organelles, other subcellular structures significant to cellular structure and function include the plasma membrane and cytoskeleton.

In order to carry out their diverse functions, each organelle maintains a distinct chemical microenvironment, where the analytes within must be kept at optimal levels [1]. Disruption to the chemical composition of an organelle can disturb its normal function, leading to abnormal cellular function. Similarly, when a cell becomes diseased, the chemical composition inside an organelle can change and disrupt its function, affecting overall cellular health [2]. Consequently, visualising and studying the chemistry within specific organelles is vital in investigating the physiology and pathology of cells.

Due to its high spatial and temporal resolution [3], fluorescence microscopy has become an important tool for understanding the structure and molecular dynamics of organelles. Unlike destructive techniques like cell fractionation, which requires the lysis of cells and cannot capture live cellular dynamics, fluorescent staining is non-invasive and allows the imaging of intact organelles within live cells. The two most common types of fluorophores are fluorescent dyes and fluorescent proteins. While fluorescent proteins can have exquisite

Molecular Fluorescent Sensors for Cellular Studies, First Edition. Edited by Elizabeth J. New.
© 2022 John Wiley & Sons Ltd. Published 2022 by John Wiley & Sons Ltd.

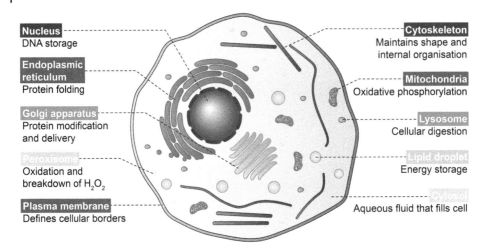

Nucleus
DNA storage

Endoplasmic reticulum
Protein folding

Golgi apparatus
Protein modification and delivery

Peroxisome
Oxidation and breakdown of H₂O₂

Plasma membrane
Defines cellular borders

Cytoskeleton
Maintains shape and internal organisation

Mitochondria
Oxidative phosphorylation

Lysosome
Cellular digestion

Lipid droplet
Energy storage

Cytosol
Aqueous fluid that fills cell

Figure 3.1 Schematic of the key organelles and subcellular structures.

subcellular localisation due to their straightforward fusion to a protein of interest, their expression within the cell necessitates genetic modification, which is incompatible with the analysis of clinical samples or unmodified cell lines.

Small-molecule fluorescent dyes present a number of advantages over other types of fluorophores. They generally have a low molecular weight (<1 kDa), allowing them to readily pass through cell membranes and access intracellular environments. Due to the sensitivity of fluorescence microscopy, fluorophores can be used at very low concentrations, allowing minimal perturbation of the cellular environment [4]. Furthermore, sample preparation is relatively easy and straightforward, especially compared to protocols that require genetic modification. Fluorescent dyes can be classed as stains, which label specific subcellular structures with a constant fluorescent output, or responsive sensors. The same design principles apply, whether for targeting of dyes or sensors to specific organelles. Small-molecule fluorophores vary greatly in structure and charge, and the resulting diverse range of cargoes with different physical and chemical properties complicates the development of robust organelle-targeting strategies.

Targeting of fluorescent sensors is generally achieved by attachment of a targeting group to the sensor *via* a linker. Some organelles have relatively robust targeting strategies (*e.g.* triphenylphophonium as a targeting group for the mitochondria [5]), but there are many organelles without robust targeting strategies. In order to develop fluorescent tools that target the organelles, it is important to have robust targeting strategies that matches the diversity of the cargoes. In this chapter, we provide an overview of the current targeting strategies used to direct fluorescent sensors to organelles and other subcellular structures, including challenges and issues. We aim to provide the mechanism of targeting where possible, as well as methods for achieving accumulation or trapping of sensors within cells.

3.2 Targeting the Nucleus

The nucleus is responsible for both cellular reproduction and gene expression, two essential functions of the cell. Due to its critical role, there are many mechanisms that prevent potential mutagens from reaching the DNA, which complicates targeting of exogenous molecules to the nucleus [6]. The nuclear envelope is a highly regulated membrane consisting of two lipid bilayers. Spanning the nuclear envelope are nuclear pore complexes (NPCs) that mediate the transport of all molecules between the nucleus and cytoplasm [7, 8]. Small molecules and proteins can freely diffuse through protein channels in the NPC, while the uptake of larger proteins is prevented.

Perhaps the most successful class of targeting group for the nucleus encompasses minor groove binders of DNA, crescent-shaped molecules that bond non-covalently via directed hydrogen-bond interactions with base pairs in this shallow helical space. Hoechst dyes, such as **Hoechst 33258** [9], are a classic example of fluorescent minor groove binders, which are as cell-permeable DNA nuclear stains in their own right (Figure 3.2). A Hoechst tag has also been covalently attached to a number of fluorophores to ensure nuclear localisation [10–12], such as the sulforhodamine **hoeSR** [10] (Figure 3.2). As a Hoechst tag is relatively large, smaller variants have been synthesised, such as the coumarin dye **CQPP** (Figure 3.2), although this dye also showed some lipid droplet localisation [13]. A less commonly used minor groove binder is pyrrole polyamide, which has been successfully used to target the fluorescein **F-DisT** to the nucleus (Figure 3.2) [14]. The key issue with the use of minor groove binders as targeting groups are their potentially mutagenic and cytotoxic effects arising from DNA binding [15].

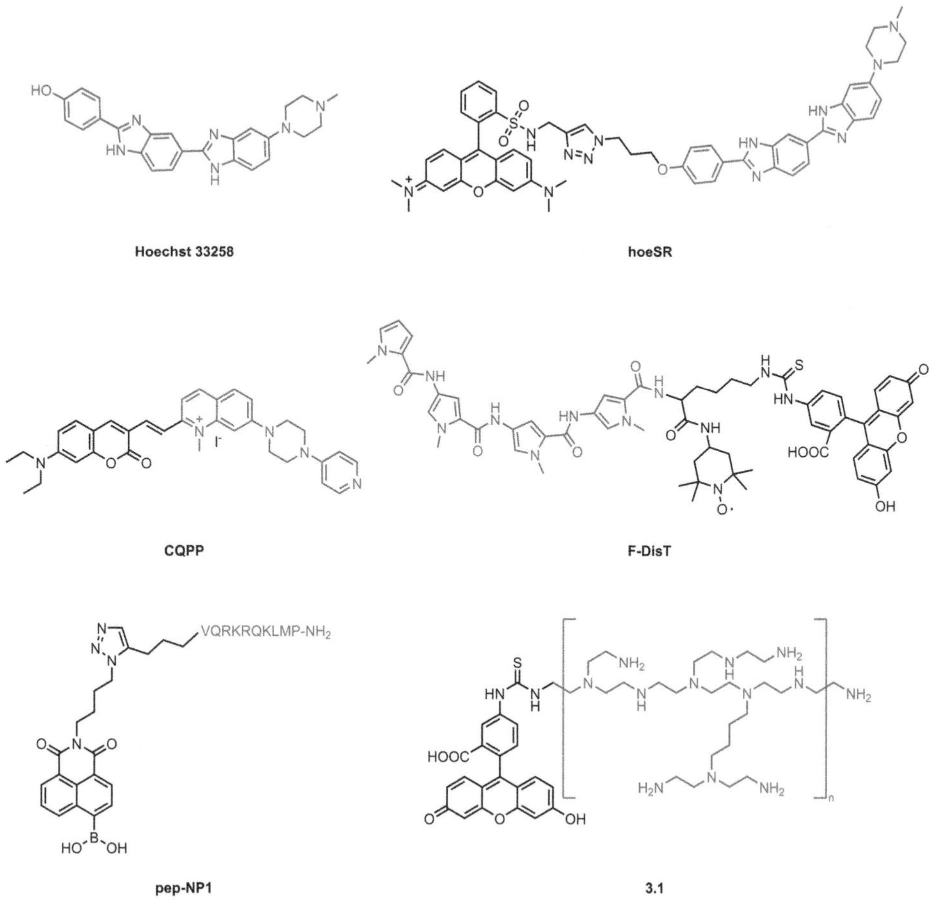

Figure 3.2 Structure of nuclear-targeted fluorescent sensors, with targeting groups highlighted in red. Hoechst 33258 is a DNA intercalator [9], while a Hoechst tag used in hoeSR [10] and a Hoechst mimic is used in CQPP [13]. Another DNA intercalator, pyrrole polyamide, is used in F-DisT [14]. pep-NP1 uses the NLS50 sequence [16] while **1** has a polyethyleneimine tag [17].

Alternatively, small molecules may be targeted to the nucleus *via* the cell's native protein-sorting machinery. The nuclear localisation sequence (NLS) describes a peptide sequence that tags a native protein for nuclear import and has been covalently attached to a number of fluorophores [14, 16, 18, 19], such as in the naphthalimide-based hydrogen peroxide sensor **pep-NP1** [16] (Figure 3.2). Unfortunately, due to their hydrophilic nature and high molecular weight, cellular uptake requires long incubation times, microinjection, or some degree of cell permeabilisation. Long incubation times may also result in lysosomal localisation [20, 21]. Other large molecules have been explored for nuclear targeting. Polyamines are a polyvalent cation that have been explored for nuclear targeting due to their association with the negatively charged DNA backbone. In **3.1**, a synthetic polyamine analogue, polyethyleneimine, was used for nuclear targeting of fluorescein isocyanate [17] (Figure 3.2), but the large size of this targeting group meant that cell permeabilisation was required to allow uptake.

3.3 Targeting Mitochondria

The mitochondria are commonly referred to as the 'powerhouses' of the cell as they are responsible for energy production. The mitochondria have two sets of phospholipid bilayers; the outer membrane is relatively porous to the cytoplasm, while the import of molecules to the inner membrane is more tightly regulated [22]. It is the inner membrane that is the site of cellular respiration and oxidative phosphorylation, being responsible for ATP synthesis. The presence of the electron transport chain and proton pumps result in an electrochemical gradient across the inner membrane, with a negative potential as high as $-180\,mV$ [23].

The overwhelming majority of targeting strategies for the mitochondria involve the use of delocalised lipophilic cations (DLCs). Due to their positive charge, DLCs are attracted to the negatively charged inner mitochondrial membrane and accumulate there. As the charge is delocalised over a relatively large area and the cation itself is lipophilic, the DLC can freely pass through the plasma and mitochondrial membranes. Some fluorescent scaffolds are inherently positively charged and have been used for mitochondrially targeted dyes and sensors; these include rhodamine derivatives such as **Rhodamine 123** [24], rosamine derivatives like **MitoTracker Orange** [25], and cyanine derivatives like **MitoTracker Deep Red** [26] (Figure 3.3a).

For non-cationic fluorophores, the DLC triphenylphosphonium (TPP) is the most popular targeting group, having been used with a variety of cargoes, including small molecules (*e.g.* naphthalimide [27], BODIPY [28], fluorescein [29], flavin [30]; Figure 3.3b), as well as nanomaterials [31, 32]. TPP is advantageous due to its lack of fluorescence and its compatibility with biological systems, as well as the relatively straightforward methods for covalent attachment [5]. In recent times, TPP has been modified to enhance aspects of mitochondrial uptake and targeting. A methyl-functionalisation on the phenyl ring of TPP was shown to lead to higher uptake of a fluorescein derivative **2** (Figure 3.3c) [33], while *para*-substitution with a trifluoromethyl group in a rhodamine derivative **3** prevented dissipation of the mitochondrial membrane potential compared to the unsubstituted compound (Figure 3.3c) [34].

A drawback of the use of DLCs is that their mechanism of uptake and retention is entirely dependent on the mitochondrial membrane potential. If this negative membrane potential is disrupted, mitochondrial localisation and retention is not possible. To allow for permanent mitochondrial retention, some dyes like the commercial **MitoTracker Orange** contain a

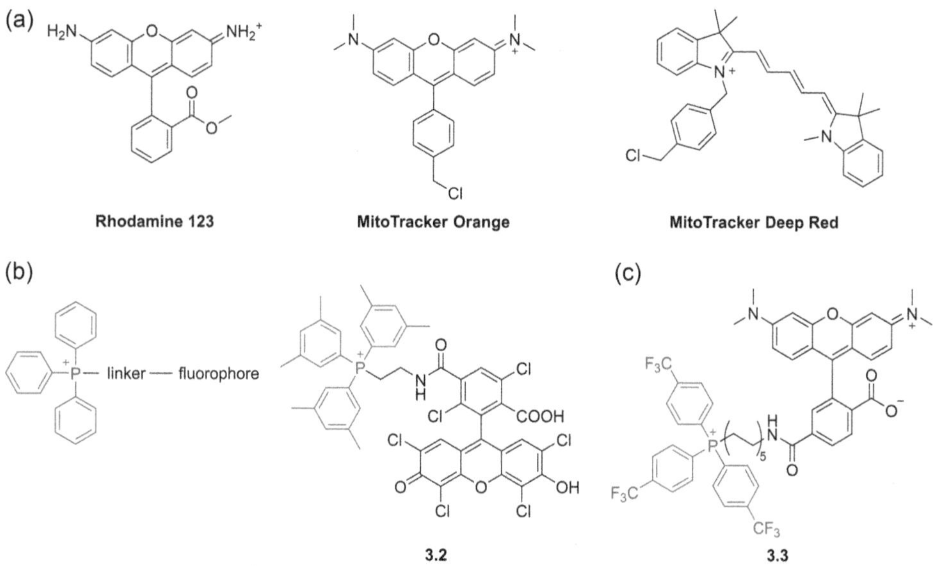

Figure 3.3 Structures of mitochondria-targeted fluorescent dyes. (a) Rhodamine 123, MitoTracker Orange, and MitoTracker Deep Red are examples of rhodamines, rosamines, and cyanines, respectively, examples of inherently fluorescent DLCs. (b) General structure for fluorescent sensors using TPP (red) as a targeting group. (c) TPP derivatives (red) include fluorescein derivative 3.2 tagged with methyl-functionalised TPP [33] and rhodamine derivative 3.3 tagged with CF$_3$ substituted TPP [34].

thiol-reactive chloromethyl group that can form covalent bonds with mitochondrial proteins after localisation is achieved [25]. It should also be noted that high concentrations of DLCs can neutralise and depolarise the mitochondrial membrane potential, resulting in disruption of mitochondrial function [35].

It is also possible to target cargo to the mitochondria using short peptide sequences, termed mitochondria-targeted peptides (MPPs). These are generally four to eight amino acids in length, consisting of alternating cationic and aromatic [36] or cationic and hydrophobic [37] residues. The exact mechanism of uptake is yet to be fully understood, but MPPs are known to localise in different compartments of the mitochondria. As with other peptide-based targeting mechanisms, longer incubation times are generally required (>1 hour), compared with relatively short times for DLCs (<15 minutes).

3.4 Targeting Lysosomes

The lysosomes are the most acidic sub-cellular organelle (pH 4–6) [38] and are responsible for the breakdown and recycling of cellular waste [39]. In addition to roles in cellular digestion, lysosomes play a role in cell signalling and energy metabolism [39]. Lysosomes are heterogeneous organelles, with differences in physical and chemical properties such as size and pH [40]. Further complicating the selective targeting of the lysosome are the membrane-bound vesicles of the endocytic and autophagic pathways that deliver waste material to the lysosome, as these can have similar properties to the lysosome.

The most widely used lysosomal targeting groups are small, lipophilic tertiary amines, which act *via* pH partitioning. At neutral pH, small molecules tagged with these tertiary amines are unprotonated and can freely diffuse through the plasma and lysosomal membranes. Once in the acidic environment of lysosome, the amine becomes protonated, and the resultant charged species can no longer pass through the lysosomal membrane and is effectively trapped. This method of lysosomal localisation has been used successfully with most classes of neutrally charged fluorophores. The most commonly used classes of tertiary amines include morpholines (*e.g.* in the commercially available **LysoSensor Green** [26], Figure 3.4) and dimethylamines (*e.g.* in the commercially available **LysoTracker Red** [26], Figure 3.4).

There are some caveats associated with the use of lipophilic tertiary amines. Their mechanism of action means that they are selective for all acidic organelles of a certain pH; this can result in the staining of endosomes, while less acidic lysosomes may remain unstained. It should also be noted that the mechanism of accumulation means that the targeting group is not inert. As lipophilic amines become protonated in the lysosome, this simultaneously results in some alkalisation of the lysosomal environment, known as lysosomotropism [43]. The increase in pH caused by lysosomotropic agents can alter lysosomal morphology and function, potentially leading to cell death. Lipophilic amines can also lead to photoinduced electron transfer (PET) quenching of fluorophores in different pH environments [44–46], which must be controlled for fluorophores detecting analytes of interest.

In recent times, new targeting strategies have been used for lysosomal targeting and retention. However, these have not been tested robustly and the mechanism of action has

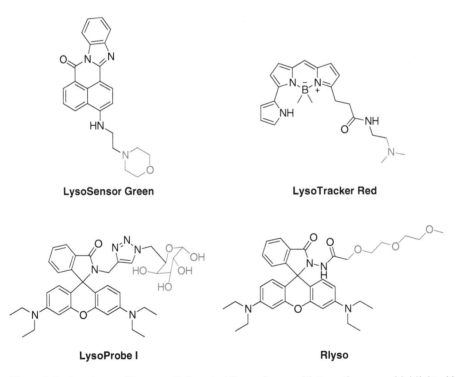

LysoSensor Green

LysoTracker Red

LysoProbe I

Rlyso

Figure 3.4 Structures of lysosomally targeted fluorophores, with targeting group highlighted in red. Most sensors use lipophilic tertiary amines, such as the morpholine in LysoSensor Green and dimethylamine in LysoTracker Red [26]. Novel targeting groups include N-linked glycans in LysoProbe 1 [41] and methylcarbitol used in Rlyso [42].

not been fully reported. These include N-linked glycans, used as a targeting group for rhodamine spirolactam derivatives such as **LysoProbe 1** (Figure 3.4) [41, 47]. Methylcarbitol has emerged as a lysosomal targeting group that does not cause alkanisation, though its mechanism of action is not understood. To date, it has been used for rhodamine and carbazole derivatives, [42, 48], such as **Rlyso** [42] (Figure 3.4).

As large macromolecules enter the cell *via* endocytosis, it is possible to fluorescently label macromolecules to follow the progress of endocytosis and track the membrane-bound vesicles in the pathway, including lysosomes. Perhaps the best-known example is FITC-dextran [49], but other combinations of macromolecule (*e.g.* albumin) and fluorophore are also possible.

3.5 Targeting Endosomes

Endosomes are digestive vesicles that are involved in endocytosis, a process by which cells internalise external materials [50]. While there are different types of endocytic pathways, the general process involves the external material being engulfed by a section of the plasma membrane. This section forms a vesicle that then breaks off and fuses with early endosomes. Here, some biomolecules may be sorted for exocytosis. Early endosomes may fuse with or transform into late endosomes, and late endosomes fuse with the lysosome [50]. In addition to early and late endosomes, there are also recycling endosomes, which are involved in the transferral of materials from the early endosome to the plasma membrane or Golgi apparatus [51].

It is relatively difficult to target the endosomes with small molecules due to the heterogeneous environments of the organelles and flux of the endocytic pathway. There are no conclusive or robust targeting groups for the endosomes, but endosomally targeted fluorescent sensors with novel targeting strategies have been reported. Some take advantage of the process of endocytosis; for example, Sampedro *et al.* targeted late endosomes with BODIPY and fluorescein-based sensors tagged with macrocyclic di(cyclosquaramides), such as **3.4** (Figure 3.5) [52]. This targeting group binds to phosphates on the plasma membrane, which are taken up by the cell *via* receptor-mediated endocytosis.

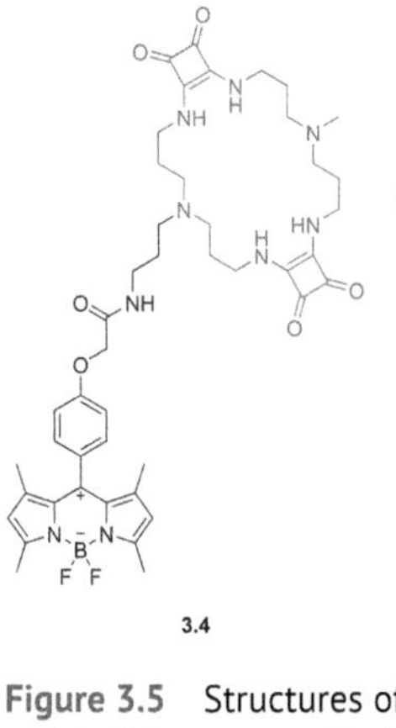

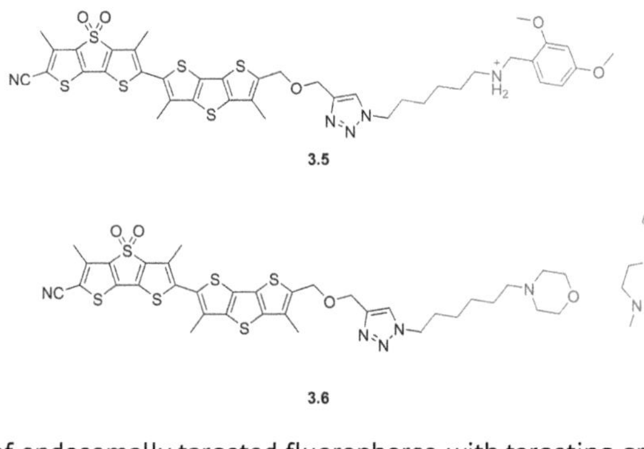

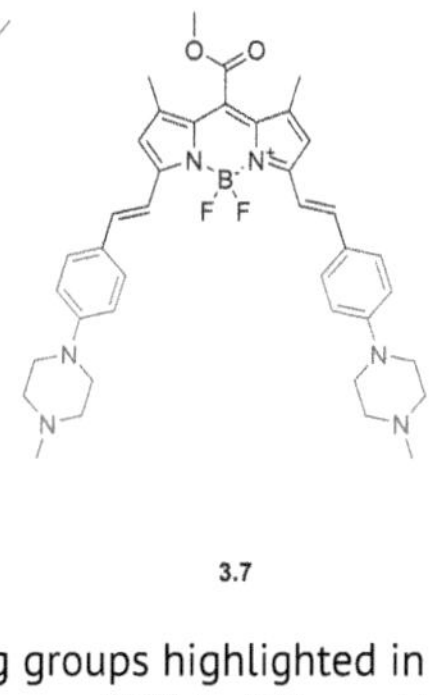

Figure 3.5 Structures of endosomally targeted fluorophores, with targeting groups highlighted in red. In **3.4**, macrocyclic di(cyclosquaramides) were used to target late endosomes [52]. In **3.5**, weakly acidic benzyl substituted amines targeted early endosomes and while in **3.6**, weakly acidic amines targeted late endosomes and lysosomes. [53]; **3.7** is a (4-methyl-1-piperazinyl)phenyl-substituted BODIPY that can stain pH flux of late endosomes [54].

Other targeting groups take advantage of the pH environment of endosomes. Piazzolla *et al.* tested a range of weakly acidic benzyl substituted amines, hypothesizing that they would be selectively protonated in the environment of the endosome to allow retention [53]. Targeting groups with higher pK_a values, such as the (2,4-dimethoxyphenyl)methanamine group in **3.5**, were found to localise fluorescent membrane tension sensors to the less-acidic early endosomes (Figure 3.5). In contrast, targeting groups with lower pK_a values, such as the morpholine moiety, can localise sensors to the more acidic late endosomes and lysosomes, such as in **3.6** (Figure 3.5). In an alternative approach, Miao *et al.* found a pH-sensitive piperazine-based group that enabled the BODIPY dye **3.7** to stain only the flux from the late endosome to lysosome (Figure 3.5) [54].

3.6 Targeting Autophagic Compartments

In contrast to endocytosis, autophagy is a process that removes damaged or unnecessary cellular components [55]. The major pathway is macroautophagy. This begins with phago-phores (also known as isolation membranes) enveloping waste material, producing autophagosomes. Autophagosomes later fuse with the lysosome, forming a vesicle often referred to as the autolysosome [56]. While there are several other pathways for autophagy, fluorescent sensors for autophagic compartments most commonly target the distinctive vesi-cles of the macroautophagic pathway.

As for endosomal targeting, there are no consensus targeting groups for the autophagic compartments. The commercial dye Cyto-ID is a cationic lipophilic tracer that labels autophagic compartments, but no further information about the chemical structure or mechanism of action has been reported [57]. **DALGreen** and **DAPGreen** are fluorescent 1,8-naphthalimides reported to stain the autophagic compartments [58]. **DAPGreen** is a pH insensitive sensor that stains all autophagic compartments, with an aminopentyl moiety in the four-position and pentyl group at the imide (Figure 3.6). While **DALgreen** also has a pentyl group at the imide, it is a pH-sensitive sensor due to the pH-sensitive aminoethylpip-erazine group at the four-position (Figure 3.6) and therefore shows greater fluorescence in the acidic environments of late-stage autophagy. The mechanism of action has not been elucidated, but the amphiphilic, detergent-like structure is thought to allow for uptake in autophagic membranes.

Some sensors bearing traditional lysosomal targeting groups (*e.g.* lipophilic tertiary amines) can examine changes in the chemical environment during the transition from the lysosome to autolysosome. For example, the coumarin-based sensor **Lyso-OC** was reported to detect the change in polarity after fusion between lysosomes and autophagosomes (Figure 3.6) [59]. Similarly, a ratiometric p-methoxyphenylacetylene-substituted carbazole, **Lyso-MPCB**, was reported to measure pH change after this fusion (Figure 3.6) [60].

3.7 Targeting Peroxisomes

Peroxisomes are involved in several biologically important oxidative processes, including the oxidation of fatty acids and the conversion of reactive oxygen and nitrogen species [61]. Molecular oxygen (O_2) is a common co-substrate, producing hydrogen peroxide (H_2O_2) as a product, which is toxic in high quantities [62]. The peroxisomal enzyme catalase can use H_2O_2 to oxidise other substrates, catalysing the decomposition of H_2O_2 in the process [63].

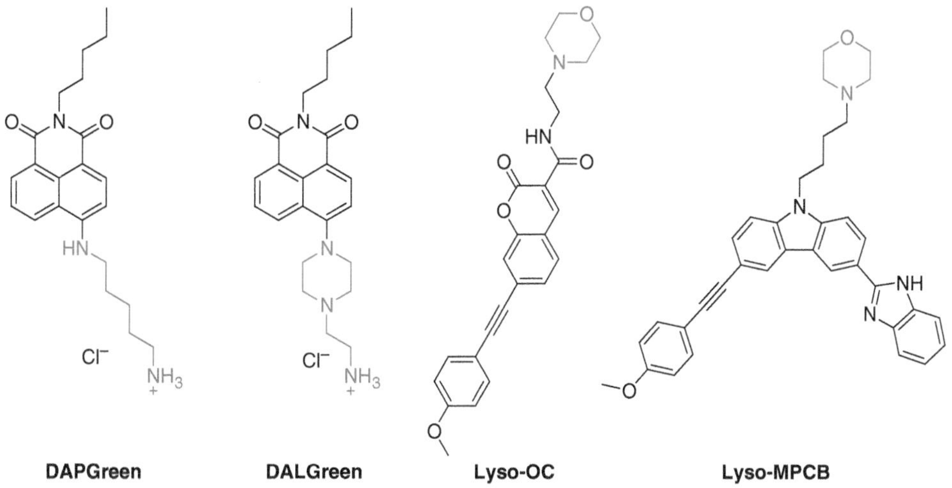

Figure 3.6 Structures of fluorescent sensors targeted to autophagy-related compartments, with targeting groups highlighted in red. DAPgreen and DALGreen contain a hydrophobic alkyl chain as imide substituent of both groups, with DAPGreen an aminopentyl moiety in the four-position and DALGreen containing an aminoethylpiperazine [58]. Lyso-OC [59] and (D) Lyso-MPCB [60] both contain the lysosome-targeted morpholine, but are sensitive to changes in the flux from lysosome to autolysosome.

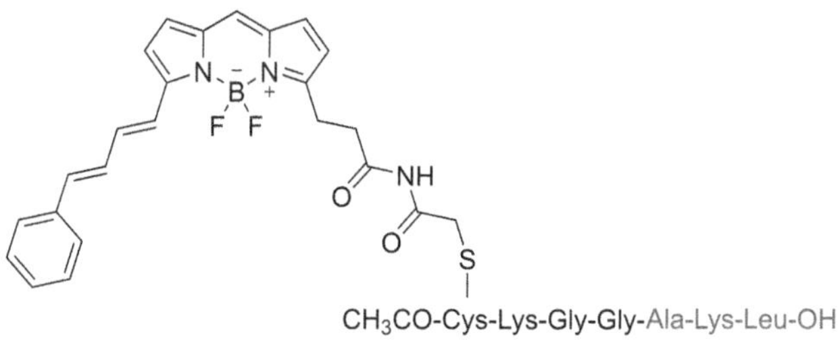

CH₃CO-Cys-Lys-Gly-Gly-Ala-Lys-Leu-OH

3.8

Figure 3.7 Structure of BODIPY tagged with C-terminal PTS-1 peptide, with targeting group highlighted in red [18].

Peroxisomal targeting signals (PTS) are short peptide sequences that allow the sorting of native proteins to the peroxisome. The most common sequences are C-terminal PTS-1 tripeptides, with consensus sequence Ser/Cys/Ala-Lys/Arg/His-Leu [64]. These peptides can be used as delivery vectors for fluorescent dyes. For example, various dyes have been targeted to the peroxisome using the PTS-1 tag Ala-Lys-Leu, such as the BODIPY **3.8** (Figure 3.7) [65].

3.8 Targeting the Endoplasmic Reticulum

The endoplasmic reticulum (ER) is a continuous, folded membrane network involved in several cellular functions, including protein folding and transport. The two major regions are the smooth ER and rough ER, with the latter containing bound ribosomes for protein

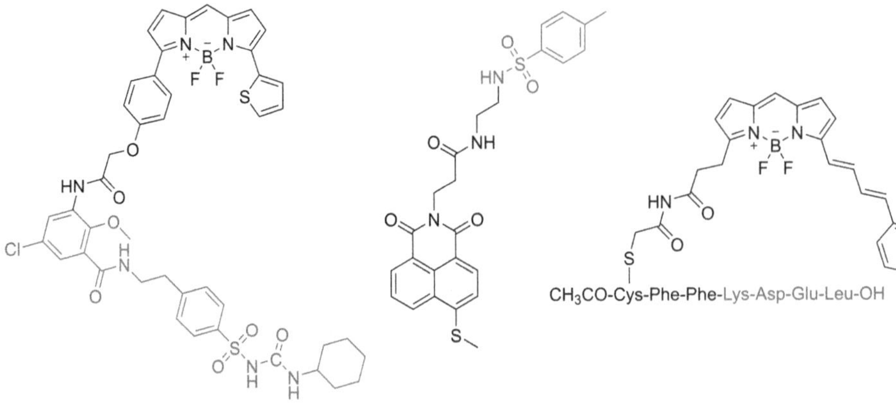

Figure 3.8 Structures of ER-targeted fluorophores, with targeting groups highlighted in red. Glibenclamide is used by ER-Tracker Red [26], NSp uses a sulfonamide [68], and **3.9** is targeted with a Lys-Asp-Glu-Leu sequence [18].

translation [66]. Some post-translation modification, such as N-linked glycosylation, also occurs in the rough ER.

The majority of ER targeting groups are small molecules that bind tightly and specifically to ER-localised proteins. A classic example is glibenclamide, used in the commercial sensor **ER-Tracker Red** [26]. Glibenclamide is a sulfonylurea-based medication traditionally used to treat diabetes and binds to sulfonylurea receptors (Figure 3.8) [67]. Smaller sulfonamides are thought to have a similar mechanism of action and are more synthetically accessible; for example, *para*-toluenesulfonamide was used in an ER-targeted naphthalimides-based polarity sensor, **NSp** (Figure 3.8) [68].

Targeting using peptide-based sequences has also been demonstrated. ER-targeted proteins often contain long N-terminal signal sequences, and ER resident proteins generally also contain a Lys-Asp-Glu-Leu retention signal [69]; **3.9** comprises a BODIPY tagged with a seven-amino-acid peptide containing the Lys-Asp-Glu-Leu retention sequence and was demonstrated to exhibit ER localisation (Figure 3.8) [18].

3.9 Targeting the Golgi Apparatus

The Golgi apparatus is responsible for the post-translational modification of proteins, including glycosylation, sulfonation, phosphorylation, or lipidation [70]. Proteins are subsequently sorted into membrane-bound vesicles and delivered to the cellular destination or secreted from the cell [70]. The structure of the Golgi apparatus is a series of flattened membrane-enclosed sacs known as cisternae, with cargo entering *via* the *cis* face of the stack and leaving *via* the *trans* face [71]. These two sections are sometimes referred to as the *cis*- and *trans*-Golgi networks.

The most common Golgi-targeting strategy involves the use of sphingolipids such as ceramide, which has successfully been used to target a range of fluorophores to the Golgi [72–74]; a classic example is the commercially available sensor **C5-DMB-Cer** (Figure 3.9) [72]. Biological ceramides are transported from their site of synthesis in the ER to the Golgi *via* the ceramide protein [75].

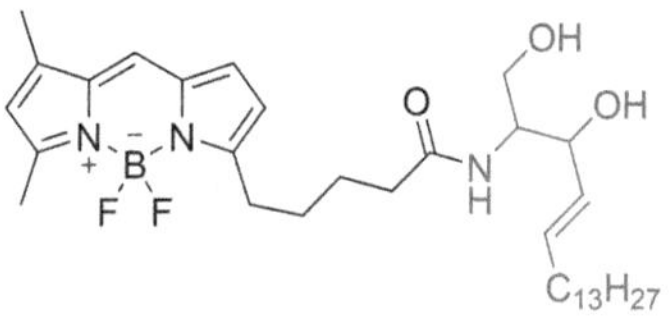

C₅-DMB-Cer

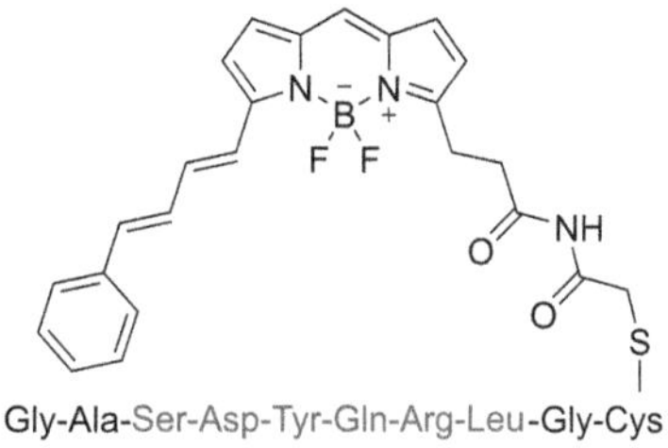

3.10

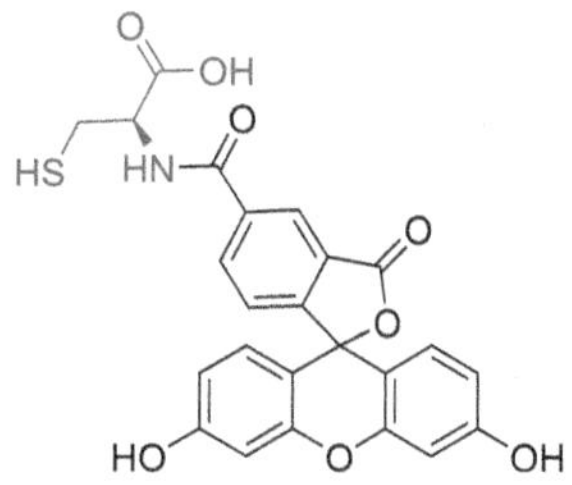

Gly-Ala-Ser-Asp-Tyr-Gln-Arg-Leu-Gly-Cys

Fluorescein-cysteine-1

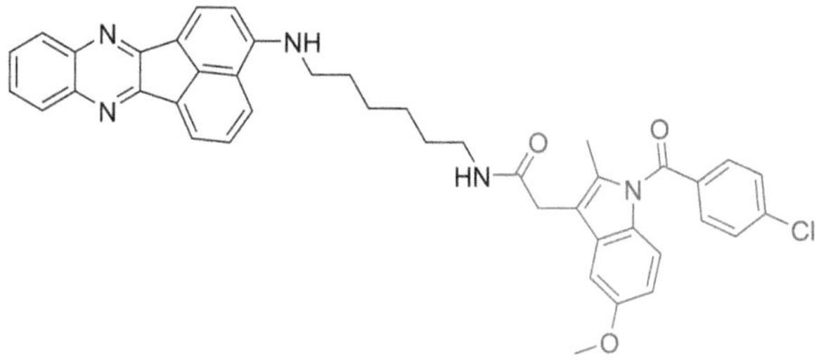

ANQ-IMC-6

Figure 3.9 Structures of Golgi-targeted fluorophores, with targeting groups highlighted in red. Ceramide is used in C5-DMB-Cer (also known as BODIPY FL C5 ceramide) [72], **3.10** is tagged with Ser-X-Tyr-Gln-Arg-Leu sequence [18], L-cysteine is used in fluorescein-cysteine-1 [76], and ANQ-IMC-6 uses indomethacine, a COX-2 inhibitor, is used in ANQ-IMC-6 [77].

A variety of novel approaches have also been reported for Golgi targeting. A variant of the Golgi signal peptide Ser-X-Tyr-Gln-Arg-Leu was used to localise a BODIPY variant **3.10** to the *trans*-Golgi network (Figure 3.9) [18]. L-cysteine was successfully used to target a range of cargo, including nanomaterials and small-molecule fluorophores (Figure 3.9) [76]; the mechanism of uptake is thought to mimic that of the proteins galactosyltransferase and protein kinase D, which anchor in Golgi apparatus *via* cysteine residues or cysteine-rich domains. Indomethacin, a drug that binds to the protein COX-2, has also been used to target the fluorophore acenaphtho[1,2-b]quinoxaline to the Golgi apparatus (Figure 3.9) [77]; however, significant uptake was only seen in cancer cell lines that overexpressed COX-2.

3.10 Targeting Lipid Droplets

Lipid droplets are the organelles responsible for energy storage, consisting of a core of neutral lipids and a phospholipid monolayer [78]. While traditionally considered inert bodies, they are now known to interact dynamically with all major organelles [79]. While found in virtually all cells, lipid droplets differ greatly in size and distribution, which depend strongly on cell type and nutrient availability [80].

Most lipid droplet sensors do not have a specific binding mechanism or targeting moiety and may not necessarily be localised only in the lipid droplet. Rather, they tend to be highly

(a)　　　　　　　　　(b)　　　　　　　　　(c)

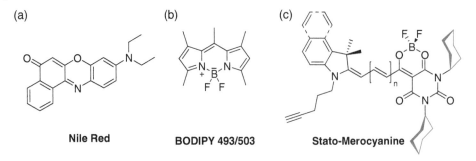

Nile Red　　　　　**BODIPY 493/503**　　　**Stato-Merocyanine**

Figure 3.10 Structures of lipid droplet-targeted fluorophores, with targeting groups highlighted in red. (a) Nile Red and (b) BODIPY 493/503 are both examples of highly conjugated, lipophilic structures. (c) Two cyclohexyl rings are grafted onto the core of a class of meracyanine derivatives [84].

conjugated, lipophilic structures that exhibit a photophysical response (*e.g.* fluorescence turn-on, emission wavelength shift) in the non-polar, hydrophobic environment of lipid droplets compared to the aqueous environment of the rest of the cell. Commercial markers such as **Nile Red** (Figure 3.10) [81] and **BODIPY 493/503** (Figure 3.10) [82] operate *via* this mechanism. However, these dyes can also accumulate in the lipophilic environment of intracellular membranes and display some non-specific staining [81, 83].

Few selective targeting groups for the lipid droplets have been developed. In one example, in order to target lipid droplets over other lipophilic cellular structures, two cyclohexyl rings were attached to the core of **Stato-Merocyanines**, a class of merocyanine derivatives (Figure 3.10) [84]. This increased hydrophobicity and bulkiness while preventing quenching *via* π-stacking. Two cyclohexyl rings have also been attached to a Nile Red derivative to increase lipid droplet selectivity [85].

3.11 Targeting the Plasma Membrane

While not an organelle, the plasma membrane is an important cellular structure responsible for separating the interior contents from the outside environment. The membrane consists of a phospholipid bilayer embedded with sterols that assist with maintaining structure and fluidity [86]. The major function of the plasma membrane is the regulation and movement of materials entering and exiting the cell. While uncharged small molecules can pass through the phospholipid bilayer *via* passive diffusion, other species must pass through transmembrane protein channels and transporters.

The most common method of achieving plasma membrane localisation is through the utilisation of fatty acid or alkyl chain appendages. These can insert directly into the hydrophobic core of the lipid bilayer, while the fluorophore is situated among the phospholipid head groups. Examples include the di-alkyl indocarbocyanine dye DiI (Figure 3.11) [87] and the lauric acid moiety attached to Laurdan (Figure 3.11) [88], itself a derivative of the membrane stain Prodan [89]. It is also possible for membrane probes to respond selectively to cellular events. For example, Zwicker *et al.* produced **P-IID**, a fluorogenic probe for apoptosis [90]. In addition to a stearate anchor that inserts into the membrane, the fluorogenic response only occurs when the bis(zinc-dipicolylamine) moiety opens in the presence of phosphatidylserine, an apoptosis marker. Alternatively, fluorescent analogues of membrane

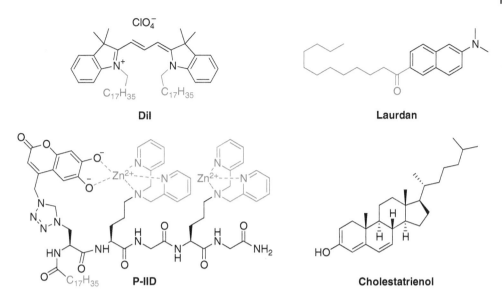

Figure 3.11 Structure of plasma membrane-targeted fluorophores, with targeting groups highlighted in red. Dil (or Dil$_{18}$(3)) [87] and Laurdan [88] use long, hydrophobic alkyl chains. P-IID also contains a long hydrophobic alkyl chain but also has a bis(zinc-dipicolylamine) moiety that opens selectively in the presence of the apoptosis marker phosphatidylserine [90]. Cholestatrienol is a fluorescent cholesterol analogue [91].

lipids may be used to track their cellular activity, though this may not be necessarily suitable for plasma membrane staining in general. An example is cholestatrienol, a UV-emitting analogue of cholesterol (Figure 3.11) [91].

In recent times, non-traditional targeting strategies have been employed to image events near and outside the plasma membrane. Most involve the use of a small-molecule ligand that can bind to a plasma membrane protein. For example, Fujishima *et al.* reported ligand-directed acyl imidazole (LDAI) chemistry as an approach to modify endogenous proteins, allowing the selective fluorescent labelling of the folate receptor. One sensor involved alkyloxyacyl imidazole derivatives containing the ligand methotrexate, as well as a fluorescein. Another example is NR-PEG-CBT, consisting of the fluorophore Nile Red conjugated to the peptide ligand carbetocin *via* a polyethylene glycol linker, allowing for the labelling of a membrane-localised G-protein-coupled receptor [92].

3.12 Targeting the Cytoskeleton

The cytoskeleton is a network of fibres and filaments that provide structure and mechanical support for the cell. The main components are microtubules, microfilaments, and intermediate filaments [93]. Microtubules are polymers of the protein tubulin and are rigid hollow rods approximately 25 nm in diameter, the largest cytoskeletal fibres. Microfilaments are polymers of the actin protein and the smallest cytoskeletal fibres at approximately 6 nm in diameter. Notably, they interact with myosin, which allows for changes in cell shape and movement. Intermediate filaments are around 10 nm in diameter and are composed of numerous protein subunits.

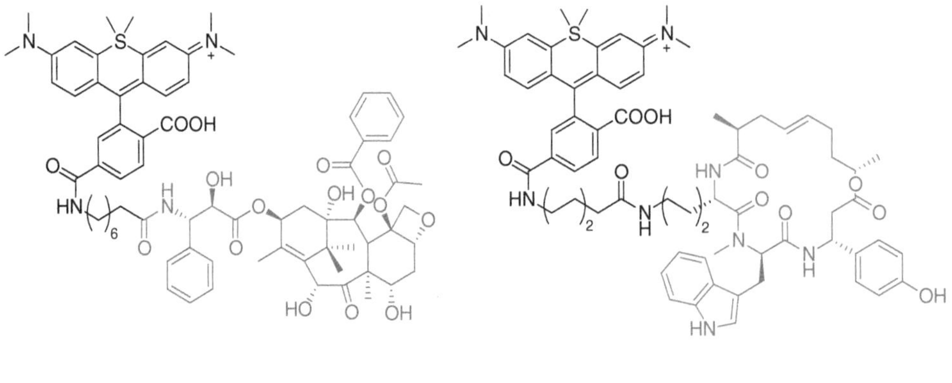

SiR-tubulin　　　　　　　　　**SiR-actin**

Figure 3.12 Structures of fluorescent sensors for the cytoskeleton, targeting groups highlighted in red. SiR-tubulin uses a docetaxel derivative to target tubulin [94], while SiR-actin uses jasplakinolide to target actin [94].

The cytoskeleton is not an organelle, and the proteins that compose the fibres and filaments must be directly targeted. The primary targeting method is through drugs known to bind to components of the cytoskeleton. Docetaxel binds to tubulin and can be attached to fluorophores to label microtubules, such as in the silicon-rhodamine derivative **SiR-tubulin** (Figure 3.12) [94]. Similarly, jasplakinolide binds to actin and can be used to label microfilaments, such as in the stain **SiR-actin** (Figure 3.12) [94]. Small-molecule fluorophores may also be directly appended onto tubulin or actin, but it can be difficult to control the point of fluorophore attachment, and cellular uptake is a major issue due to the large size of the protein [95, 96].

3.13 Targeting the Cytosol

The cytosol refers to the intracellular fluid matrix that envelops the organelles [97]. It is the major component of the cytoplasm, which is defined as all the cellular contents enclosed by the plasma membrane, excluding the nucleus. The cytosol is a complex solution of ions and other biomolecules that are involved in various cellular processes, including metabolism and transport [98].

A general, diffuse staining pattern throughout the cytoplasmic or cytosolic regions is often considered sufficient evidence for localisation. However, incubation time and dose concentration should be considered before any conclusions are drawn; low incubation times may not allow for sufficient time for the sensor to reach the final cellular destination, while an excessive concentration of sensor may mask the true localisation.

3.14 Trapping and Accumulation of Sensors

For ideal use, targeting groups for fluorescent sensors should allow for selective localisation and retention with minimal disruption to the biological cellular environment. Furthermore, the targeting group should not affect the photophysical response of the fluorescent sensor, especially if the sensor is designed to respond to an analyte of interest. However, there is no 'one-size-fits-all' approach for the selection of appropriate targeting groups in organelle-targeted sensors and their subsequent use in biological experiments. For novel sensors, even relatively robust groups may not guarantee uptake and retention, and incubation time and dosage can be highly variable considering the organelle of interest and the structure of the sensor. Considerations of the general uptake and retention mechanisms are useful for biological experimental design.

Many of the targeting strategies for small-molecule fluorophores involve passive targeting groups that rely on diffusion, with retention reliant on relatively weak non-covalent interactions. For example, the most prevalent mitochondrial targeting group, TPP, relies on the electrostatic attraction between the positive DLC and negative mitochondrial membrane potential, while for lysosomes, lipophilic amines are trapped due to their protonation in the acidic lumen. The uptake and retention of sensors with these targeting groups are thus dependent on the maintenance of the chemical environment, and long-term retention is not guaranteed. Where uptake to the organelle of interest is rapid, it is possible to attach a moiety that can attach covalently in the organelle to allow for permanent retention. An example is the thiol-reactive chloromethyl moiety in some MitoTracker stains [25]; while this moiety can form covalent attachment with all thiol-containing proteins, uptake is sufficiently rapid to ensure that selective attachment only occurs in the mitochondria.

Other targeting strategies are less reliant on the continual maintenance of the chemical environment of the organelle. Some are still reliant on passive diffusion to reach the subcellular location of interest, but bind to a target molecule; for example, docetaxel binds to microtubules and Hoechst conjugates bind to adenine-thymine-rich regions in the minor groove of the DNA. While not necessarily covalent bonds, these binding interactions are unlikely to be displaced, ensuring retention. Peptide targeting sequences are recognised by the native protein sorting machinery of the cell to allow targeting and retention in the organelle of interest, such as the PTS-1 tripeptide for peroxisomal targeting. The chief difficulty is that they can be relatively bulky and may require long incubation times, and that peptide sequences may be cleaved by cytoplasmic enzymes. Similarly, some targeting groups are non-peptide biomolecules that are targeted through cell-sorting mechanisms, such as ceramide for the Golgi apparatus.

It should be also noted that self-labelling proteins (SLPs, Chapter 7) can be used for specific targeting of a protein of interest. While the targeting can be more specific, this strategy requires genetic engineering and is not suitable for all experimental designs. SLPs are genetically engineered enzymes that form covalent attachment to a specific substrate group. The SLP can be directed to an organelle of interest *via* fusion with an appropriate signal peptide or protein. The substrate group can be attached to a fluorescent sensor, allowing for labelling of cells expressing the SLP. Examples of common SLPs are HaloTag [99], SNAP-Tag [100], and CLIP-tag [100].

In summary, there are well-established strategies for achieving the subcellular localisation of fluorescent sensors, though robust targeting groups are only available for a small number of organelles. An understanding of the uptake and retention mechanisms of these sensors is valuable for biological experimental design and provides insight for the further development of fluorescent tools.

References

1 Cohen, S., Valm, A.M., and Lippincott-Schwartz, J. (2018). *Current Opinion in Cell Biology* 53: 84–91.

2 Weisz, O.A. (2003). *Traffic* 4 (2): 57–64.

3 Specht, E.A., Braselmann, E., and Palmer, A.E. (2017). *The Annual Review of Physiology* 79 (1): 93–117.

4 New, E.J. (2016). *ACS Sensors* 1 (4): 328–333.

5 Zielonka, J., Joseph, J., Sikora, A. et al. (2017). *Chemical Reviews* 117 (15): 10043–10120.

6 Shibata, Y. and Morimoto, R.I. (2014). *Current Biology* 24 (10): R463–R474.

7 Alber, F., Dokudovskaya, S., Veenhoff, L.M. et al. (2007). *Nature* 450 (7170): 695–701.

8 Kelich, J.M. and Yang, W. (2014). *The International Journal of Molecular Sciences* 15 (8): 14492–14504.

9 Pjura, P.E., Grzeskowiak, K., and Dickerson, R.E. (1987). *The Journal of Molecular Biology* 197 (2): 257–271.

10 Zhang, X., Ye, Z., Zhang, X. et al. (2019). *Chemical Communications* 55 (13): 1951–1954.

11 Bucevicius, J., Keller-Findeisen, J., Gilat, T. et al. (2019). *Chemical Science* 10 (7): 1962–1970.

12 Nakamura, A., Takigawa, K., Kurishita, Y. et al. (2014). *Chemical Communications* 50 (46): 6149–6152.

13 Wang, K.N., Liu, L.Y., Mao, D. et al. (2021). *Angewandte Chemie International Edition* 60 (27): 15095–15100.

14 Ikeda, M., Nakagawa, H., Ban, S. et al. (2010). *Free Radical Biology and Medicine* 49 (11): 1792–1797.

15 Durand, R.E. and Olive, P.L. (1982). *The Journal of Histochemistry and Cytochemistry* 30 (2): 111–116.

16 Wen, Y., Liu, K., Yang, H. et al. (2014). *Analytical Chemistry* 86 (19): 9970–9976.

17 Saito, M. and Saitoh, H. (2012). *Bioscience, Biotechnology, and Biochemistry* 76 (9): 1777–1780.

18 Pap, E.H., Dansen, T.B., van Summeren, R., and Wirtz, K.W. (2001). *Experimental Cell Research* 265 (2): 288–293.

19 Gronewold, A., Horn, M., and Neundorf, I. (2018). *The Beilstein Journal of Organic Chemistry* 14: 1378–1388.

20 Ragin, A.D., Morgan, R.A., and Chmielewski, J. (2002). *Chemistry & Biology* 9 (8): 943–948.

21 Tkachenko, A.G., Xie, H., Coleman, D. et al. (2003). *Journal of the American Chemical Society* 125 (16): 4700–4701.

22 Giacomello, M., Pyakurel, A., Glytsou, C., and Scorrano, L. (2020). *Nature Reviews Molecular Cell Biology* 21 (4): 204–224.

23 Kuhlbrandt, W. (2015). *BMC Biology* 13: 89.

24 Johnson, L.V., Walsh, M.L., and Chen, L.B. (1980). *Proceedings of the National Academy of Sciences of the United States of America* 77 (2): 990–994.

25 Scorrano, L., Petronilli, V., Colonna, R. et al. (1999). *Journal of Biological Chemistry* 274 (35): 24657–24663.

26 Johnson, I.S. and Michelle, T. (ed.) (2010). *The Molecular Probes® Handbook*, 11e. Carlsbad, CA: Life Technologies.

27 Pak, Y.L., Li, J., Ko, K.C. et al. (2016). *Analytical Chemistry* 88 (10): 5476–5481.

28 Miao, J., Huo, Y., Liu, Q. et al. (2016). *Biomaterials* 107: 33–43.

29 Denisov, S.S., Kotova, E.A., Plotnikov, E.Y. et al. (2014). *Chemical Communications* 50 (97): 15366–15339.

30 Kaur, A., Brigden, K.W.L., Cashman, T.F. et al. (2015). *Organic & Biomolecular Chemistry* 13 (24): 6686–6689.

31 Marrache, S. and Dhar, S. (2012). *Proceedings of the National Academy of Sciences of the United States of America* 109 (40): 16288–16293.

32 D'Souza, G.G.M., Cheng, S.-M., Boddapati, S.V. et al. (2008). *Journal of Drug Targeting* 16 (7–8): 578–585.

33 Hu, Z., Sim, Y., Kon, O.L. et al. (2017). *Bioconjugate Chemistry* 28 (2): 590–599.

34 Kulkarni, C.A., Fink, B.D., Gibbs, B.E. et al. (2021). *The Journal of Medicinal Chemistry* 64 (1): 662–676.

35 Trnka, J., Elkalaf, M., and Anděl, M. (2015). *PLoS One* 10 (4): e0121837.

36 Szeto, H.H. (2006). *The AAPS Journal* 8 (2): E277–E283.

37 Jean, S.R., Ahmed, M., Lei, E.K. et al. (2016). *Accounts of Chemical Research* 49 (9): 1893–1902.

38 Johnson, D.E., Ostrowski, P., Jaumouille, V., and Grinstein, S. (2016). *Journal of Cell Biology* 212 (6): 677–692.

39 Settembre, C., Fraldi, A., Medina, D.L., and Ballabio, A. (2013). *Nature Reviews Molecular Cell Biology* 14: 283–296.

40 Xu, H. and Ren, D. (2015). *The Annual Review of Physiology* 77: 57–80.

41 Yapici, N.B., Bi, Y., Li, P. et al. (2015). *Scientific Reports* 5: e8576.

42 Zhu, H., Fan, J., Xu, Q. et al. (2012). *Chemical Communications* 48 (96): 11766–11768.

43 Lu, S., Sung, T., Lin, N. et al. (2017). *PLoS One* 12: e0173771.

44 Gan, J., Chen, K., Chang, C.-P., and Tian, H. (2003). *Dyes and Pigments* 57 (1): 21–28.

45 Zhang, J., Yang, M., Mazi, W. et al. (2016). *ACS Sensors* 1: 158–165.

46 Duan, H., Ding, Y., Huang, C. et al. (2019). *Chinese Chemical Letters* 30 (1): 55–57.

47 Chen, X., Bi, Y., Wang, T. et al. (2015). *Scientific Reports* 5: 9004.

48 Wang, X.-D., Fan, L., Ge, J.-Y. et al. (2019). *Spectrochimica Acta Part A: Molecular and Biomolecular Spectroscopy.* 221: 117175.

49 Lencer, W.I., Weyer, P., Verkman, A.S. et al. (1990). *American Journal of Physiology-Endocrinology and Metabolism* 258 (2 Pt 1): C309–C317.

50 Doherty, G.J. and McMahon, H.T. (2009). *Annual Review of Biochemistry* 78: 857–902.

51 Taguchi, T. (2013). *Journal of Biochemistry* 153 (6): 505–510.

52 Sampedro, A., Villalonga-Planells, R., Vega, M. et al. (2014). *Bioconjugate Chemistry* 25 (8): 1537–1546.

53 Piazzolla, F., Mercier, V., Assies, L. et al. (2021). *Angewandte Chemie-International Edition* 60 (22): 12258–12563.

54 Miao, F., Uchinomiya, S., Ni, Y. et al. (2016). *ChemPlusChem* 81 (11): 1209–1215.

55 Bento, C.F., Renna, M., Ghislat, G. et al. (2016). *Annual Review of Biochemistry* 85: 685–713.

56 Yu, L., McPhee, C.K., Zheng, L. et al. (2010). *Nature* 465 (7300): 942–946.

57 Guo, S., Liang, Y., Murphy, S.F. et al. (2015). *Autophagy* 11 (3): 560–572.

58 Iwashita, H., Sakurai, H.T., Nagahora, N. et al. (2018). *FEBS Letters* 592 (4): 559–567.

59 Jiang, J., Tian, X., Xu, C. et al. (2017). *Chemical Communications* 53 (26): 3645–3648.

60 Ning, P., Hou, L., Feng, Y. et al. (2019). *Chemical Communications* 55 (12): 1782–1785.

61 Wanders, R.J.A., Waterham, H.R., and Ferdinandusse, S. (2016). *Frontiers in Cell and Developmental Biology* 3 (83).

62 Beckman, K.B. and Ames, B.N. (1998). *Physiological Reviews* 78 (2): 547–581.

63 Alberts, B., Johnson, A., Lewis, J. et al. (2017). *Molecular Biology of the Cell* 1451 (1): 17–34.

64 Hettema, E.H., Distel, B., and Tabak, H.F. (1999). *Biochimica et Biophysica Acta* 1451 (1): 17–34.

65 Dansen, T.B., Pap, E.H.W., Wanders, R.J., and Wirtz, K.W. (2001). *The Histochemical Journal* 33 (2): 65–69.

66 Schwarz, D.S. and Blower, M.D. (2016). *Cellular and Molecular Life Sciences* 73 (1): 79–94.

67 Hambrock, A., Löffler-Walz, C., and Quast, U. (2002). *The British Journal of Pharmacology* 136 (7): 995–1004.

68 Song, W., Dong, B., Lu, Y. et al. (2019). *The New Journal of Chemistry* 43 (30): 12103–12108.

69 Pelham, H.R.B. (1990). *Trends in Biochemical Sciences* 15 (12): 483–486.

70 Day, K.J., Staehelin, L.A., and Glick, B.S. (2013). *Histochemistry and Cell Biology* 140 (3): 239–249.

71 Farquhar, M.G. and Palade, G.E. (1981). *Journal of Cell Biology* 91 (3): 77s–103s.

72 Pagano, R.E., Martin, O.C., Kang, H.C., and Haugland, R.P. (1991). *Journal of Cell Biology* 113 (6): 1267–1279.

73 Lipsky, N.G. and Pagano, R.E. (1983). *Proceedings of the National Academy of Sciences of the United States of America* 80 (9): 2608–2612.

74 Erdmann, R.S., Takakura, H., Thompson, A.D. et al. (2014). *Angewandte Chemie (International Edition in English).* 53 (38): 10242–10246.

75 Rao, R.P., Yuan, C., Allegood, J.C. et al. (2007). *Proceedings of the National Academy of Sciences of the United States of America* 104 (27): 11364–11369.

76 Li, R.S., Gao, P.F., Zhang, H.Z. et al. (2017). *Chemical Science* 8 (10): 6829–6835.

77 Zhang, H., Fan, J., Wang, J. et al. (2013). *Journal of the American Chemical Society* 135 (31): 11663–11669.

78 Ruggles, K.V., Turkish, A., and Sturley, S.L. (2013). *Annual Review of Nutrition* 33: 413–451.

79 Gao, Q. and Goodman, J.M. (2015). *Frontiers in Cell and Developmental Biology* 3: 49.

80 Olzmann, J.A. and Carvalho, P. (2019). *Nature Reviews Molecular Cell Biology* 20 (3): 137–155.

81 Greenspan, P., Mayer, E.P., and Fowler, S.D. (1985). *The Journal of Cell Biology* 100: 965–973.

82 Spangenburg, E.E., Pratt, S.J.P., Wohlers, L.M., and Lovering, R.M. (2011). *Journal of Biomedicine and biotechnology* 2011: 598358.

83 Gocze, P.M. and Freeman, D.A. (1994). *Cytometry* 17 (2): 151–158.

84 Collot, M., Fam, T.K., Ashokkumar, P. et al. (2018). *Journal of the American Chemical Society* 140 (16): 5401–5411.

85 Danylchuk, D.I., Jouard, P.H., and Klymchenko, A.S. (2021). *Journal of the American Chemical Society* 143 (2): 912–924.

86 Bernardino de la Serna, J., Schutz, G.J., Eggeling, C., and Cebecauer, M. (2016). *Frontiers in Cell and Developmental Biology* 4: 106.

87 von Bartheld, C.S., Cunningham, D.E., and Rubel, E.W. (1990). *Journal of Histochemistry and Cytochemistry* 38 (5): 725–733.

88 Parasassi, T., Di Stefano, M., Loiero, M. et al. (1994). *Biophysical Journal* 66 (3): 763–768.

89 Krasnowska, E.K., Gratton, E., and Parasassi, T. (1998). *Biophysical Journal* 74 (4): 1984–1993.

90 Zwicker, V.E., Oliveira, B.L., Yeo, J.H. et al. (2019). *Angewandte Chemie International Edition* 58 (10): 3087–3091.

91 Fischer, R.T., Stephenson, F.A., Shafiee, A., and Schroeder, F. (1984). *Chemistry and Physics of Lipids* 36 (1): 1–14.

92 Karpenko, I.A., Kreder, R., Valencia, C. et al. (2014). *ChemBioChem* 15 (3): 359–363.

93 Fletcher, D.A. and Mullins, R.D. (2010). *Nature* 463 (7280): 485–492.

94 Lukinavicius, G., Reymond, L., D'Este, E. et al. (2014). *Nature Methods* 11 (7): 731–733.

95 Waterman-Storer, C.M. and Salmon, E.D. (1998). *Biophysical Journal* 75 (4): 2059–2069.

96 Feramisco, J.R. (1979). *Proceedings of the National Academy of Sciences of the United States of America* 76 (8): 3967–3971.

97 Clegg, J.S. (1984). *American Journal of Physiology-Regulatory, Integrative and Comparative Physiology* 246 (2): R133–R151.

98 Verkman, A.S. (2002). *Trends in Biochemical Sciences* 27 (1): 27–33.

99 Los, G.V., Encell, L.P., McDougall, M.G. et al. (2008). *ACS Chemical Biology* 3 (6): 373–382.

100 Gautier, A., Juillerat, A., Heinis, C. et al. (2008). *Chemistry & Biology* 15 (2): 128–136.

4

Recognition-based Sensors for Cellular Imaging

Amy A. Bowyer[1], Jianping Zhu[1,2], and Elizabeth J. New[1,2,3]

[1] *School of Chemistry, The University of Sydney, NSW, Australia*
[2] *Australian Research Council Centre of Excellence for Innovations in Peptide and Protein Science, The University of Sydney, NSW, Australia*
[3] *The University of Sydney Nano Institute (Sydney Nano), The University of Sydney, NSW, Australia*

The development of responsive fluorescent sensors for cellular imaging has been predominated by recognition-based sensors. In this chapter, we define recognition-based sensing as a change in fluorescence response caused by an electronic interaction between sensor and analyte. This interaction does not chemically alter the structure of the sensor or the analyte, but rather is a reversible process (Figure 4.1). In contrast, activity-based, or reaction-based sensing results in the structural modification of the sensor upon analyte interaction, giving rise to a fluorescence response (Figure 4.1). Probes for activity-based sensing are discussed in Chapter 5. The two main forms of recognition between sensor and analyte are coordinate covalent bonds and hydrogen bonding. Both of these bond types are far weaker than organic covalent bonds, which allows analytes to bind when concentrations are high and detach when concentrations fall. In this way, recognition-based sensors are routinely used to monitor real-time fluctuating levels of labile (non-protein bound) analytes in cells. In general, recognition-based sensors are used to sense cations and anions, which will be the focus of this chapter.

4.1 Considerations for Recognition-based Sensing

In their most simple form, recognition-based small-molecule fluorescent sensors consist of three parts: a fluorophore, linker, and receptor group (Figure 4.2). The types and properties of common fluorophores are outlined in Chapter 1, while more detail on receptor groups will be given in the following sections. There are several measures of a successful recognition-based fluorescent probe for cellular imaging. First, analytes must bind spontaneously to the receptor group and remain bound once recognised. Second, upon recognition, the fluorophore component of the probe must exhibit a conspicuous change in fluorescence response such as a change in the emission/excitation intensity or colour, explained in detail in Chapter 1. Finally, in order to be used in the cellular environment, probes must be selective to the analyte of interest over other species naturally occurring in the cell.

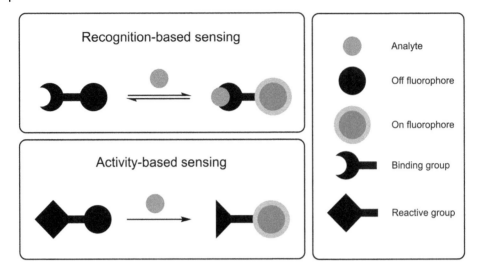

Figure 4.1 Schemes of recognition-based versus reaction-based fluorescent probes with a turn-on fluorescence response.

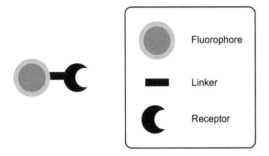

Figure 4.2 General structure of a small-molecule fluorescent probe.

Section 4.1 will focus on the interaction between the sensor receptor group and the analyte. It will outline key considerations to enhance analyte binding and improve receptor group selectivity.

4.1.1 Receptor–Analyte Recognition and Binding Affinity

In designing sensors for the cellular environment, it is important to consider how well the receptor group and the analyte interact, and the binding affinity of the interaction. This is because the relative concentration of analyte to competing species within a cell is often low. It is therefore essential for fluorescent sensors to have an analyte binding affinity that is sufficiently high to detect trace concentrations of analytes with appropriate sensitivity. Similarly, some sensors must also compete with endogenous chelators of the analyte of interest, such as proteins and peptides. Therefore, binding affinity is an important consideration for keeping probe dosing concentrations low. Conversely, if the binding affinity is too high, the sensor may remove the analyte from a biologically important role, such as sequestering the metal from a metalloprotein, which will perturb the system being analysed. This may cause stress to the system resulting in false results, artefacts, and cell toxicity. Therefore, balance in binding affinity must be struck so that it is high enough to detect labile analyte competitively, but not too high as to change the system being studied.

4.1.1.1 Defining Binding Affinity

The binding affinity between a sensor and an analyte can be quantified by a binding constant, also called stability or formation constant, when referring to a metal complex. This measure is a type of equilibrium constant, K, which reports on the binding and unbinding of receptors (hosts) and analytes (guests). In solution, the host and guest can exist separately, H and G, respectively, or bound together, HG. This interaction is formalised as:

$$H + G \rightleftharpoons HG$$

The reaction has both an on-rate constant and an off-rate constant, k_{on} and k_{off} respectively. In an equilibrium, the forward and backward reactions are balanced so that:

$$k_{on}\left[H\right]\left[G\right] = k_{off}\left[HG\right]$$

where [H] and [G] represent the concentration of unbound host and guest molecules, and [HG] represents the concentration of the host–guest complex. In this way, the binding constant K_a is defined by:

$$K_a = \frac{k_{on}}{k_{off}} = \frac{\left[HG\right]}{\left[H\right]\left[G\right]}$$

Conversely, the rate of the backwards reaction is defined by the dissociation constant, K_d, which is inversely related to the binding constant.

$$K_d = \frac{1}{K_a} = \frac{\left[H\right]\left[G\right]}{\left[HG\right]}$$

Strong probe–analyte interactions therefore have high binding constants (measured in inverse molarity, M^{-1}) and low dissociation constants (measured in molarity, M).

For studying cellular environments, it is important for sensors to have binding affinities that align with cellular concentrations of the analyte. Ideal receptor groups will have smaller dissociation constants than the concentration of the labile analyte being studied. For example, it is known that labile iron exists in relatively high (nM to μM) levels in the cell [1]. Therefore, receptor groups for iron require only relatively weak binding affinities. On the other hand, probes for analytes such as labile Cu(I), which is only present in picomolar concentrations, need to have much lower, picomolar dissociation constants [2].

4.1.1.2 Measuring Binding Stoichiometries and Binding Affinity

The above definition of binding constants is based on a 1 : 1 binding stoichiometry between host and guest, but other ratios are possible. Several analytes can bind to a single probe or vice versa. Binding constants are reliant on binding stoichiometries, and it is often therefore important to determine the binding stoichiometry before formulating a method to measure the binding constant.

Binding stoichiometries are typically determined through Job's method [3]. In this analysis, the sum of the molar concentration between sensor and analyte is kept constant, while the mole fraction, the proportion of analyte to sensor, is varied from 0 to 1. A physical property such as fluorescence intensity or UV-vis absorbance is measured against each mole fraction and plotted on the y-axis. The resulting plot generates a peak that corresponds to the

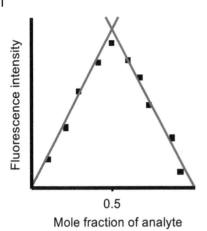

Figure 4.3 Example Job plot demonstrating 1 : 1 binding stoichiometry as the peak occurs at an analyte mole fraction of 0.5.

binding ratio of analyte to sensor [3] (Figure 4.3). Additionally, the shape of the peak gives insight into the kinetics of binding, where a sharp, triangular peak indicates large K values, while a peak with gradual curvature often indicates a smaller binding constant [4].

Once the sensor-analyte binding stoichiometry has been determined, an appropriate method of measuring the binding affinity can be employed. One of the most common methods of calculating a binding constant is *via* titration, often set up in a way that uses a known sensor-analyte system as a starting point. A simple way of determining the binding constant of systems that have 1 : 1 stoichiometries is through the Benesi-Hildebrand (B-H) method [5]. With this approach, data collected from one of a variety of spectroscopic techniques (NMR, absorption spectroscopy, fluorescence spectroscopy, *etc.*) can be used in the calculation. For fluorescent probes, fluorescence intensity is the most sensitive spectroscopic technique and is therefore an effective way to monitoring binding.

For an interaction between a host or receptor (H) and a guest or analyte (G) represented as:

$$H + G \rightleftharpoons HG$$

As the Bouguer-Beer-Lambert law can be expressed in the form: $A = \varepsilon Cl$, then a 1 : 1 complex in the Benesi-Hildebrand method has the equation in terms of K, the binding constant, as:

$$\frac{[G_0]}{\Delta A} = \frac{1}{K[H_0](\varepsilon_{HG} - \varepsilon_G)} + \frac{1}{\varepsilon_{HG} - \varepsilon_G}$$

where ΔA is the change in absorbance with binding, $[G_0]$ and $[H_0]$ are the initial concentrations of host and guest, and ε_{HG} and ε_G are the absorption coefficients of HG and G, respectively. This relationship is derived from the assumptions that ε_B is negligible and that $[H_0] \gg [G_0]$. A similar equation can be used for fluorescence intensity because it is also proportional to the concentration of the complex, HG. The Benesi-Hildebrand equation for fluorescence data is:

$$\frac{1}{\Delta FI} = \frac{1}{K[G]\Delta FI} + \frac{1}{\Delta FI}$$

where ΔFI is the change in fluorescence intensity with binding and [G] is the concentration of analyte, or guest. Therefore, the binding constant, K (M^{-1}), can be determined by taking the slope of the straight line from plotting $1/\Delta FI$ against $1/[G]$ (Figure 4.4). Advantages of

Figure 4.4 Example of B-H plot using fluorescence intensity as the spectroscopic measurement. The binding constant is obtained from the gradient of the straight line.

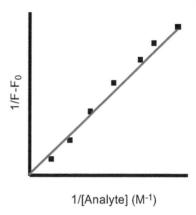

$1/F-F_0$

$1/[Analyte]$ (M^{-1})

this method include simplicity of data collection and analysis. However, concerns have arisen with this method based on its breakdown when assumptions such as $[H_0] \gg [G_0]$ are violated and to the fact that it is restricted only to investigating complexes with 1 : 1 stoichiometry [6].

An alternative approach to the Benesi-Hildebrand method is potentiometric titration. In these experiments, the concentration of sensor and analyte is kept constant, while the pH of the system is changed [7]. The presence or absence of additional H^+ ions in solution alters the relative concentration of bound analyte and receptor, which can be measured spectroscopically, commonly with NMR or with UV-vis or fluorescence spectroscopy. This data is then compared to known binding models to obtain a binding constant.

Finally, an increasingly popular method of binding constant determination is isothermal titration calorimetry (ITC). ITC was originally developed for investigating enzyme kinetics and measures subtle heat changes from interacting molecules [8, 9]. In this method, standard solutions of ligand, or receptor, are titrated into a sample cell with a known concentration of analyte. The cell is kept at a constant temperature accurate to a millionth of a degree. The output of the technique is a plot that quantifies the energy required to keep the system at a constant temperature. Upon titration, peaks in the plot decrease, then disappear when there is no longer any free analyte to bind to. The data can then be processed to calculate binding enthalpy and the binding constant and stoichiometry between sensor and analyte and compare them to common interferants to prevent unwanted cross-reactivity. The use of techniques such as these enable intelligent design of sensor receptor groups for studying complex cellular environments.

4.1.2 Key Considerations to Enhance Selective Receptor to Analyte Recognition

Binding affinity and stoichiometry of a receptor–analyte interaction are key to controlling selectivity, especially in complex cellular environments. Both binding affinity and stoichiometry can be manipulated by a number of factors to enhance probe selectivity. These include physical properties such as receptor group size and the number of receptor atoms, as well as chemical properties such as type of atom and electronic configuration of analytes. The following section will define these properties and explain how they can be used to increase sensor selectivity and sensitivity.

4.1.2.1 Size

A key consideration when designing receptor groups for analytes is how easily they will fit together. Selectivity may be achieved through tailoring the receptor group to fit the size of the analyte, much like a lock and key model. This is especially true for ionic analytes, as they can share similar chemical properties but differ in the size of their effective ionic radii. For example, Zn(II) and Cd(II) have very similar electronic properties but have differing effective ionic radii of 74 and 95 pm, respectively, in their octahedral geometries [10]. The property of size is especially important in designing receptor groups for Group I and Group II metals that lack characteristic d-block electronic interactions (see Section 4.2.1).

4.1.2.2 The Chelate Effect

The chelate effect is a phenomenon relating to metal coordination complexes, first described by Schwarzenbach in 1952 [11], based on the observation that metal complexes with multidentate ligands are more stable and form more readily than metal complexes with simple monodentate ligands. For receptor groups, this means that binding affinity can be improved through incorporating multiple atoms that can donate electrons to form coordination bonds with a metal ion analyte.

The theory behind this is based on thermodynamic principles. Increasing the number of ligand–metal bonds decreases enthalpy and favourably increases entropy as water molecules are released from aquated metal ions. In this way, once complexation has occurred, the metal is much less likely to dissociate, enhancing stability and binding affinity. For example, $[Cd(en)_2]^{2+}$, a cadmium complex with the bidentate ligand ethylenediamine (en), has a much lower free energy of $-60.67 \, kJ \, mol^{-1}$ compared to $[Cd(NH_2Me)_4]^{2-}$, a monodentate complex, with a free energy of $-37.41 \, kJ \, mol^{-1}$ [12]. This means that the multidentate coordination complex will form more spontaneously and will be less likely to dissociate. Additionally, metal complexes with en ligands are highly stable because they form five-member rings when bound (Figure 4.5). Five- and six-membered rings are favoured over smaller rings in diamine and diol ligands as bond angles are larger and steric strain is minimised [13].

The chelate effect can be enhanced with preorganisation (giving ligands more structure). The well-known multidentate ligand, ethylenediaminetetraacetic acid (EDTA), forms hexadentate, octahedral metal binding through two tertiary amines and four carboxylic acid groups (Figure 4.5). Despite being a very successful metal chelator, being readily employed to sequester heavy metals [14, 15], its carbon chain between amines is flexible, which means that the carboxylic acid groups are not always in position to form the hexadentate complex. To address this, cyclohexylenediaminetetraacetic acid (CDTA) was created [16]. CDTA has a

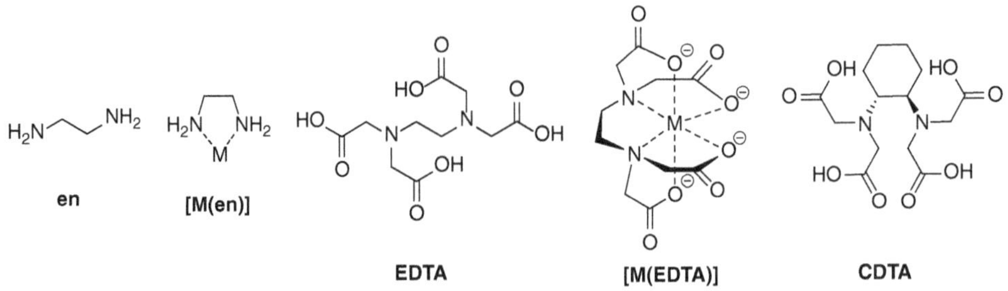

Figure 4.5 Structures of multi-dentate ligands and complexes.

trans-cyclohexyl ring, which is inflexible, ensuring that the carboxylic acid groups are always in a favourable direction for forming octahedral complexes with metals (Figure 4.5) [17]. Preorganisation is also a key contributing factor to the success of macrocyclic ligands such as crown ethers and porphyrins in complexing metals.

In summary, increasing the number of binding atoms and preorganisation of ligands increases complex stability. These principles can be employed when designing receptor groups for fluorescent sensors, as incorporating multiple coordinating atoms and some structure rigidity will enhance analyte recognition and sensing ability.

4.1.2.3 Hard–Soft Acid–Base Theory

The hard–soft acid–base (HSAB) theory was devised by Pearson in 1963 [18]. It states that Lewis acids can be classified as hard or soft based upon their polarisability (how easily their electron cloud is distorted), and that their classification will predict what types of Lewis bases they will favourably interact with. Hard, non-polarisable acids will more strongly interact with hard, non-polarisable bases, while soft, polarisable acids will react more readily with soft, polarisable bases. With this classification system, receptor groups on sensors can be selected to match the hardness/softness of the analyte of interest to increase selectivity and sensitivity.

Properties of hard Lewis acids include low polarisability, small ionic radii, and high oxidation states in the cases of metal ions. Conversely, soft Lewis acids are highly polarisable, have large ionic radii, and low oxidation states. Bonds between hard acids and bases are highly ionic in nature, while bonds between soft acids and bases involve more sharing of electrons, making them more covalent. Lewis acids that fall in-between the two categories are classified as 'borderline'. Examples of hard, soft, and borderline Lewis acids and bases are provided in Table 4.1.

Tailoring receptor groups to match the hardness/softness of analytes will greatly improve selectivity. For example, heavy metal ion receptor groups tend to include sulfurous functional groups, while sensors for halides (hard bases) take advantage of hydrogen bonding because protons are hard Lewis acids.

4.1.2.4 Crystal and Ligand Field Theory

While HSAB theory uses electron cloud size and polarisability to determine the likelihood of atoms forming coordination bonds, it cannot predict the number of bonds or bond geometries between receptor groups and analytes. Ligand field theory (LFT), a combination of molecular orbital theory (MOT) and crystal field theory (CFT), is used to describe the bonding and orbital arrangement of transition metal complexes.

In MOT, electrons from atoms in a molecule are not treated as belonging to specific bonds, but rather are under the influence of all nuclei in the molecule. The theory is used to predict bonding characteristics of molecules by using linear combinations of atomic orbitals to assign molecular orbitals as bonding, anti-bonding, and non-bonding [20]. Simply put, molecular orbital theory provides an explanation for why certain molecules can exist while others cannot.

CFT concerns coordination bonding between ligands and elements with d and f atomic orbitals. It is based on the electrostatic attraction between the positive charge of a metal cation and the negative charge of a non-bonding pair of electrons on a ligand, or receptor group. As a ligand approaches, the degenerate (equal energy) d orbitals of the metal will be repelled by the non-bonding electrons of the ligand. Some of these d electrons will be closer

Table 4.1 Classification of Lewis acids and bases according to hard–soft acid–base theory [19].

Lewis acids	Lewis bases
Hard	
H(I), Li(I), Na(I), K(I), Be(II), Mg(II), Ca(II), Mn(II), Al(III), Sc(III), Cr(III), Fe(III), Co(III), Ga(III), As(III), La(III), Si(IV), Ti(IV), Zr(IV), Gd(III)	H_2O, NH_3, OH^-, F^-, Cl^-
	R–NH_2 (Primary amines) R–OH (Alcohols) R–COO$^-$ (Carboxylates) R–O–R (Ethers)
O=C=O (Carbon dioxide) SO_3 (Sulfur trioxide)	Carbonate Sulfate Nitrate Phosphate
Borderline	
Fe(II), Co(II), Ni(II), Cu(II), Zn(II), Ru(II), Sn(II), Pb(II), Au(II), Bi(III)	Br^-
SO_2 (Sulfur dioxide)	Sulfite N≡N (Nitrogen) Teriary amines Azides
	Aniline Pyridine Imidazole
Soft	
Cu(I), Ag(I), Au(I), Tl(I), Pd(II), Pt(II), Cd(II), Hg(II), Tl(III), Pt(IV)	H^-, I^-
Br–Br (Bromine) I–I (Iodine)	Benzene R–SH (Thiols) Thioethers Organophosphorus
	R—≡N (Nitriles) N≡C$^-$ (Cyanide) S=C=N$^-$ (Thiocyanate) $^+$O≡C$^-$ (Carbon monoxide)

to the ligand than others, causing a loss of degeneracy as the electrons that are closer to the ligand will now have a higher energy causing the orbitals of the d-block to 'split'. This splitting is highly influenced by the ligand type, and so a spectrochemical series of ligands has been developed to predict which ligand will cause larger splitting. Ligands such as halides, hydroxides, and water cause weak splitting, while ligands including ammine, carbon monoxide, triphenylphosphine, and cyanide induce a large degree of splitting [21]. Similarly, metal ions can also be arranged from causing lowest to highest splitting: Mn(II) < Ni(II) < Co(II) < Fe(II) < V(II) < Fe(III) < Cr(III) < V(III) < Co(III) [22]. The general pattern is that splitting is increased with oxidation number and down a group of the periodic table, although this is extremely dependent on the ligand. It is this splitting in the metal d-orbitals that gives rise to an assortment of bonding preferences between metals and ligands.

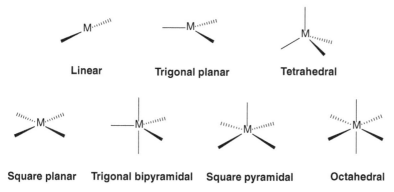

Figure 4.6 Some coordination geometries of metal (M) complexes.

CFT has been criticised for simplifying the bonding between ligands and metals as purely electrostatic. In response, LFT was developed by building upon concepts of both MOT and CFT. It incorporates both the orbital arrangement and bonding of metals and ligands to predict which ligands a *d*-block element may complex with, as well as their coordination geometries.

Coordination geometries describe the three-dimensional shape that ligands form around a central metal ion. The most widely studied geometry is octahedral where the metal ion is positioned on the middle of six ligand atoms (Figure 4.6). However, many more exist, including linear, trigonal planar, tetrahedral, square planar, trigonal bipyramidal and square pyramidal (Figure 4.6) [23].

Design of receptor groups of fluorescent sensors for transition metal ions should take into account coordination geometry preferences (Table 4.2) [24, 25]. For example, soft Lewis acids such as Ag(I) and Hg(II) prefer to coordinate in tetrahedral and linear geometries, respectively [24]. Therefore, selectivity for Hg(II) over Ag(I) may be achieved by only incorporating two soft Lewis bases into a receptor group, because Ag(I) generally prefers to bind to four electron donors. Incorporating rigidity and preorganisation into the structure will also enhance selectivity, as some flexible multi-dentate ligands can bind to metal ions in stoichiometries greater than 1 : 1. However, this may compromise the reversibility of a recognition-based sensor, if the metal ion is encapsulated too successfully.

Taking into account the above considerations, highly selective recognition-based fluorescent sensors can be designed. The following sections highlight specific examples of fluorescent sensors for cations and anions.

4.2 Recognition-based Cation Sensing

All living organisms require metals to survive as they have essential roles in biology as signalling molecules, catalysts, osmotic regulators, and molecular transporters. In humans, group I and II metal ions such as sodium and potassium are responsible for controlling nerve impulses, while iron in haemoglobin is used to transport oxygen around the body [26]. Detecting metal cations in biological systems is important for many reasons including enhanced understanding of cellular processes, diagnosing disease and disorders, and the development of therapies.

Table 4.2 Common metal coordination numbers, coordination geometries, and effective ionic radii including high spin (HS) and low spin (LS) variants where applicable.

Metal ion	Preferred coordination [24]	Preferred coordination geometry [24]	Effective ionic radius (pm) [10]
Na(I)	6	Octahedral	102
K(I)	4–12	Flexible	137–164
Mg(II)	6	Octahedral	72
Ca(II)	6	Octahedral	100
Mn(II)	6	Octahedral	LS = 67 HS = 83
Mn(III)	6	Octahedral	LS = 58 HS = 65
Fe(II)	6	Octahedral	LS = 61 HS = 78
Fe(III)	6	Octahedral	LS = 55 HS = 65
Co(II)	6	Octahedral	LS = 65 HS = 75
Ni(II)	4	Square planar	55
	6	Octahedral	69
Cu(I)	4	Tetrahedral	60
Cu(II)	4	Square planar	57
	5	Square pyramidal	57
Zn(II)	4	Tetrahedral	60
Ag(I)	4	Tetrahedral	100
Cd(II)	4	Tetrahedral	78
	6	Octahedral	95
Hg(II)	2	Linear	69
Pb(II)	4	Tetrahedral	98

Metal ions are not the only important cations in biology. Sensing cellular concentrations of hydrogen ions, or pH level, is also of great use (see Chapter 6). Polyatomic cationic species such as ammonium and nicotinamide adenine dinucleotide play vital metabolic roles and can also be associated with some pathologies.

Small-molecule recognition-based sensors for polyatomic cations are rare, while activity-based sensors are generally favoured for studying these cations in biology. Consequently, Section 4.2 will focus on recognition-based sensing of metal ions including group I and II metals (Section 4.2.1), essential transition metals (Section 4.2.2), and toxic heavy metals (Section 4.2.3). The specific focus will be on receptor group design and selection, and how it has been applied in successful examples of recognition-based metal sensors for cellular studies.

4.2.1 Group I and II Metal Sensing

4.2.1.1 The Biological Significance of Group I and II Metals
Group I and II metals, also called alkali and alkaline earth metals, respectively, are the most abundant metal species found in the human body. These include sodium (0.2% of total body mass), potassium (0.4%), magnesium (0.1%), and calcium (1.5%) [27]. Group I metals, sodium and potassium, are essential nutrients. Sodium is mainly located outside the cell,

while potassium is intracellular. This difference in distribution forms the cell-membrane potential, which moderates key bodily functions such as muscle contraction, nerve impulses, and organ function [27].

Lithium is also present in trace concentrations in the body as an essential metal, but its biological role is less well understood. Li(I) is routinely prescribed for patients suffering bipolar disorders and manic-depressive psychosis [28]. While monitoring Li(I) in patient sera is highly important, few fluorescent probes have been developed for cellular imaging of Li(I).

Group II metals, calcium and magnesium, are essential to all known organisms. Like sodium and potassium, they are involved in cellular ion pumps and each have their own roles. For example, calcium is an essential building block in connective tissues, such as bone [29], while magnesium is an important cofactor in many enzymes [30].

4.2.1.2 Receptor Group Design for Group I and II Metals

All group I and II metal ions are classified as hard under hard–soft acid–base theory, so they have a preference for binding to oxygen and nitrogen atoms over sulfur [18]. As they are not d block elements, their binding geometries cannot be easily manipulated by ligand type. Na(I), Mg(II), and Ca(II) all generally prefer to coordinate to ligands in an octahedral geometry [24], while K(I) forms six to eight coordination bonds with flexible geometries [31]. As a result, the most used method of achieving selectivity between these metal ions is size, as each ion has a unique effective ionic radius (Table 4.1). This section details types of receptor groups used for the selective sensing of group I and II metals.

4.2.1.2.1 Sodium and Potassium   The most widely used type of receptor group for sensing group I metals ions, Na(I) and K(I), is the crown ether. In their simplest form, crown ethers are heterocycles with the repeating unit of ethyleneoxy ($-CH_2CH_2O-$), first described by Pedersen in 1967 [32]. The smallest crown ether is arguably dioxane, with two ethyleneoxy repeating units. As the heterocycles increase in size, they are named according to the total number of atoms, and the number of oxygen atoms (Figure 4.7) [33]. For example, a crown ether with 6 ethyleneoxy repeating units is known as 18-crown-6 because it is an 18-member ring with 6 oxygen atoms. Pedersen proposed that the stability of the coordination between the guest ion and the crown ether was based on a number of factors, particularly their relative sizes. For example, 18-crown-6 forms a more stable complex with K(I) over Na(I) because the estimated pore size of the crown ether (4.0 Å) is too large relative to the size of Na(I) (1.90 Å) [32]. Despite being a good foundation for achieving selectivity, pore size alone is not sufficient. This has resulted in a large expansion in the field of crown ethers, leading to more elaborate designs.

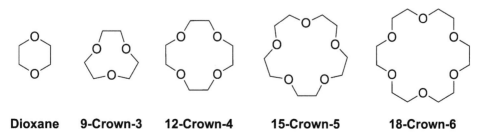

Dioxane   **9-Crown-3**   **12-Crown-4**   **15-Crown-5**   **18-Crown-6**

Figure 4.7   Structures of crown ethers with 1–6 ethyleneoxy repeating units.

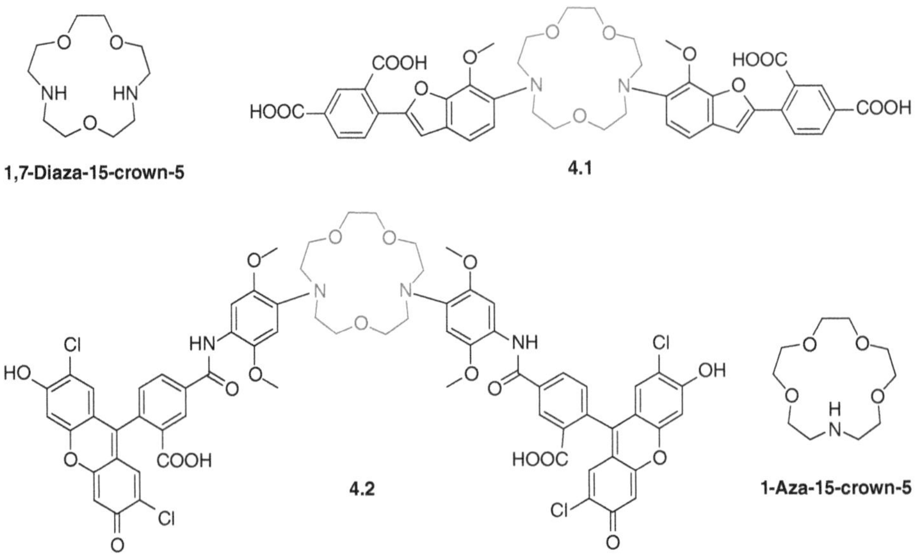

Figure 4.8 Structures of simple crown ether analogues and commercial dyes SBFI and Sodium Green. Metal receptor groups are coloured in red.

Simple analogues of crown ethers are aza-crown ethers, where some oxygen atoms in the heterocycle are replaced by nitrogen atoms. These compounds are named according to the position of the nitrogen atoms and the parent crown ether. For example, 1,7-diaza-15-crown-5, a widely used selective receptor group for Na(I), has 15-crown-5 as its parent structure with two NH groups as the first and seventh atom in the ring instead of two of the original oxygen atoms (Figure 4.8).

The addition of amine groups into the ring has two functions. Firstly, it alters selectivity and sensitivity as amines can be protonated in various conditions, while ethers cannot. Secondly, as amines can form up to three covalent bonds, the aza-crown can be tethered to a fluorophore to create a sensor. For example, the two nitrogen atoms in 1,7-diaza-15-crown-5 allow for up to two fluorophore units to be attached, a property that has been exploited in commercial sodium dyes, SBF1 (**4.1**) and Sodium Green (**4.2**) (Figure 4.8). SBFI and Sodium Green are fluorescent turn-on probes that emit light at 499 and 535 nm, respectively, in response to Na(I) [34]. Other fluorescent sensors with this receptor group report excellent selectivity for Na(I) over K(I) [35–37]. A similar aza-crown with a single amine, 1-aza-15-crown-5, is also routinely used as a receptor group for sodium ions (Figure 4.8) [38].

Cryptands are another variation on crown ethers, routinely used in group I metal recognition-based fluorescent sensors. Cryptands are composed of a diaza-crown with two nitrogen atoms that are opposite each other, acting as attachment points for another ethylen-exoxychain to bridge the heterocycle. The most common cryptand is 1,10-diaza-18-crown-6 with an ethylene glycol diethyl ether linker, which is named [2.2.2]cryptand because each chain contains two oxygen atoms (Figure 4.9) [33].

Cryptands are regularly employed as K(I) sensors, owing to their large cavity size, although they do suffer strong pH interferences as their bridgehead nitrogen atoms are easily protonated under neutral pH conditions. Cryptand selectivity for alkali metals can therefore be achieved through the introduction of aromatic nitrogens on the backbone, which decreases interference from protons [39]. As a result, a benzo-annelated analogue of [2.2.2]cryptand (**4.3**), first

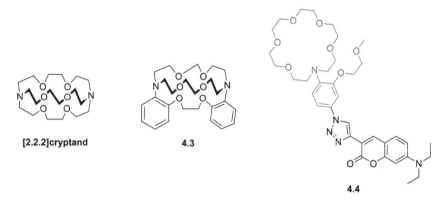

[2.2.2]cryptand   4.3

4.4

Figure 4.9 Structures of receptor groups and sensors for K(I). Metal receptor groups are coloured in red.

developed by de Silva *et al.* [40] (Figure 4.9), has become the most widely used receptor group in recognition-based fluorescent potassium sensors [41].

Ast *et al.* developed a highly selective potassium sensor for intra and extracellular environments. The sensor, **4.4**, uses a phenylaza-18-crown-6 with a 2-methoxyethoxy lariat chain as a receptor group linked *via* a triazole moiety to a 7-diethylaminocoumrin fluorophore (Figure 4.9) [42]. Lariat crown ethers are attractive alternatives to cryptands as they have similar binding properties but are much easier to synthesise. The coumarin sensor exhibited a fluorescent intensity enhancement at 493 nm in the presence of potassium ions in *in vitro* studies with simulated physiological conditions. The authors further demonstrated this turn-on response in a normal rat kidney cell line, quantifying intracellular and extracellular potassium concentrations; **4.4** was adapted for membrane sensing by converting it into a hydrogel form.

4.2.1.2.2 Magnesium Most commercially available fluorescent sensors for magnesium ions contain multiple carboxylic acids as their receptor groups. For example, mag-fura-2 (**4.5**) and Magnesium Green (**4.6**) both utilise the popular *ortho*-aminophenol-*N,N,O*-triacetic acid (APTRA) moiety in their structures (Figure 4.10). However, these structures, and others using the APTRA receptor group, can suffer from strong interference from Ca(II) [43], because both Mg(II) and Ca(II) are classified as strong Lewis acids under HSAB theory [18] and have a preference for octahedral coordination geometries [24]. For example, mag-fura-2 has dissociation constants of 1.9 mM and 25 µM for Mg(II) and Ca(II) complexes, respectively [44]. Despite the tighter binding of APTRA to Ca(II) over Mg(II), APTRA is still effective for quantifying intracellular Mg(II) because cellular concentrations of Mg(II) are in the low mM range [45], while intracellular Ca(II) is closer to 100 nM [27].

An advantage of recognition-based fluorescent sensors is their adaptability to a variety of cellular environments. Compared to activity-based sensing, analyte receptor groups on recognition-based sensors are more robust and can therefore be incorporated into numerous fluorescent sensing platforms. A good example of the adaptability of recognition-based sensing is demonstrated through the work of Gruskos *et al.* In their studies, they modified their previously reported Mg(II) sensor, **4.7** [46] with a tetrazine moiety to give **4.8**, and demonstrated its ability to sense Mg(II) in various sub-cellular organelles of HEK293T cells (Figure 4.10) [47].

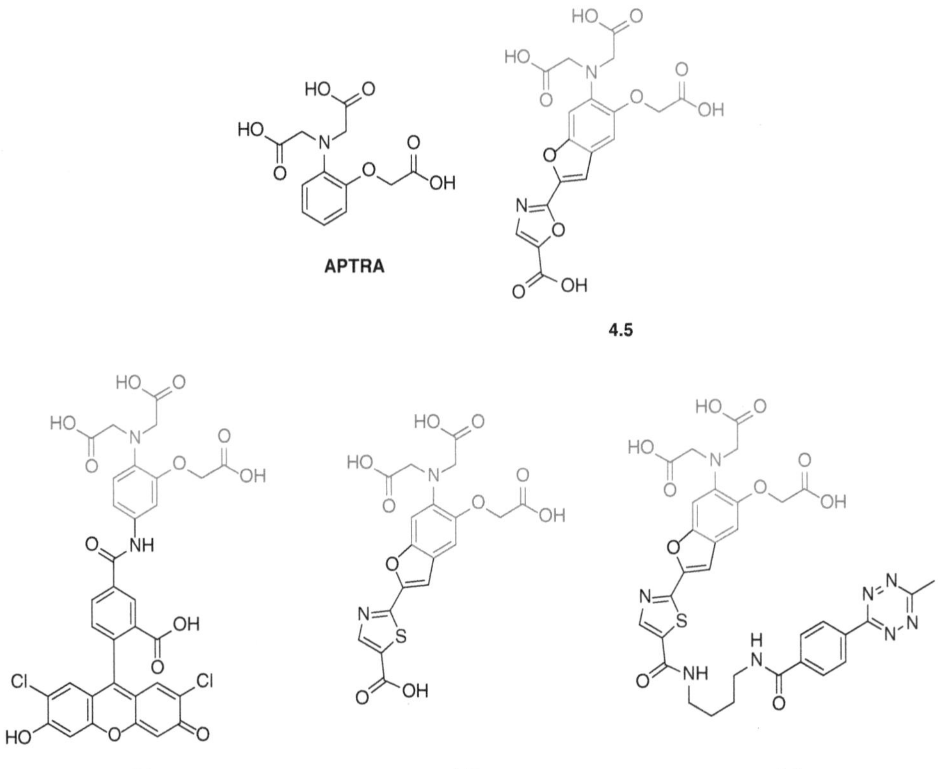

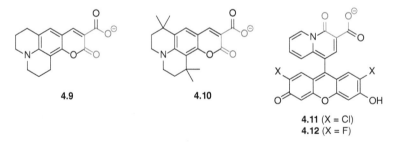

APTRA

4.5

4.6   **4.7**   **4.8**

Figure 4.10 Structures of APTRA and associated Mg(II) sensors. Metal receptor groups are coloured in red.

4.9   **4.10**

4.11 (X = Cl)
4.12 (X = F)

Figure 4.11 Structures of Mg(II) sensors using β-diketoester receptor group. Metal receptor groups are coloured in red.

An alternative approach to selective Mg(II) receptor groups is using a charged β-diketone/ketodiester. It is proposed that Mg(II) forms coordination bonds between an aromatic ketone and deprotonated carboxylic acid. Using this approach, Suzuki and co-workers developed two fluorescent magnesium sensors, **4.9** and **4.10** (Figure 4.11), that exhibited a 200-fold enhancement of selectivity compared to commercial dyes, mag-fura-2 (**4.5**) and Magnesium Green (**4.6**) (Figure 4.10) [48]. The same group also demonstrated that including an electron-donating nitrogen group next to their β-diketoester receptor group further increased the binding affinity of Mg(II) in their sensors, **4.11** and **4.12** (Figure 4.11) [49]. Aryl Schiff bases with *ortho*-phenol groups have also been used as Mg(II) receptor groups in fluorescent sensors [41, 43]. However, these systems lack selectivity, suffering interference from other metal ions such as mercury [50] and zinc [51].

4.2.1.2.3 *Calcium*

The overwhelming majority of small-molecule fluorescent Ca(II) sensors use 1,2-bis(2-aminophenoxy) ethane-*N,N,N′,N′*-tetraacetic acid (BAPTA) as their receptor group [52]. BAPTA was first reported by Tsien as an analogue of ethylene glycol-bis(β-aminoethyl ether)-*N,N,N′,N′*-tetraacetic acid (EGTA) (Figure 4.12) [53]. Both groups have good selectivity for Ca(II) over Mg(II), but BAPTA has superior resistance to pH changes. The aromatic nitrogen atoms in BAPTA have a significantly lower pK_a than the amino groups in EGTA, meaning that BAPTA is less likely to be protonated in neutral pH conditions, which favours metal binding [53]. Other receptor groups for Ca(II) sensing are usually analogues of BAPTA, tailored for specific biological studies. For example, BAPTA-AM has acetoxymethyl (AM) esters in the place of carboxylic acids to allow it to cross the cell membrane and is therefore commonly used as an intracellular Ca(II) chelator (Figure 4.12).

Selective receptor groups can be used in combination to construct recognition-based fluorescent sensors to answer important biological questions. For example, monitoring cellular sodium and calcium levels can increase understanding of how calcium homeostasis is achieved, or how disease occurs upon calcium imbalances. Kim *et al.* combined their previously reported intracellular Na(I) sensor, **4.13** [35], with a newly designed Ca(II) sensor, **4.14**, to simultaneously monitor concentrations of both species using two-photon microscopy [54]. The Na(I) sensor, **4.13**, uses the highly selective 1,7-diaza-15-crown-5 receptor group with a 2-acetyl-6-(diethylamino)naphthalene turn-on fluorophore, with red fluorescence indicating the presence of sodium ions; **4.14** uses a BAPTA-analogue called

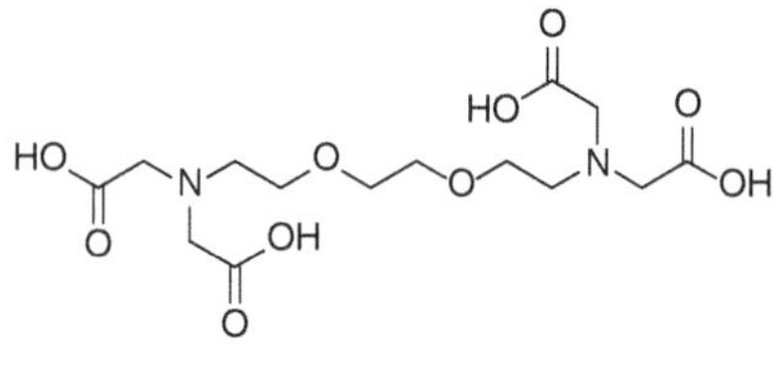

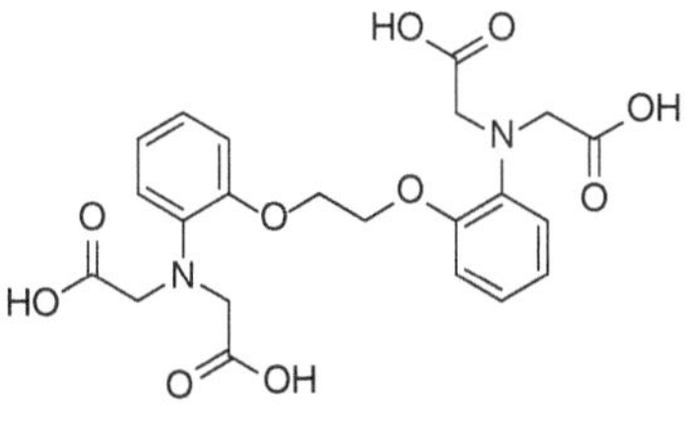

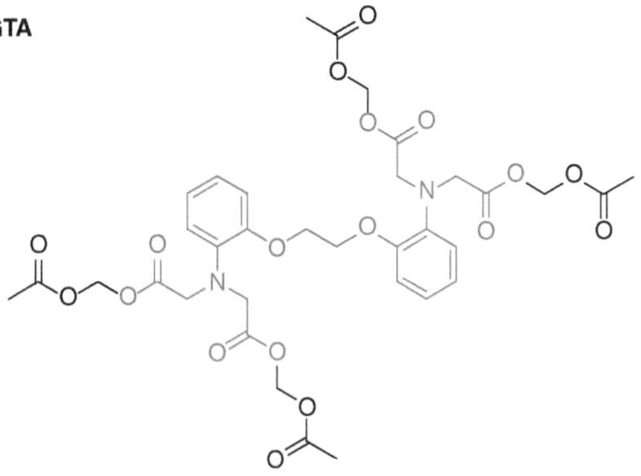

Figure 4.12 Structures of Ca(II) receptor groups and analogues. Metal receptor groups are coloured in red.

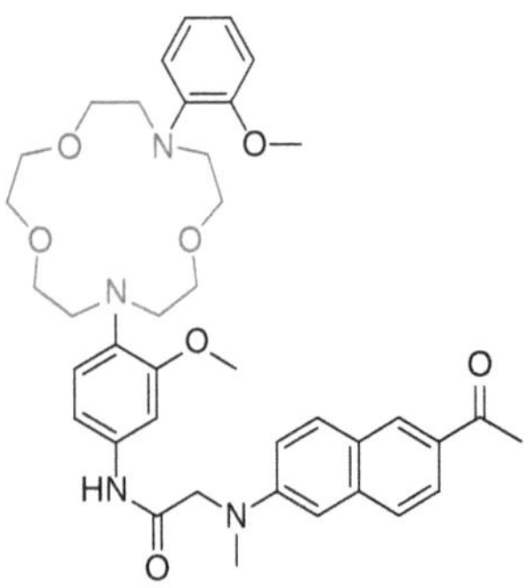

4.13

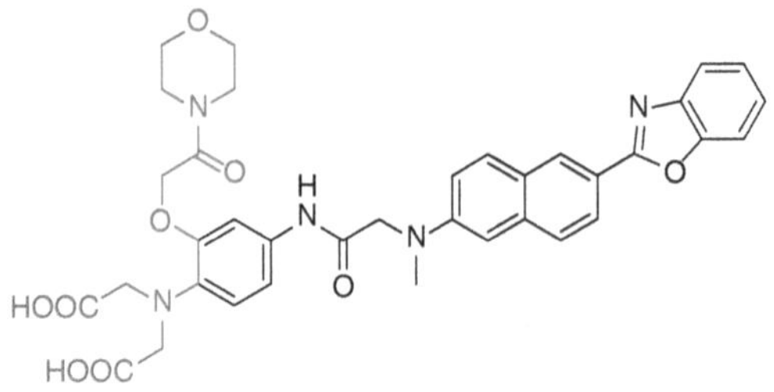

4.14

Figure 4.13 Structures of recognition-based sensors used in the experiments of Kim *et al.* Metal receptor groups are coloured in red.

2-(2'-morpholino-2'-oxoethoxy)-*N,N*-bis(hydroxycarbonylmethyl)-aniline (MOBHA) and has a benzoxazole-functionalised (diethylamino)naphthalene for its fluorescent reporter group, and emits green fluorescence when chelated to calcium ions (Figure 4.13). The authors co-labelled HeLa cells with both sensors, followed by treatment with histamine to induce Ca(II) release. The fluorescence response of both sensors was recorded as a function of time and revealed that the rate of fluorescence increase for Na(I) in the cell membrane was faster than for Ca(II) in the cytosol (Figure 4.14). Such results suggest that Na(I) influx occurs after Ca(II) through the sodium/calcium exchanger in the plasma membrane, enhancing understanding of cellular Na(I)/Ca(II) exchange. The return to a baseline fluorescence intensity after histamine treatment (Figure 4.14e) for both dyes demonstrates the advantage of recognition-based sensors to measure transient levels of analytes for real-time elucidation of cellular processes.

4.2.2 Essential Transition Metal Sensing

4.2.2.1 The Biological Significance of Essential Transition Metals

Transition metals are those that occupy the central block of the periodic table, named for their ability to transition between multiple oxidation states. Many of these metals are essential for human biological processes, although they are considered as trace elements because collectively they make up less than 0.1% of total body mass [55]. Essential transition metals include iron, copper, zinc, manganese, cobalt, and nickel. All these metals, apart from zinc, have much higher binding constants to organic ligands compared to group I and II metals. As a result, transition metals predominantly occur bound to proteins in the cell, with fewer labile ions present.

All these metals perform essential biological tasks. Briefly, iron is mainly responsible for bodily oxygen transport in haeme proteins, haemoglobin, and myoglobin. Copper is involved in neuromodulation and the development of new blood vessels. Zinc is essential for DNA synthesis and immunity, and it is proposed that 10% of the human proteome is zinc metalloproteins [56]. Manganese is important in metabolic processes and the antioxidant system, being involved in mitochondrial superoxide dismutase for oxidative stress control. Cobalt is usually bound tightly in a heterocyclic ring called a corrin. Most notably, cobalamin (Vitamin B$_{12}$) is an important cofactor in DNA synthesis and fatty acid and

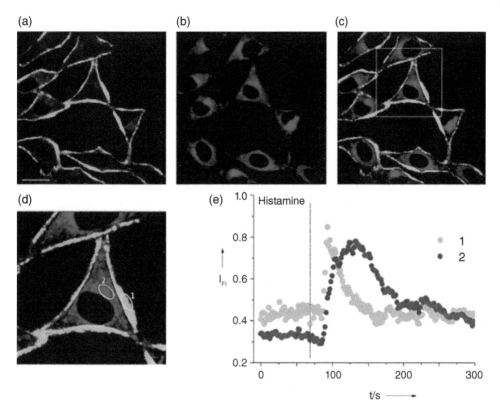

Figure 4.14 (a–d) Dual-channel two-photon microscopy images of HeLa cells co-labelled with **4.13** and **4.14** collected at (a) 390–450 nm (**4.14**) and (b) 500–560 nm (**4.13**). (c) Merged image of (a) and (b); (d) enlargement of the white box in (c). (e) Time course of two-photon excited fluorescence at designated positions (1) and (2) in (d) after stimulation with 100 mM histamine in nominally calcium-ion-free buffer. *Source:* From Kim et al. [54]/with permission from John Wiley & Sons.

amino acid metabolism. Nickel occurs less frequently in humans than other transition metals, but is more vital in microorganisms and plants for energy production and nitrogen metabolism [57].

Despite these metals being naturally present and useful in the human body, an excess or imbalance of any one of them has severe adverse consequences [58]. The chronic brain and liver disorder, Wilson's disease, is caused by excess copper accumulation, giving patients progressive neurological dysfunction. Excess iron, such as levels that occur with haemochromatosis, can cause liver disease, diabetes, and heart conditions. It is therefore highly important to monitor quantities and the movement of labile transition metals in the body, particularly at a cellular level to enhance understanding of their biological roles.

4.2.2.2 Receptor Group Design for Essential Transition Metals

4.2.2.2.1 Zinc Small molecular fluorescent sensors are very useful tools for understanding the role of transition metals in cellular processes. The first fluorescent transition metal sensor to study live cells was devised by Zalewski and co-workers in 1993 to quantify intracellular labile Zn(II) [59]. Their compound, **4.15**, was based on 6-methoxy-8-*p*-toluenesulfonamido-quinoline (TSQ), which had been previously noted by Frederickson

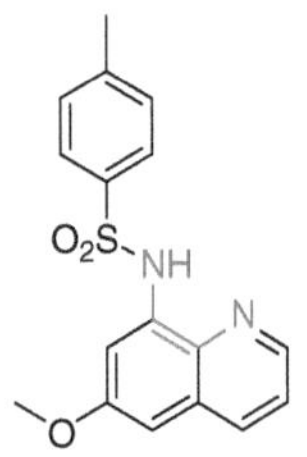

TSQ

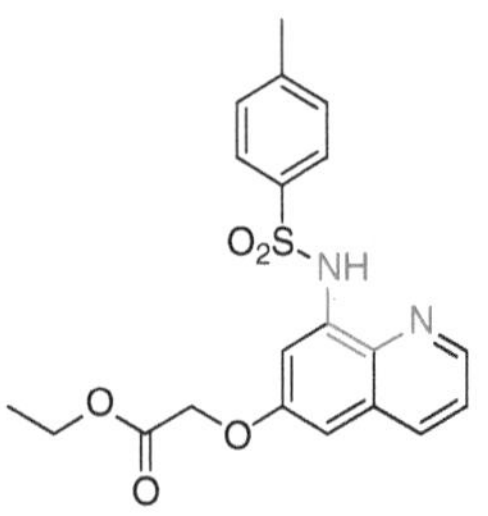

4.15

Figure 4.15 Structures of sulfonamide quinoline-based Zn(II) probes. Receptor groups are coloured in red.

and co-workers to have an excellent fluorescence response to Zn(II) in *in vitro* screening and tissue studies [60] (Figure 4.15). Both structures bind to Zn(II) through the secondary amine and aromatic nitrogen of the quinoline, in 1 : 1 or 2 : 1 (probe:Zn(II)) binding stoichiometries. Unlike TSQ, **4.15** contained a methyl ester conjugated to the methoxy group at the 6-position, which enhanced both solubility and cellular retention (Figure 4.15).

Since **4.15**, fluorescent probes for cellular Zn(II) have been extensively developed. Across the literature, the current predominant binding group for Zn(II) ions is the di-2-picolylamine (DPA) moiety (Figure 4.16) [57]. This receptor group has two pyridines and a single aliphatic nitrogen atom, which bind to borderline hard–soft acids such as Zn(II). It is based on the

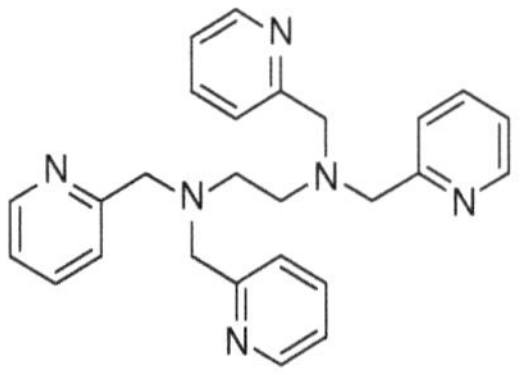

TPEN

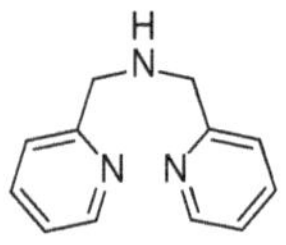

DPA

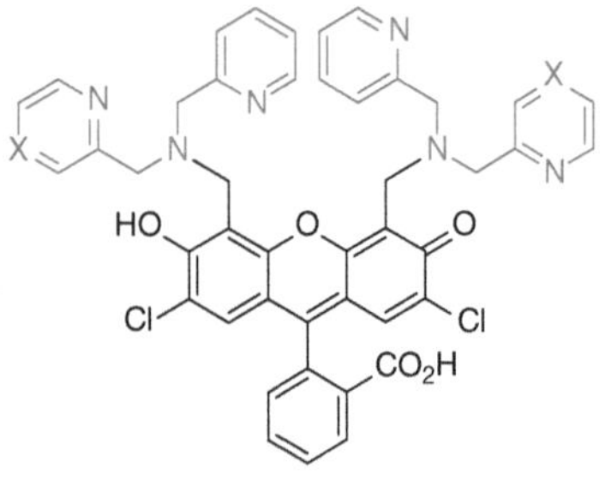

4.16 X = CH
4.18 X = N

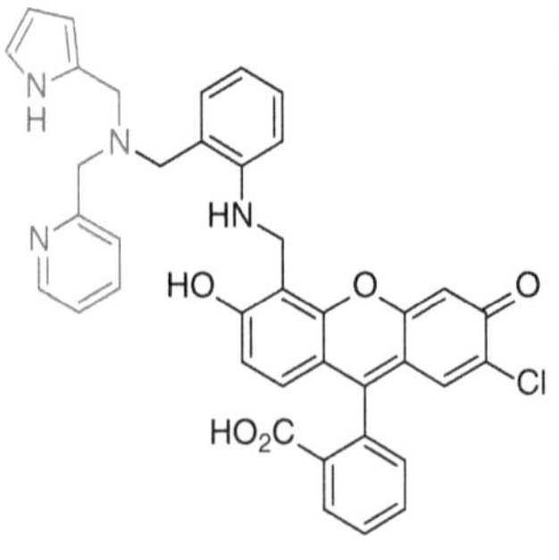

4.17

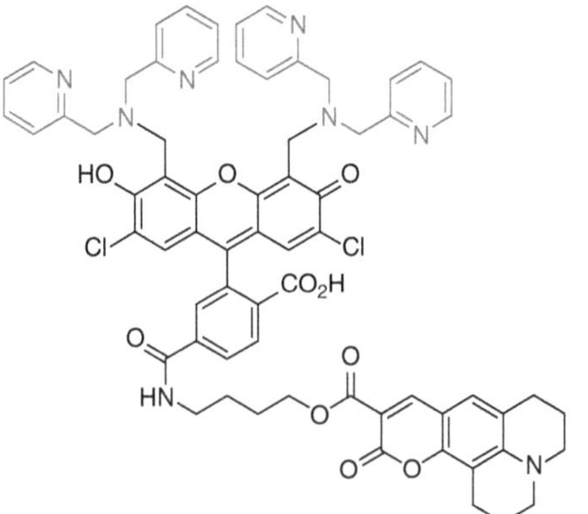

4.19

Figure 4.16 Structure of DPA-based Zn(II) chelators and probes. Zn(II) receptor groups are coloured in red.

well-known transition metal chelator, *N,N,N′,N′*-tetrakis(2-pyridylmethyl) ethylenediamine (TPEN), which contains two DPA moieties (Figure 4.16) and has an extremely high affinity for Zn(II) ($K_d = 2.6 \times 10^{-16}$ M) (Figure 4.16) [61]. A prominent family of fluorescent Zn(II) sensors, the Zinpyr or ZP family, comprise a fluorescein fluorophore with one or more DPA receptor groups.

The first ZP probe, **4.16**, features two DPA receptor groups and a dichlorofluorescein fluorophore (Figure 4.16) [62]. This probe exhibits a fluorescence turn-on response around 530 nm upon Zn(II) coordination [62]. The ZP family has been extensively studied and refined over the past couple of decades to optimise their use for cell studies [63, 64]. For example, Nolan and co-workers replaced one of the pyridines of a single DPA receptor group with a pyrrole in their compound, **4.17**, to increase Zn(II) binding affinity. Other studies have modified the DPA group with methyl groups [65], thiophenes [66], thioethers [67], and aminoquinolines [68], all with varying levels of success with regards to increasing Zn(II) selectivity.

Zhang *et al.* modified **4.16** by replacing one pyridine at each DPA group with a pyrazine to give the novel Zn(II) probe, **4.18** (Figure 4.16) [69]. Subsequent studies using **4.18** showed significantly lower fluorescence in a prostate cancer cell line (RWPE2) versus a healthy prostate cell line (RWPE1) [70]. The authors also reported the same, lower fluorescence result in a mice model of prostate cancer, suggesting that low Zn(II) could be a hallmark of prostate cancer. Furthermore, the DPA receptor group has been successfully applied to fluorescent scaffolds other than fluorescein such as rhodamines [71] and cyanines [72], demonstrating its versatility. DPA has also been incorporated onto a fluorescein-coumarin dual-fluorophore sensor, **4.19**, to achieve a ratiometric fluorescence response in cells through an esterase-cleavage mechanism (Figure 4.16) [73].

4.2.2.2.2 *Copper*

Copper, another essential transition metal, exists in two oxidation states in biology. Cu(I) predominates in the intracellular, reducing environment of the cytosol, while Cu(II) is generally only present in the extracellular matrix. Cu(I) and Cu(II) have different chemical properties, including their Lewis acidity under HSAB theory. Cu(I) is the only soft metal species in the cell, while Cu(II) is classified as a borderline Lewis acid. As a result, Cu(I) prefers to bind to soft receptor atoms such as sulfur, while Cu(II) readily binds to nitrogen and oxygen electron donors.

The vast majority of biological probes for copper are Cu(I) sensors. The first small-molecule probe for studying labile copper in cells, **4.20**, was reported in 2005 by the Fahrni group [74]; **4.20** consists of an azatetrathia-crown ether receptor group and pyrazoline-based fluorophore (Figure 4.17) and undergoes a 4.6-fold fluorescence emission enhancement at 485 nm upon the addition of one molar equivalent of Cu(I). Cell microscopy and X-ray fluorescence studies with **4.20** demonstrated that labile Cu(I) appears to localise in the Golgi region and mitochondria in acute copper overload conditions [74]. The same group developed another Cu(I) sensor, **4.21**, with just three sulfur atoms in the azathia ring and with four hydroxymethyl groups as ring substituents (Figure 4.17) [75]. This analogue had greater water solubility and did not dimerise as easily, enhancing probe–metal interactions by decreasing probe–probe interactions.

All subsequent Cu(I) probes for cell studies also use thioethers as the receptor group motif. For example, the Chang group recognised an azatetrathia dipodal ligand as a more synthetically accessible Cu(I) chelator and has since incorporated it into their Cu(I) sensors, **4.22**, **4.23**, and **4.24** (Figure 4.17) [76–78]. The use of these sensors in cell microscopy studies

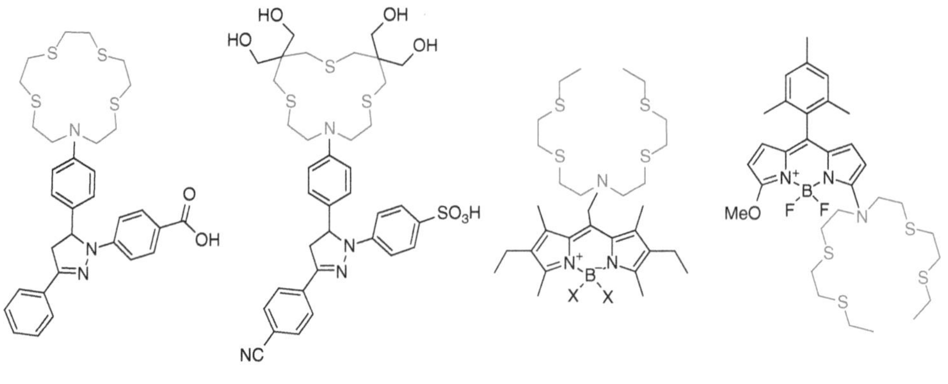

Figure 4.17 Structures of fluorescent probes for Cu(I). Cu(I) receptor groups are coloured in red.

has yielded some important biological results. For example, a study of **4.23**, a turn-on sensor for Cu(I), revealed that during cell depolarisation, hippocampal neurons redistribute endogenous copper to peripheral regions, linking calcium release to copper mobilisation [77].

Fluorescent probes for Cu(II) with application in microscopy studies are far less common in the literature. This is could be a result of the lower prevalence of Cu(II) in the cell, and therefore a decreased need for Cu(II) probes compared to Cu(I) probes. However, another factor is the increased difficulty in designing selective Cu(II) probes. Unlike Cu(I), Cu(II) does not have a unique soft Lewis acidity, but rather a very common classification of a borderline Lewis acid, which is shared with other essential metal ions such as Fe(II), Co(II), and Zn(II). Borderline Lewis acids such as Cu(II) interact with a large variety of Lewis bases, ranging from hard ethers to soft thioethers. This is demonstrated through the work of the Yoon group, who presented 20 fluorescent chemosensors for Cu(II) in a review of their own work [79]. Unfortunately many of these sensors suffered considerable interferences from other metal ions, particularly Co(II) and Ni(II), limiting their use for cell studies.

As a result, other strategies, such as coordination geometry manipulation, need to be employed for receptor groups to achieve selectivity for Cu(II). This was achieved by Pathak *et al.* in their calix[4]-arene-based probe, **4.25**, which selectively coordinates to Cu(II) through two benzimidazole and two triazole ligands in a square planar geometry (Figure 4.18) [80]. The structure can accomodate a second Cu(II) ion that coordinates to the two triazole nitrogrens and two phenolate oxygens, somewhat distorting the coordination geometry (Figure 4.18). Zn(II) and Fe(II) also readily bond to aromatic nitrogen groups such as these [18], but prefer tetrahedral coordination geometries [24], giving **4.25** selectivity for Cu(II).

4.2.2.2.3 Iron Despite being the most abundant transition metal in the human body, fluorescent chemosensors to study cellular iron are less common than those for zinc and copper. Like copper, ionic iron predominantly exists in two oxidation states, Fe(II) and Fe(III), while other oxidation states exist only briefly within catalytic cycles. In the cytosol, Fe(II) is the major oxidation state of iron due to the cellular reduction potential and the poor solubility of Fe(III) in neutral pH conditions [57]. Probes for Fe(II) are reasonably prevalent in the literature, though many of these are activity-based sensors. The first report of labile iron pool visualisation used the recognition-based sensor, calcein (**4.26**) [81]. This probe

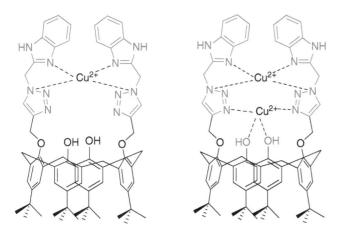

Figure 4.18 Structure of **4.25** complexing one (left) and two (right) Cu(II) ions. Receptor groups are coloured in red.

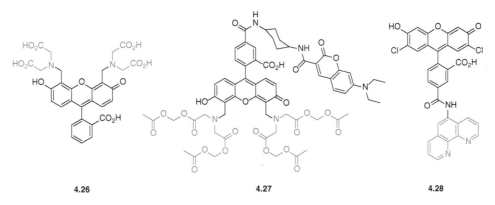

4.26	**4.27**	**4.28**

Figure 4.19 Structures of early iron probes. Receptor groups are coloured in red.

utilises fluorescein as its fluorophore and contains two iminodiacetate receptor groups as iron chelators (Figure 4.19).

The main limitations of calcein are that it is not selective to a particular oxidation state of iron, and it has a fluorescence quenching response. In attempt to remedy the latter issue, Carney *et al.* devised a calcein analogue, **4.27**, containing an additional coumarin fluorophore that gave a more desirable ratiometric fluorescence response. AM esters were also installed for cell membrane permeability, a routine probe adjustment (Figure 4.19) [82]. Another early iron probe, **4.28**, uses a 1,10-phenanthroline receptor group on a dichlorofluorescein scaffold (Figure 4.19) [83]. It has a greater quenching response that of calcein (96% *vs.* 46%) but is still unable to differentiate between Fe(II) and Fe(III).

As recognition-based iron sensing progressed, different approaches for selectively sensing Fe(II) and Fe(III) have been adopted. Fe(II) is a borderline Lewis acid under HSAB theory [18] and prefers to bind to six donor atoms in an octahedral geometry. As a result, one reported receptor group for Fe(II) featured a terpyridine (Tpy) [84], while another used a 2-amino-2-(hydroxylmethyl)propane-1,3-diol (Tris) chelating group [85] (Figure 4.20). The disparity between these receptor groups for Fe(II) highlights the lack of an established, widely adapted receptor group for Fe(II), unlike those described for other metals thus far.

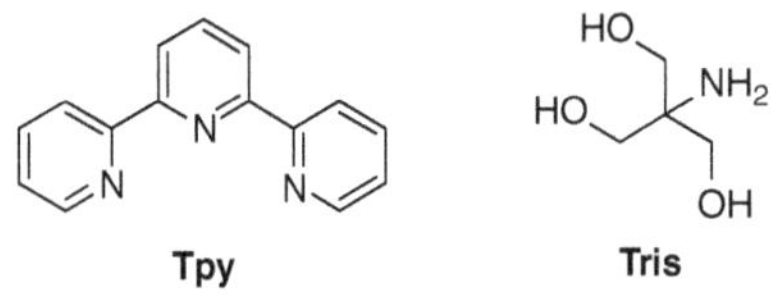

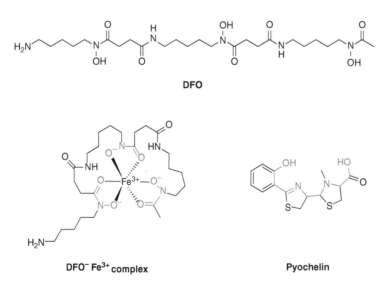

Figure 4.20 Fe(II) chelating groups used in reported Fe(II) probes.

A common approach to developing receptor groups for Fe(III) is to design them from naturally occurring organic molecules called siderophores, which are secreted by soil microbes to chelate and transport Fe(III) [57]. One such siderophore, desferrioxamine B (DFO), forms hexadentate complexes with Fe(III), coordinating to six oxygen atoms from three hydroxamic acid functional groups (Figure 4.21) [86]. Lytton *et al.* successfully attached a 7-nitrobenz-2-oxa-1,3-diazole (NBD) fluorophore onto DFO to create a probe (**4.29**) for monitoring the removal of Fe(III) from ferriproteins (Figure 4.22) [87]. However, the extremely high stability constant of Fe(III)-DFO complexes, $10^{31}\,\mathrm{M}^{-1}$ [86], limits its use for studying the lower concentrations of the labile iron pool [1].

Pyochelin is another siderophore that has been used to develop fluorescent probes for Fe(III). Pyochelin is a mixed-ligand siderophore, consisting of a phenol, a thiazoline, and a 4-thiazolinecarboxylic acid (Figure 4.21), that coordinates to Fe(III) in a 2 : 1 binding stoichiometry [88, 89]. Noël *et al.* used pyochelin with an NBD fluorophore for their Fe(III) probe, **4.30** (Figure 4.22). This probe displayed an enhancement in fluorescence intensity upon Fe(III)-chelation. While the lower binding constant of $6.2\times10^{10}\,\mathrm{M}^{-2}$ for probe **4.30** makes it better suited to studying labile iron, other essential metal ions such as Cu(II) and Zn(II) were not tested for their response and are likely to interfere with Fe(III) chelation [90].

Other Fe(III) binding motifs inspired by siderophores include catechols and 3-hydroxy-4-one; both of these groups coordinate to Fe(III) through oxygen atoms alone and have been used as receptor groups in fluorescent iron sensors **4.31** [91] and **4.32** [92], respectively (Figure 4.22). Notably, **4.32**, a quenching probe, was used in cell studies to monitor the

Figure 4.21 Siderophores used in Fe(III) probes. Chelating groups are coloured in red.

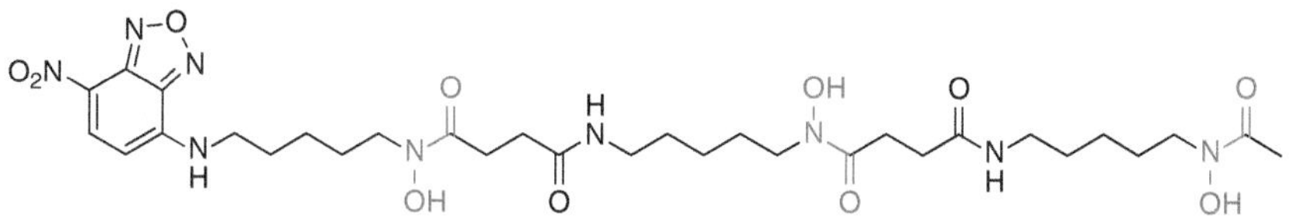

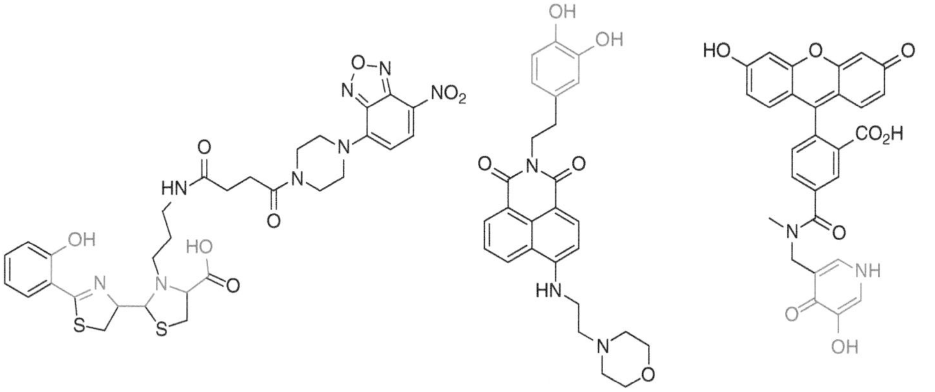

Figure 4.22 Structures of fluorescent probes for Fe(III). Receptor groups are coloured in red.

effects of iron chelators on iron homeostasis in the lysosome and endosome [92]. This probe likely has a lower binding affinity for iron than NBD-DFO, allowing it to measure labile, rather than protein-bound iron.

4.2.2.2.4 Other Essential Transition Metals: Manganese, Nickel, and Cobalt

Essential transition metals such as manganese, nickel, and cobalt are less extensively studied than zinc, copper, and iron. As a result, there are significantly fewer recognition-based fluorescent sensors reported for them, compared to other biologically relevant metals already discussed in Section 4.2. This is partly due to their lower prevalence in human biology but may also be attributed to the difficulty in developing selective sensors for these metal ions. Mn(II), Ni(II), and Co(II) are all paramagnetic metal ions, making them more likely to quench rather than enhance fluorescence signals of probes [93]. These ions are also notoriously difficult to distinguish from each other and other metals using recognition-based sensing. Ni(II) and Co(II) are both classified as borderline Lewis acids, so these ions often interfere with each other, while Mn(II) is classified as a hard Lewis acid and has a $3d^5$ electron configuration, so it often suffers interference from Ca(II) and Mg(II) [57]. Furthermore, manganese, nickel, and cobalt are all close neighbours on the periodic table, making their effective ionic radii very similar (Table 4.2).

Despite the challenges of detecting these ions with recognition-based sensing, some selective probes have been achieved. For example, Liang and Canary developed **4.33** as a selective turn-on difluorescein probe for Mn(II) by modifying the BAPTA Ca(II) receptor group with two pyridines in the place of two carboxylates (Figure 4.23) [94]. Alternatively, Mao *et al.* used a combination of secondary amines and ethers in the receptor group design of their bipyrene fluorescent sensor, **4.34**, which also displayed adequate selectivity to Mn(II) (Figure 4.23) [95].

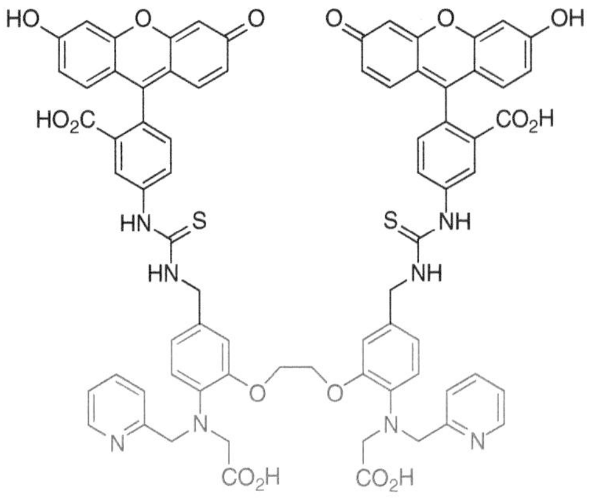

Figure 4.23 Structures of recognition-based probes for Mn(II). Receptor groups are coloured in red.

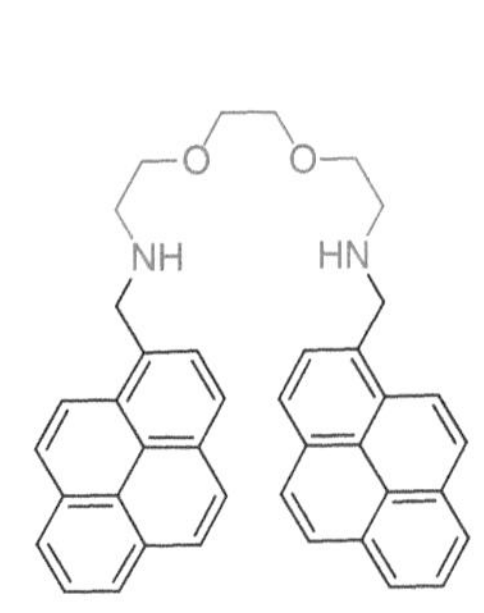

Notably, the above two reported chelating groups for Mn(II) differ considerably in structure, which is also true for Ni(II) receptor groups in the literature. The only example of the same Ni(II) receptor group being used on multiple fluorescent scaffolds is a dipodal thioether-carboxylate motif used by Dodani *et al.* and Kang *et al.* on their BODIPY and 2-acetyl-6-dialkyl-aminonaphthalene (acedan) Ni(II) probes, **4.35** and **4.36**, respectively (Figure 4.24) [96, 97]. Many other fluorescent probes for Ni(II) use receptor groups with *N*-heterocycles and imines, which are theoretically better suited to the borderline acidic nature of Ni(II), although many of these suffer interference from Cu(II) and Co(II) [98].

Cobalt is another challenging transition metal to sense using a recognition-based strategy. This is reflected in the lack of reported sensors in the literature for Co(II), the predominant oxidation state of cobalt in biology. Ghazali *et al.* created a probe, **4.37**, with a BODIPY reporter and phenol-amide receptor group that exhibited a turn-on fluorescence response in the presence of Co(II) ions (Figure 4.25) [99]. Zhang *et al.* used a different approach to the receptor group with their naphthalimide probe, **4.38** (Figure 4.25) [100]. They used a combination of secondary amines and pyridines that chelated to Co(II), leading to ratiometric fluorescence emission response. However, it is important to note that significant concentrations of organic solvents needed to be used in the sample matrix for both **4.37** and **4.38**, limiting their utility for biological experiments.

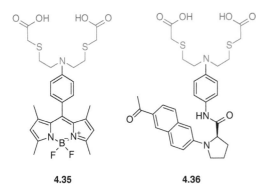

4.35   **4.36**

Figure 4.24 Structures of recognition-based probes for Ni(II) using the same receptor group coloured in red.

Figure 4.25 Structures of recognition-based probes for Co(II). Receptor groups are coloured in red.

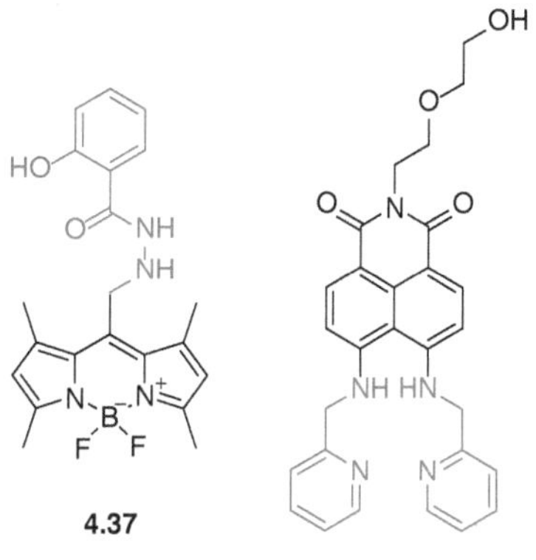

An alternative reported approach to sensing Co(II) involves oxidising it to Co(III) *in situ* with exogenous hydrogen peroxide, then applying a probe for Co(III) recognition [101]. While the addition of excess hydrogen peroxide is valid for detecting cobalt in environmental water samples, it is not suitable for biological experiments owing to its detrimental effect on cells [102].

It is evident that receptor groups for Cu(II), Mn(II), Ni(II), and Co(II) are highly variable, compared to ions such as Na(I), Ca(II), Zn(II), and Cu(I) that employ one receptor group class that functions on many fluorescent scaffolds. This highlights the challenging nature of designing selective probes for paramagnetic transition metals, especially those that are borderline Lewis acids. As a result, this is a highly active and important current area of research.

4.2.3 Toxic Metal Sensing

4.2.3.1 The Biological Significance of Toxic Metals

In this section, toxic metals are defined as those that do not naturally occur in organisms and cause adverse effects to cells and biological systems upon exposure, even in trace concentrations. These include, but are not limited to, mercury, cadmium, and lead. Mercury is a potent neurotoxin to humans, causing ataxia and muscle weakness, as well as damage to sight, hearing, and speech [103]. Cadmium and lead are toxic as they have been found to mimic or antagonise the biological roles of Ca(II) and Zn(II). Cadmium is carcinogenic and exposure through inhalation in industrial settings causes symptoms ranging from flu-like symptoms to pulmonary oedema and kidney and liver disorders [104]. Chronic lead exposure through polluted water from lead pipes or mining activities is dangerous as it causes high blood pressure and reduced renal function in human blood lead concentrations below $2\,\mu M$, and anaemia and lead encephalopathy in concentrations over $4\,\mu M$ [105].

These toxic metals share common characteristics such as higher atomic weights and more diffuse electronic orbitals than essential transition metals described in Section 4.2.2. As a result, mercury, cadmium, and lead are more covalent in nature and are more readily able to form metal-carbon bonds to make organometallic molecules. Organomercury, organocadmium, and organolead compounds are significantly more toxic to humans than the Hg(II), Cd(II), and Pb(II) metal ions because their lipophilicity enables them to cross the blood–brain barrier. However, they are less stable in biological systems and can degrade and reduce into their ionic forms. Therefore, the majority of small-molecule florescent probes for toxic metals are developed for studying the ionic forms. While activity-based probes for toxic metal ions exist, the following section will focus on recognition-based approaches for studying toxic metal ions in biology.

4.2.3.2 Receptor Group Design for Toxic Metals

4.2.3.2.1 Lead   A major strategy for developing probes for toxic metals is to adapt probes for the metals that they commonly mimic in biology. For example, many small-molecule Pb(II) probes utilise binding units inspired by Ca(II) probes. One example of this is the early Pb(II) probe, **4.39**, which contains a dipodal pseudo crown ether receptor group with terminal carboxylates (Figure 4.26) [106]. This receptor group is structurally similar to calcium chelator, BAPTA, as it also contains ethers, aliphatic amines, and carboxylates to coordinate to Pb(II) in an octahedral geometry. The sensor displays a favourable turn-on fluorescent enhancement in HEK cells incubated with Pb(II), though high micromolar concentrations of Pb(II) were required. Pseudo crown ether receptor groups form a common pattern across the literature, appearing in several other probes for Pb(II) [107, 108].

Owing to the borderline Lewis acid nature of Pb(II), many other strategies for Pb(II) receptor groups have been reported including hydroxyl groups [109], crown ethers [110], aza-crown ethers [111], and DPA [112–114]. However, one of the main challenges of using non-specific chelating groups for Pb(II) is that probes suffer interferences from other metals. For example, in most cases, Pb(II) sensors using DPA receptor groups also respond significantly to Zn(II). This creates a challenge for applying these sensors in biological cell studies, where Zn(II) is an essential metal, occurring in higher concentrations than the toxic metals being studied. Zhu *et al.* addressed this issue by designing a DPA-carboxylate receptor group for their cellular Pb(II) rhodamine-based probe, **4.40** (Figure 4.26) [115]. The probe demonstrated a 100-fold fluorescence emission enhancement upon Pb(II) chelation due to the opening of the rhodamine spirolactam, while Cd(II) addition resulted in a slight response and Zn(II) elicited no fluorescence change. The probe was used to visualise intracellular, exogenous Pb(II) in DLD-1 cells and to monitor the effect of Pb(II) in K562 cell differentiation into red blood cells.

4.2.3.2.2 Cadmium   Cd(II) is very chemically similar to Zn(II) because both ions have d^{10} electronic configurations, are classified as borderline Lewis acids, and prefer tetrahedral geometries. As a result, the predominant strategy for developing receptor groups for cadmium is using DPA or other known Zn(II) chelating groups. For example, Peng *et al.* used a single

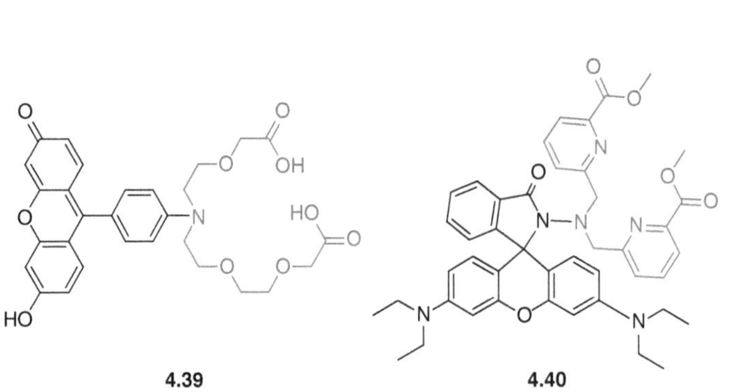

Figure 4.26   Structures of fluorescent Pb(II) probes. Receptor groups are coloured in red.

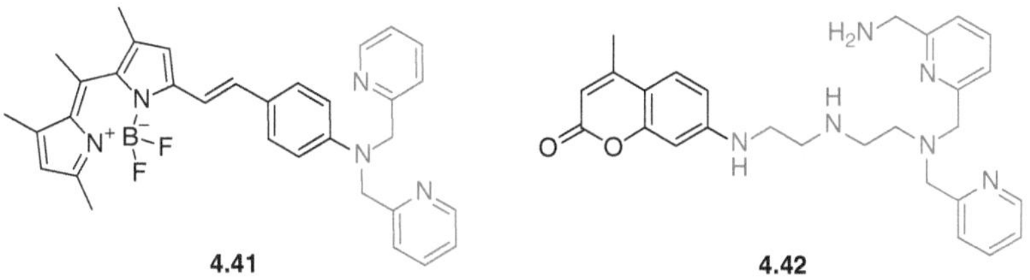

Figure 4.27 Structures of fluorescent Cd(II) probes. Receptor groups are coloured in red.

DPA receptor group on their BODIPY sensor, **4.41**, which displayed a ratiometric fluorescence emission response to cadmium ions, with little interference from other metal ions including Zn(II) (Figure 4.27) [116].

While DPA-based Cd(II) are common [57, 104, 117–120], many unsurprisingly have unfavourable responses with Zn(II) also. A successful example of a selective sensor for Cd(II) using a DPA receptor group was developed by Taki *et al.* [121]. Their coumarin sensor, **4.42**, contained a secondary amine ICT donor, an alkyl-amine linker, and a DPA group decorated with a single primary amine (Figure 4.27), which all bind to Cd(II) in an octahedral geometry, causing a unique blue-shift in the fluorescence excitation spectrum. While Cu(II), Co(II), Ni(II), and Hg(II) caused a decrease in fluorescence intensity, Cd(II) was the only metal ion that evoked a ratiometric response. This probe may have displayed a selective response to Cd(II) over Zn(II), because it has six coordinating atoms, which are more suited to bind Cd(II) as it more readily forms octahedral complexes and has a larger ionic radius. Other strategies for designing receptor groups for recognition-based Cd(II) sensors include using both oxygen and nitrogen atoms such as in a tetraamide moiety [122, 123], which displayed an improved selective response to Cd(II).

4.2.3.2.3 Mercury A challenge for designing fluorescent probes for toxic heavy metals such as lead and mercury is that they tend to quench fluorescence due to the heavy atom effect. As a result, the majority of small-molecule fluorescent probes for Hg(II) use rhodamine fluorophores. Like other xanthene-core dyes, rhodamines dramatically increase in fluorescence intensity upon the opening of the spirolactam ring, which can be triggered by the presence of a metal analyte [57, 124]. The spirolactam ring-opening is a reversible process, although many probes have been designed in such a way that Hg(II) cleaves/reacts with the ring in a non-reversible manner [57], making them as activity-based probes.

Those that are reversible and open the ring with coordination bonding often use sulfur atoms to chelate Hg(II), as it is classified as a soft Lewis acid. For example, Lin *et al.* used soft Lewis bases of a thioketone and an alkene in the probe, **4.43**, to reversibly monitor Hg(II) in HeLa cells (Figure 4.28) [125]. Another common approach to Hg(II) receptor groups is using a combination of amines and thioethers, similar to receptor groups employed for Cu(I) sensing. Although these sensors show impressive responses to Hg(II), many do not screen their probes against Cu(I), creating potentially confounding results in a cell biology context. For example, a reported mercury probe, **4.44** [126], uses an identical azatetrathia-crown ether receptor group to copper probe, **4.20**, so is very likely to also produce a response to Cu(I) (Figure 4.28).

Like the paramagnetic essential transition metals detailed in Section 4.2.2, toxic metals are notoriously difficult to develop selective sensors for because many of them quench fluorescence and are borderline Lewis acids. As a result, robust selective receptor groups for Pb(II),

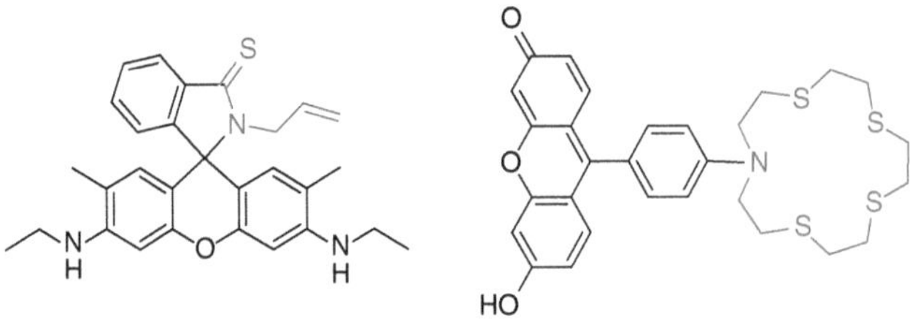

4.43   **4.44**

Figure 4.28   Structures of fluorescent Hg(II) probes. Receptor groups are coloured in red.

Cd(II), and Hg(II) are yet to be discovered. This has resulted in the use of existing essential metal receptor groups such as DPA and azatetrathia-crown ethers for toxic metal sensors, leading to limited sensor selectivity. The development of selective chelating groups for toxic metals is an ongoing area of research.

4.3   Recognition-based Anion Sensing

Anions participate in a variety of physiological processes and are crucial to many biological functions. For instance, one of the most important biological anions, phosphate, is involved in many cellular metabolic processes, such as skeletal development, ATP synthesis, and protein phosphorylation [127, 128]. Therefore, dysregulation of cellular anion levels can induce diverse diseases. Fluorescent sensors play an important role in detecting and quantifying specific anions, functioning as diagnostic tools for some disease states. Fluorescent probes sense specific anions through a change in electronic properties that are induced by interactions with receptors. In addition to electrostatic interactions between the negatively charged anions and protonated (or positively charged) receptors, hydrogen bonding, metal displacement, and metal coordination are widely used to recognise anions.

4.3.1   Anion Sensing Approaches

4.3.1.1   Hydrogen Bonding

Hydrogen bonding interactions are widely used in anion recognition. A hydrogen bond occurs between an electron rich atom and a hydrogen atom that is connected to a more electronegative atom [129]. This interacting system is denoted as X–H···A, where X and A can be fluorine, oxygen, or nitrogen. In certain cases, sulfur, carbon, chlorine, bromine, and iodine can also form weak hydrogen bonds [130]. Commonly used hydrogen bond donors in the design of anion receptors contain amide, urea, thiourea, indole, pyrrole, carbazole, and phenol groups, which coordinate to anions with different binding affinities (Figure 4.29) [131, 132].

There are several strategies to increase the selectivity of receptors for anions. Introducing more hydrogen bond donors that are well preorganised is a useful approach to increase receptors' ability to bind with anions [133, 134]. Due to the basic nature of anions, increasing the acidity of hydrogen bond donors by including electron-withdrawing groups in the receptor scaffolds is a useful way to enhance the interaction between the receptors and

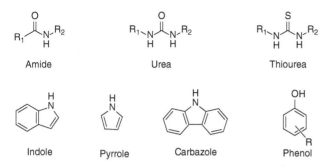

Figure 4.29 Common hydrogen bond donor species.

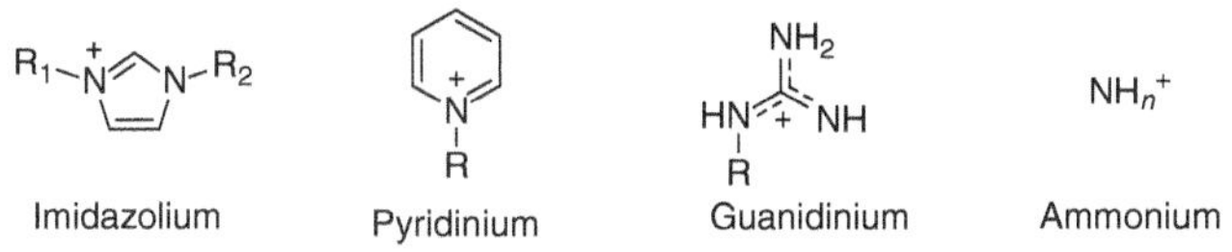

Figure 4.30 Positively charged binding sites.

anions [135]. Introducing positively charged binding sites is also an effective way to increase binding affinities to anions due to the enhanced electrostatic interactions between negatively charged anions and the positively charged receptors [136]. Guanidinium, ammonium, imidazolium, and pyridinium are commonly used positively charged receptors (Figure 4.30) [137–139]. Both hydrogen bonding and electrostatic interactions govern binding behaviours, leading to strong binding affinities to anions.

A number of anion receptors have been developed but have not been widely adopted in fluorescent sensor scaffolds due to their low water solubility and lack of selectivity for specific anions in aqueous solution [140]. Even though anions bind with a variety of moieties such as amide, urea, thiourea, and phenol groups *via* hydrogen bonding, a very limited number of fluorescent sensors for anions based on hydrogen bonding have been reported. This is likely due to the heavily hydrated nature of anions in aqueous solutions that weaken hydrogen bonding to receptors, resulting in poor selectivity [141].

4.3.1.2 Displacement Approach

Metal displacement is a widely used strategy to generate fluorescence responses to detect anions. In this approach, the fluorescent sensor consists of four major units: a fluorophore that provides fluorescent signals, a receptor that binds to a metal ion with low binding affinities, a spacer that connects the fluorophore to the receptor, and a metal ion that dissociates from the receptor once the metal interacts more strongly with targeted anion (Figure 4.31a,b) [142, 143].

Turn-on and turn-off fluorescent sensors have been developed for the recognition of anions *via* metal displacement approach. A turn-on fluorescent sensor can be built by a fluorescent molecule coordinated to a quenching metal ion. The fluorescence is quenched by the metal ion either due to the paramagnetic nature of some metals or heavy metal effects (Figure 4.31a) [144]. In the presence of anions, the quenching metal ion is displaced from the sensor to form a more stable complex, restoring the fluorescence of the sensor.

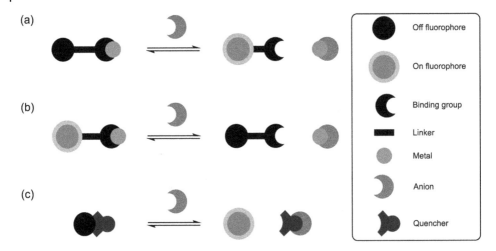

Figure 4.31 General structure of metal displacement-based sensors for anions with turn-on (a) or turn-off response (b), as well as fluorophore-quencher ensembles for recognition of anions (c).

For turn-off anion sensors, the organic probe is normally more fluorescent when bound to a metal ion, and when the metal is displaced by anions, the fluorescence is quenched (Figure 4.31b) [145]. Cu(II), Hg(II), and Fe(III) are widely used to construct sensors with turn-on responses to anions due to their fluorescent quenching properties. Zn(II) and Al(III) are effective for developing turn-off sensors for anions. These metal ions exhibit strong interactions with fluoride (F^-), iodide (I^-), phosphate (PO_4^{3-}), dihydrogen phosphate ($H_2PO_4^-$), and pyrophosphate ($P_2O_7^{4-}$) by either forming stable complexes with high association constants or generating stable products with low solubility product constant [143]. Examples of organic molecule-metal ensembles and corresponding sensing anions will be shown in Sections 4.3.2, 4.3.3 and 4.3.4.

Other than molecule-metal combination, a fluorophore-quencher ensemble can also operate as a displacement sensor for anions (Figure 4.31c). Similar to metal displacement sensors, the fluorescence of the probe is reduced by a non-covalently linked quencher [146]. The combination of fluorophore and quencher is chosen to ensure that the targeted anion has higher affinity with the quencher than it does with the fluorophore to ensure that the fluorophore will be released, and its fluorescence restored, upon anion binding. Representative sensors for anions adopting fluorophore-quencher ensembles will be presented in Sections 4.3.3 and 4.3.4.

4.3.1.3 Metal Coordination

Complexes of some transition metal ions (*e.g.* Zn(II), Cu(II)) can serve as receptors for anions since their unsaturated coordination provides free binding sites with strong electrostatic interactions. When anions bind to these metals, metal-sensor bonds weaken, triggering a change in fluorescence response (Figure 4.32) [147]. This metal coordination approach is commonly adopted for the recognition of pyrophosphate due to its strong bidentate binding to metal centres, as detailed in Section 4.3.3.3.

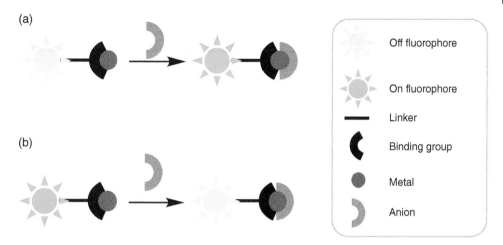

Figure 4.32 General structure of a metal coordination-based sensors for anions with turn-on (a) or turn-off (b) response.

4.3.2 Halogen Ions Sensing

4.3.2.1 The Biological Role of Halogen Ions

Halogen ions play significant roles in physiological activities and medical treatment. Fluoride (F^-) is an additive in toothpaste to prevent tooth decay [148]. Chloride (Cl^-) is the most abundant anion in living systems and its transport across cellular membranes is essential to maintain cellular osmotic balance [149]. Iodide (I^-) is also a commonly used additive in table salt as iodide is critical to the generation of thyroid hormones T_3 and T_4 [150]. A high level of bromide (Br^-) in blood can influence exchangeable iodide pool in the thyroid [151]. Due to their biological relevance, great efforts have been made to develop fluorescent sensors for detecting halogen ions.

4.3.2.2 Recognition-based Fluorescent Sensors for Halogen Ions

4.3.2.2.1 Displacement Motifs Hydrogen bonding recognition-based sensors for halides are uncommon in the literature, as they act as bases, deprotonating receptor groups, rather than binding with them. Alternatively, metal displacement is a widely adopted approach to detect halogen ions because the complexations of some metals (*e.g.* Al(III), Ag(I), Hg(II)) with halogen ions leads to insoluble products or stable complexes. For example, F^- complexes the Al(III) ion in **4.45**, liberating a phenolic oxygen of the fluorescein, enhancing fluorescence (Figure 4.33) [152]. This interaction is highly selective for F^- over Cl^-, Br^-, and I^-, since the Al^{3+}-F^- complex is less soluble than the complexes with Cl^-, Br^-, and I^-. In another example, **4.46** is built by complexation of thiourea ligands with Hg(II), in which Hg(II) acts as a quencher for fluorescence due to the heavy atom effect (Figure 4.33) [153]. I^- is capable of displacing Hg(II) from **4.46**, forming insoluble HgI_2, thus turning on the fluorescence of the sensor with high selectivity. Ko and co-workers developed **4.47**, a Cl^- and Br^- sensor using a triazole-Ag(I) complex (Figure 4.33) [154]. The addition of Cl^- or Br^- assists the release of Ag(I), leading to enhanced fluorescence. Sensor **4.47** does not respond to I^- because it is too large to fit into the binding cavity. In a slightly different displacement mechanism, a coumarin fluorophore is displaced from **4.48** when F^- binds to the Gd(III), producing a turn-on fluorescence response (Figure 4.33) [155]. This interaction with Gd(III) is selective to F^-, as it is more electronegative than other halides.

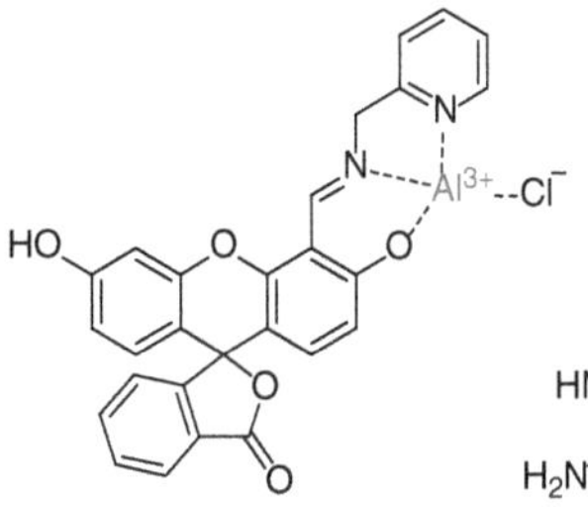

4.45 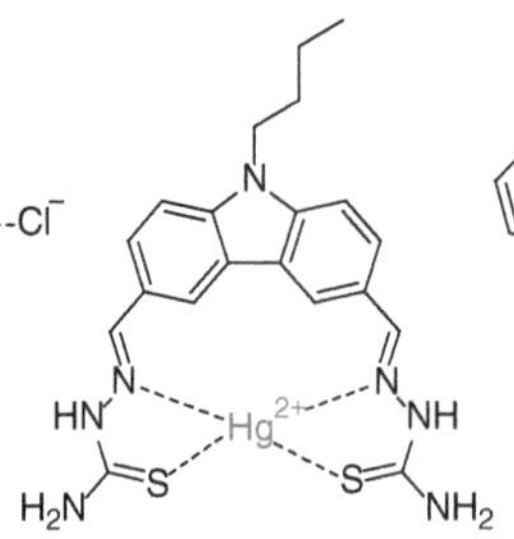   **4.46** 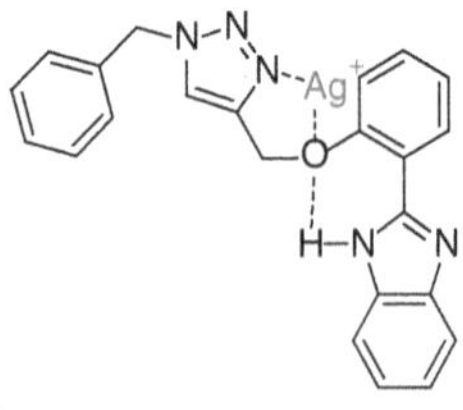   **4.47**

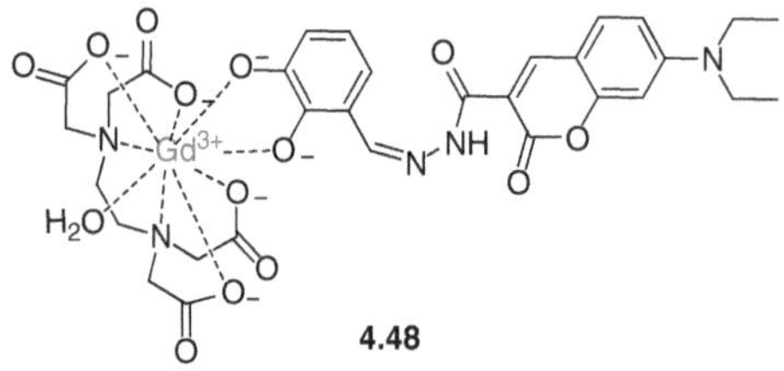

4.48

Figure 4.33 Sensors for halogen ions developed by displacement approach. Anion-binding displacement motifs are coloured in red.

4.3.3 Inorganic Phosphates and Pyrophosphates

4.3.3.1 The Biological Role of Inorganic Phosphates and Pyrophosphates

Phosphates and pyrophosphates have significant roles in living organisms. Phosphates are involved in many cellular metabolic processes, such as skeletal development, bone integrity maintenance, ATP synthesis, and protein phosphorylation [127, 128]. In body fluids, phosphate exists in two forms, dihydrogen phosphate ($H_2PO_4^-$) and hydrogen phosphate (HPO_4^{2-}). In tissues and cells, phosphates are present as various organic substances, including phosphoproteins, sugar phosphates, phospholipids, and nucleic acids [156]. Approximately 80–85% of phosphates are found in bone, 10% in skeletal muscle, and less than 1% in extracellular fluids [157]. The deficiency of inorganic phosphate results in skeletal mineralisation, abnormal function of platelets and leukocytes, increased erythrocyte membrane rigidity, and cardiac dysfunction [156]. Inorganic pyrophosphate (PPi, $P_2O_7^{4-}$) is the dimeric form of inorganic phosphate and a by-product of the hydrolysis of adenosine triphosphate (ATP) under cellular conditions, which makes pyrophosphate a research target for disease diagnosis [158]. Several diseases, such as arthritis and Mönckeberg's arteriosclerosis, are associated with abnormal levels of pyrophosphate [159].

Developing fluorescent sensors that are capable of detecting inorganic phosphate and pyrophosphate in living cells and tissues is of great interest. At physiological pH, inorganic phosphates are present in both $H_2PO_4^-$ and HPO_4^{2-} forms in a near 1 : 2 ratio [160]. As a result, reported sensors for inorganic phosphate do not specify which form of phosphate they target. Alternatively, in organic solvents, phosphate can exist in one specific form, and therefore many fluorescent sensors are reported to detect $H_2PO_4^-$ in organic solvent rather than in aqueous solution. In this section, fluorescent sensors for recognising inorganic phosphate and pyrophosphate will be introduced.

4.3.3.2 Sensors for Inorganic Phosphates

4.3.3.2.1 Hydrogen Bonding Motifs Inorganic phosphate (HPO_4^{2-} or $H_2PO_4^-$) possesses hydrogen atoms that can serve as hydrogen bonding donors and oxygen atoms that can serve as hydrogen bonding acceptors. The existence of both hydrogen bonding donors and acceptors within one molecule enables phosphate to readily bind to both hydrogen bonding acceptors or donors. As a result, amides have been employed as phosphate sensors, as seen in probes **4.49** and **4.50** (Figure 4.34) [161, 162]. In sensor **4.49**, phosphate quenches fluorescence, while sensor **4.50** experiences a shift in fluorescence emission upon phosphate binding due to pyrene excimer formation. The fluorescence change in **4.50** is selective for $H_2PO_4^-$ over HPO_4^{2-}, but it suffers interference from HSO_4^-. A limitation of both of these probes is that studies were conducted purely in organic solvents, limiting their use in biological applications.

Other hydrogen bonding motifs for phosphate include imidazolium and benzimidazolium (Figure 4.34). These groups are positively charged, forming strong electrostatic interactions with phosphate. These motifs have been used to construct phosphate probes, **4.51** and **4.52**, that contain anthracene and acridine fluorophores, respectively (Figure 4.34) [163, 164]. The anthracene of **4.51** is PET-quenched by phosphate, resulting in a turn-off response. Alternatively, phosphate binding of **4.52** causes a new emission band to form, allowing for ratiometric detection. Both **4.51** and **4.52** have good selectivity for $H_2PO_4^-$ over HSO_4^- in acetonitrile and a mixture of acetonitrile and DMSO.

4.3.3.2.2 Displacement Motifs Zn(II), Cu(II), Al(III), and Fe(III) have high binding affinities for phosphate and are therefore commonly used for constructing displacement-based sensors

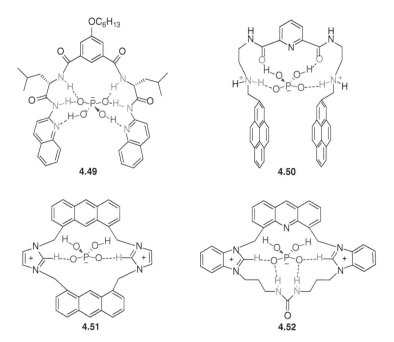

4.49

4.50

4.51

4.52

Figure 4.34 Sensors for phosphate (coloured in blue) based on hydrogen bonding motifs (coloured in red).

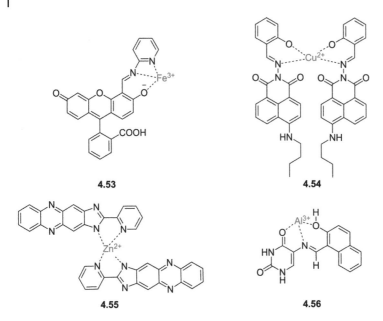

Figure 4.35 Sensors for phosphate based on metal displacement motifs. The displaced metal is coloured in red.

(Figure 4.35). For example, Fe(III) is displaced from fluorescein-based sensor **4.53** upon phosphate binding, which results in a turn-on fluorescence response (Figure 4.35) [165]. In a similar way, Cu(II) is released from **4.54** when bound to phosphate, enhancing naphthalimide fluorescence (Figure 4.35) [166]. Zn(II) was used as a displacement element in **4.55** (Figure 4.35) [167]. Upon $H_2PO_4^-$ bonding, Zn(II) is liberated and **4.55** can become twisted, which is accompanied by a new fluorescence emission band through a twisted internal charge transfer (TICT) mechanism. In a final example, **4.56** is a Schiff base Al(III) complex that is fluorescent in the absence of phosphate (Figure 4.35) [168]. As Al(III) has a stronger binding affinity to phosphate than the Schiff base, it is displaced, enabling the free sensor to isomerise, quenching the fluorescence signal. Most of these displacement probes exhibit good selectivity for $H_2PO_4^-$ over other anions, especially HSO_4^-, which often interferes the detection of $H_2PO_4^-$ in aqueous solutions. However, **4.56** suffers from the interference of HSO_4^-, which is likely because HSO_4^- also has strong affinity for Al(III).

4.3.3.3 Sensors for Inorganic Pyrophosphate

4.3.3.3.1 Hydrogen Bonding Motifs Inorganic pyrophosphate (PPi, $P_2O_7^{4-}$) has a greater negative charge and more hydrogen bonding acceptors than inorganic phosphate. As for the recognition of pyrophosphate, urea and thiourea are effective motifs that provide hydrogen bonding donors for binding with pyrophosphate. An anthracene-based probe, **4.57**, binds pyrophosphate *via* two thiourea motifs (Figure 4.36) [169]. The hydrogen bonding donors from urea tend to associate with oxygen atoms from pyrophosphate, which quenches fluorescence emission due to PET. However, the alignment of two urea groups at opposite sides of the anthracene in **4.57** makes hydrogen binding sites accessible to both pyrophosphate and phosphate, resulting in poor anion selectivity.

Preorganisation of hydrogen binding sites in a pseudo-cavity that fits pyrophosphate is an effective way to improve sensing selectivity. The urea moieties of **4.58** are arranged at 2- and

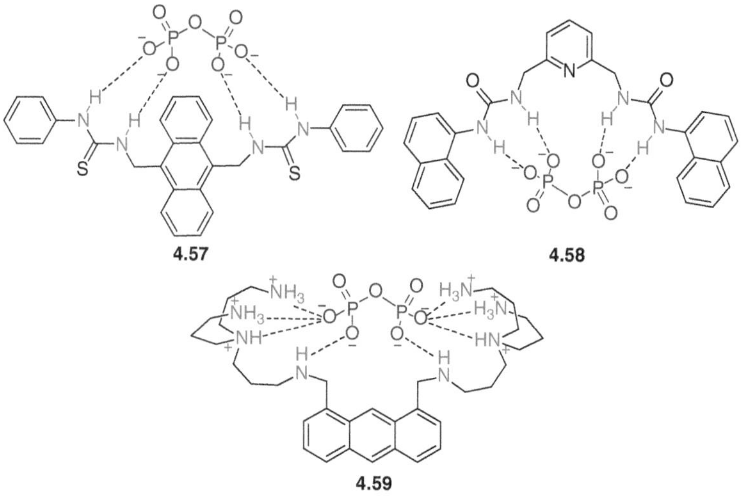

Figure 4.36 Sensors for pyrophosphate based on hydrogen bonding motifs. Receptor groups are coloured in red.

6-positions of pyridine and are tethered to two naphthalene moieties as fluorescence emitters (Figure 4.36) [170]. Flexibility of the pseudo-cavity on **4.58** allows phosphate and pyrophosphate enter and form hydrogen bonds, while only the binding of urea with pyrophosphate induces formation of naphthalene excimer, leading to a ratiometric fluorescence response. Both **4.57** and **4.58** recognise pyrophosphate in a pure organic solvent, dimethyl sulfoxide, which limits their application in biological environments. To solve this issue, a positively charged ammonium motif can be introduced to improve recognition efficiency in aqueous solution through hydrogen bonding and electrostatic interactions. This was applied in **4.59**, which possesses both amines and ammoniums as hydrogen bonding sites, exhibiting high charge density and thus forming strong electrostatic interactions with pyrophosphate (Figure 4.36) [171]. Besides high charge density, the geometry of the anion receptors allows the probe to bind to both sides of the pyrophosphate, contributing to high selectivity towards pyrophosphate over phosphate. Therefore, **4.59** is one of the few fluorescent sensors capable of selectively recognising pyrophosphate in 100% aqueous solution.

4.3.3.3.2 Displacement Motifs As a dimer of phosphate, pyrophosphate also tends to coordinate with hard metals with high binding affinity. This nature can be used to build fluorescence sensors for recognition of pyrophosphate *via* metal displacement approach. Al(III) and Cu(II) are hard and borderline metals, respectively, therefore readily associate with pyrophosphate. For example, Al(III) is displaced by pyrophosphate from the rhodamine-based probe, **4.60** (Figure 4.37) [172]. The displacement of Al(III) causes the spirolactam ring to close, which is accompanied by a decrease in fluorescence intensity. In contrast, a turn-on pyrophosphate probe, **4.61**, was developed by Zhu and co-workers (Figure 4.37) [173]. In the absence of pyrophosphate, **4.61** is quenched by paramagnetic Cu(II). When Cu(II) is bound by pyrophosphate, it is released from the probe, turning on fluorescence. Although Cd(II) is a soft metal [19], the Cd(II) displacement probe, **4.62**, binds to pyrophosphate with high affinity, liberating Cd(II) from the anthracene-terpyridine core and creating a ratiometric fluorescence response (Figure 4.37). All three of these probes, **4.60**, **4.61**, and **4.62**, exhibit

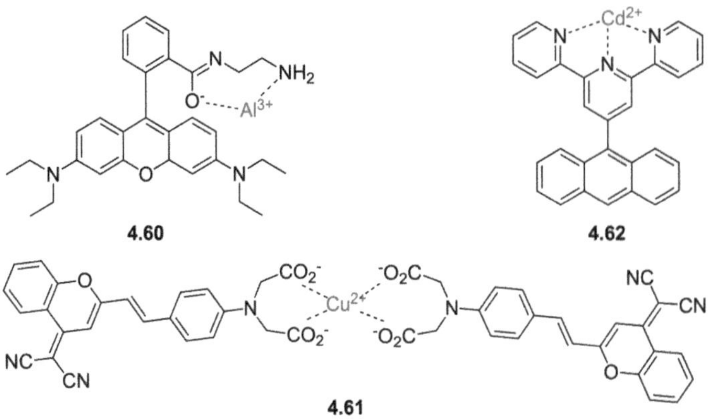

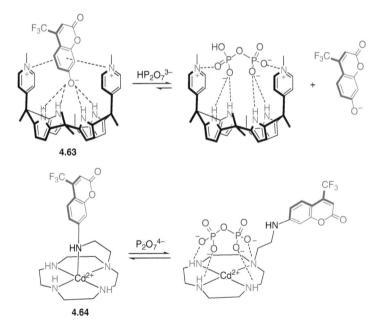

Figure 4.37 Sensors for pyrophosphate-containing metal-fluorophore combinations. Displaced metals are coloured in red.

excellent selectivity for pyrophosphate over phosphate, which should be attributed to the high charge density of pyrophosphate. Moreover, pyrophosphate could be detected in solutions with high water content, especially for **4.60** and **4.61**, whose high selectivity for pyrophosphate was achieved in 100% aqueous solutions.

In addition to metal ions, macrocyclic anion receptor groups can also be displaced from fluorophores to generate fluorescence turn-on responses in the presence of analytes. One such example, **4.63**, uses bis-pyridinium calix[4]pyrrole macrocycle coordinated to a coumarin fluorophore (Figure 4.38) [174]. When pyrophosphate binds to the macrocycle, it is displaced from the coumarin, enhancing fluorescence. Another macrocycle fluorescence quencher is 1,4,7,10-tetraazacyclododecane (cyclen). A Cd(II)-cyclen complex with a

Figure 4.38 Sensors for pyrophosphate-containing fluorophore-quencher ensembles. Receptor groups are coloured in red.

quenched coumarin fluorophore was used in probe, **4.64** (Figure 4.38) [175]. The fluorescence excitation spectrum of the complex changes upon pyrophosphate binding to cyclen, which releases the ICT donor amine from its coordination to Cd(II).

4.3.3.3.3 Metal Coordination Apart from the displacement recognition approach, metal ions can be used as anion sensing motifs themselves. This logic has been applied to a number of molecular sensors for pyrophosphate. Zn(II) complexed to dipicolylamine (DPA) is the most widely used motif for recognising pyrophosphate (Figure 4.39). For example, two preorganised Zn(II)-DPA units can be linked to a fluorophore, which creates an ideal binding site for one pyrophosphate molecule with its two phosphate units. As a result, this binding motif provides excellent selectivity for pyrophosphate over other forms of phosphate and other anions. This motif is used in sensor **4.65**, where two Zn(II)-DPA units are attached to a phenoxide of 2-(2-hydroxyphenyl)-1,3-benzoxazole (HBO) group (Figure 4.39). HBO is a typical fluorophore exhibiting excited-state intramolecular proton transfer (ESIPT) [176]. When the phenoxide oxygen is coordinated to Zn(II), ESIPT is turned off, leading to fluorescence emission at 425 nm. The coordination of pyrophosphate with two Zn(II) sites removes Zn(II) from the HBO phenoxide, restoring the fluorescence emission at 525 nm. The fluorescence switching induced by the presence of pyrophosphate makes ratiometric detection of pyrophosphate in HEPES buffer possible. Similarly, in sensor **4.66**, two Zn(II)-DPA units are attached to two coumarin moieties with a benzene linker (Figure 4.39) [177]. The probe alone exhibits strong fluorescence

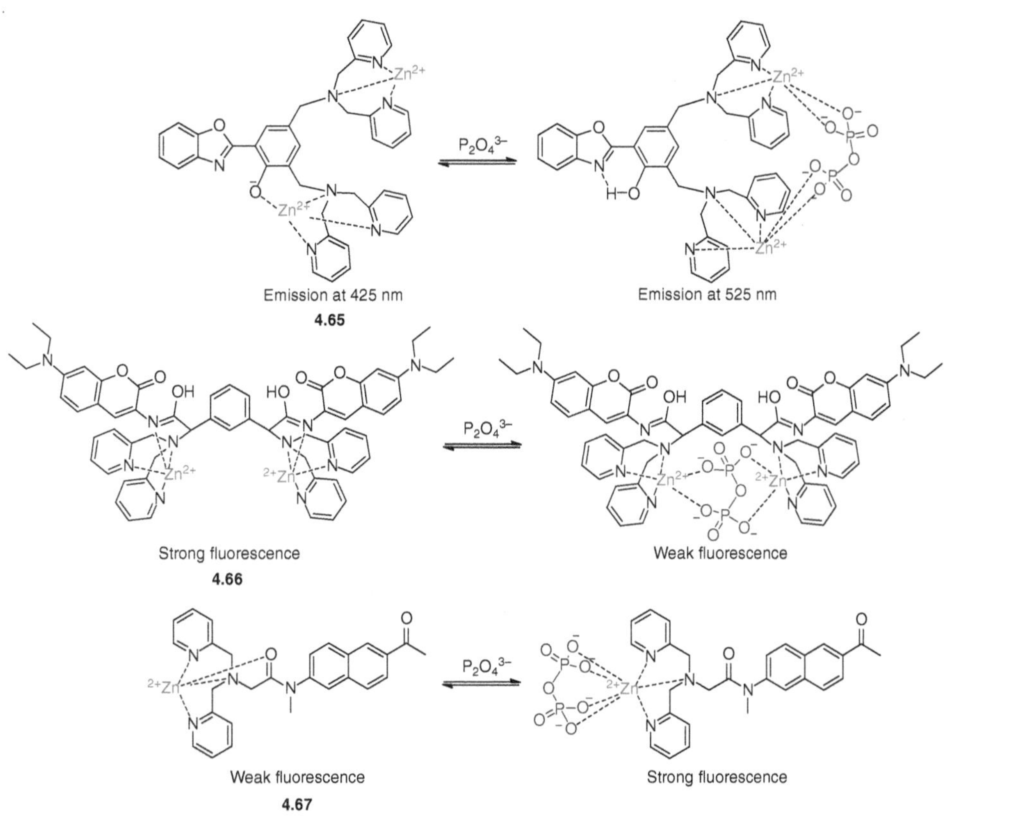

Figure 4.39 Sensors for pyrophosphate based on the Zn(II)-DPA binding motif.

due to the blocking of photoinduced electron transfer (PET) quenching of DPA by Zn(II). The association of pyrophosphate with Zn(II) weakens the coordination between DPA and Zn(II), restoring the PET quenching.

Intramolecular charge transfer (ICT) is another effective way to generate a pyrophosphate recognition signal with the Zn(II)-DPA motif. The acedan-based probe, **4.67**, contains one Zn(II)-DPA unit tethered to the fluorescent core *via* a carboxamido group (Figure 4.39) [178]. The electron donating ability and hence ICT efficiency of the carboxamido group are weakened by Zn(II) binding. As pyrophosphate coordinates with Zn(II), the carboxamido group is liberated from Zn(II) and enhances ICT, accompanied by a strong fluorescence increase. These three metal-coordination probes show excellent selectivity towards pyrophosphate over phosphate in HEPES buffer, demonstrating that Zn(II)-DPA is an effective recognition motif for pyrophosphate.

Although the Zn(II)-DPA unit is most prevalent for pyrophosphate recognition, examples with other metals/complexes exist. Polymeric sensor **4.68** uses a Cu(II)-DPA complex as the metal-receptor group and operates by a similar sensing mechanism to **4.67** (Figure 4.40) [179]; **4.69** is a pentacoordinate Sn(IV) complex with the fluorophore Alizarin red S, which exhibits strong fluorescence (Figure 4.40) [180]. In the presence of pyrophosphate, the Sn(IV) complex of **4.69** loses an aqua ligand and forms two new coordination bonds to pyrophosphate, resulting in a hexacoordinate Sn(IV) complex and a decrease in the fluorescence signal. Since phosphate and other anions can only provide one binding site to Sn(IV), **4.69** has good selectivity for pyrophosphate over other anions. Zn(II) has been used in complexes other than Zn(II)-DPA for pyrophosphate recognition. For example, **4.70** is a turn-on fluorescence sensor for pyrophosphate, with terpyridine-Zn(II) as the recognition motif, which has been successfully applied to image pyrophosphate in HeLa cells (Figure 4.40) [181].

4.3.4 Bicarbonate, Hydrogen Sulfate, and Nitrate

4.3.4.1 The Biological Roles of Bicarbonate, Hydrogen Sulfate, and Nitrate

Oxyanions, especially bicarbonate (HCO_3^-), play an important biological role in animals and plants. Bicarbonate is essential for maintaining physiological intracellular and extracellular pH, which is achieved by controlling the equilibrium between carbonic acid (H_2CO_3), bicarbonate, and carbon dioxide (CO_2) [182]. The kidneys regulate the excretion and recovery of bicarbonate to maintain normal physiological status [183]. Furthermore, the ratio of bicarbonate and carbon dioxide affects vesicle-mediated calcification processes in mammals [184]. Bicarbonate induces the production of a second messenger molecule, cAMP, and directly stimulates mammalian soluble adenylyl cyclase activity [185]. Bicarbonate also participates in the carboxylation reaction of pyruvate to oxaloacetate in mitochondria [186]. The level of bicarbonate in cells and organs are classic indicators for respiratory and metabolic disorders [187].

Nitrate (NO_3^-) is a precursor of nitric oxide (NO) and is mainly absorbed in the body through green vegetable consumption. NO_3^- is converted to NO, an important signalling molecule, by symbiotic bacteria in the oral cavity and stomach [188]. Therefore, nitrate is indirectly involved in many physiological functions. Finally, hydrogen sulfate (HSO_4^-) is used to produce primary and secondary sulfur-containing metabolites that are essential for plant growth [189]. Monitoring levels of bicarbonate, nitrate, and hydrogen sulfate in organisms is helpful for disease diagnosis. Therefore, generating fluorescent sensors for recognition of bicarbonate, nitrate, and hydrogen sulfate is an ongoing area of research.

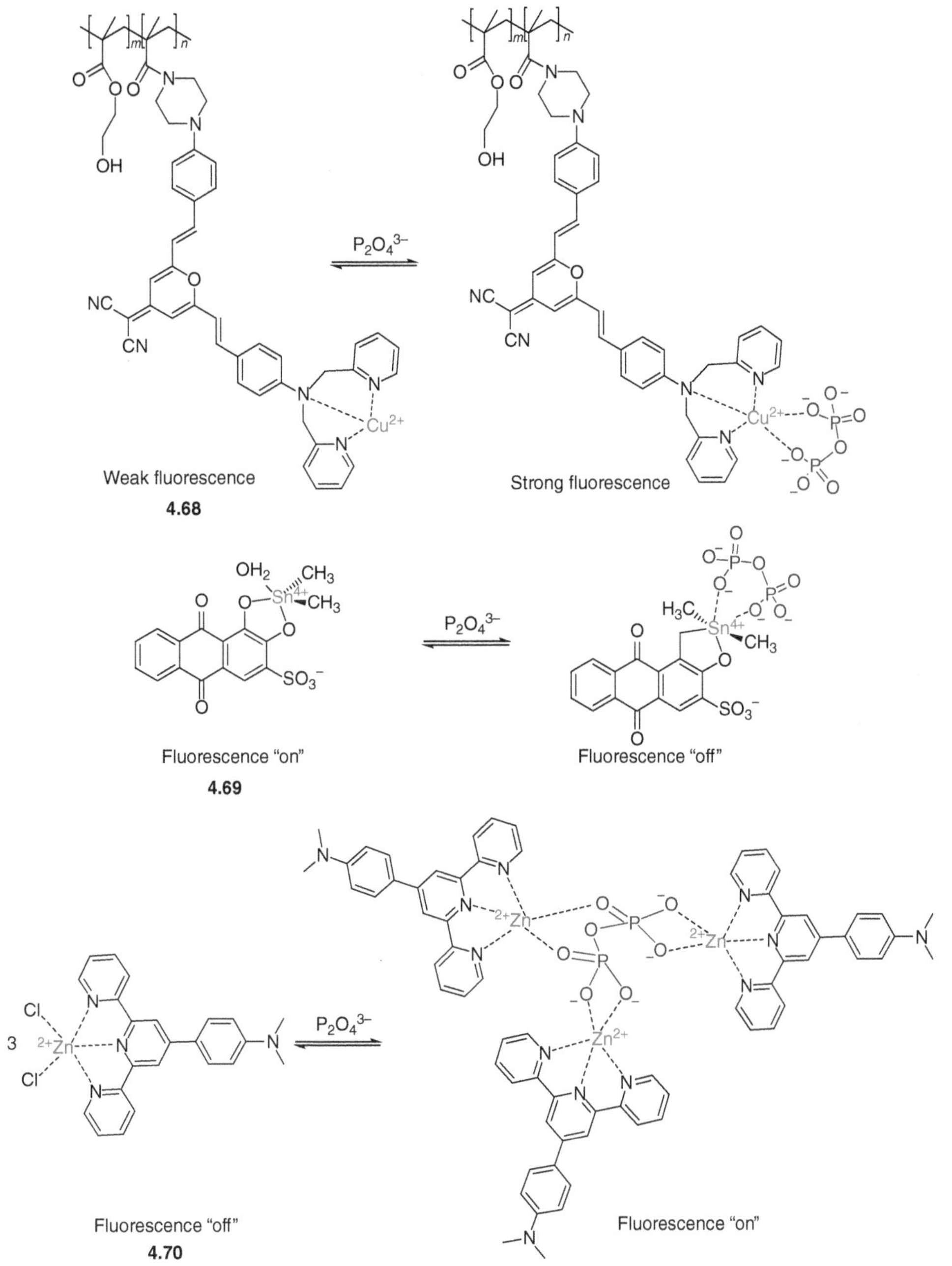

Figure 4.40 Sensors for pyrophosphate based on metal coordination.

4.3.4.2 Sensors for Bicarbonate, Hydrogen Sulfate, and Nitrate

Fluorescent sensors for bicarbonate, hydrogen sulfate, and nitrate are very rare, creating a large space for researchers to develop selective sensors for recognising these oxyanions. Here, two examples for detecting bicarbonate are described. Sensor **4.71** is composed of a calix[4]pyrrole derivative as the anion recognition unit and a trifluoromethyl coumarin fluorophore (Figure 4.41) [187]. HCO_3^- displaces coumarin from the cavity of calix[4]pyrrole

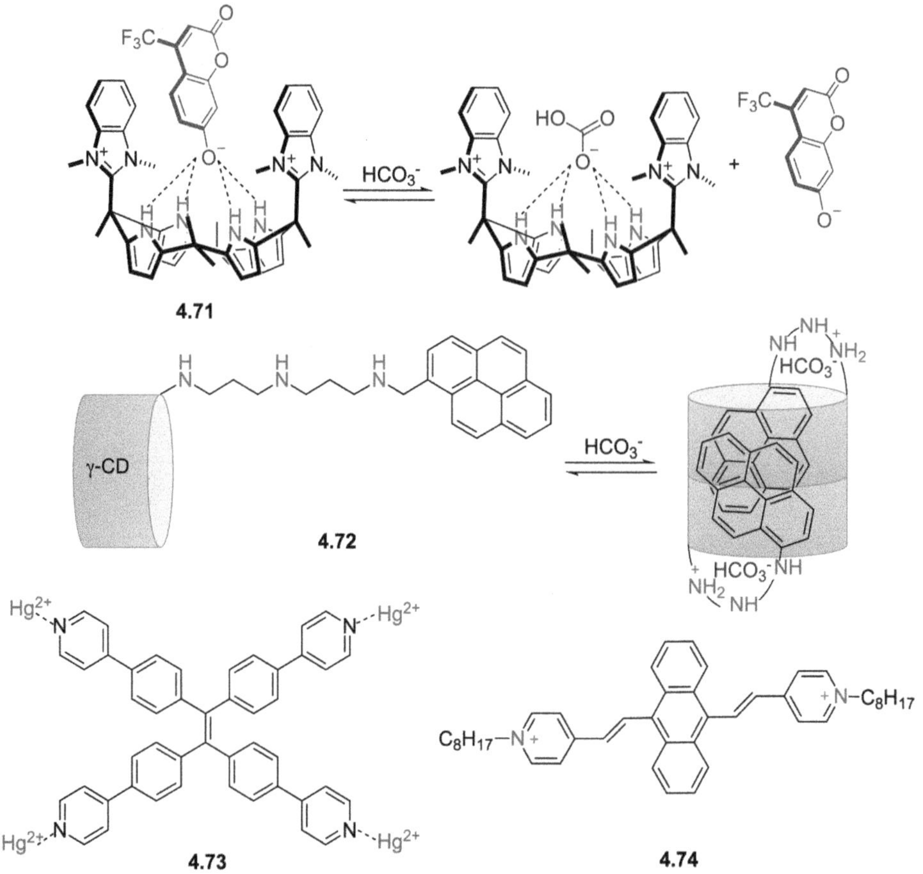

Figure 4.41 Sensors for bicarbonate, hydrogen sulfate, and nitrate. Receptor groups are coloured in red.

derivative, resulting in restored coumarin fluorescence. The binding affinity of **4.71** with HCO_3^- is on the order of $10^7 M^{-1}$, but **4.71** also binds with Pi, PPi, SO_4^{2-}, and NO_3^- with affinities around $10^4 M^{-1}$, limiting its selectivity. In sensor **4.72**, the pyrene fluorophore is appended to cyclodextrin (CD) through a triamine linker (Figure 4.41) [190]. At pH 8.6, two pyrene units form an association dimer within the cavity of two CDs, and simultaneously the triamine linker bends and forms a pseudo azacrown ring with one charged ammonium group and two neutral amino groups. The pseudo ring size, as well as the charge and hydrogen bonding sites from amino and ammonium groups, are suitable for HCO_3^- binding, which leads to a new fluorescence band. Studies of **4.72** were conducted in buffer solution (pH 8.6) and showed excellent selectivity towards HCO_3^- over other anions, making it an attractive sensor for HCO_3^- in biological settings.

Recognition-based fluorescent sensors for hydrogen sulfate and nitrate using common methods such as hydrogen bonding are rare. An alternative approach is aggregation-induced emission [191], which is becoming increasingly popular [192]. Probe **4.73** is a tetra(4-pyridylphenyl)ethylene-Hg(II) complex, which exhibits relatively weak fluorescence alone (Figure 4.41) [193]. In the presence of hydrogen sulfate, **4.73** aggregates *via* HSO_4^--Hg(II) bonds, leading to enhanced fluorescence. However, this response is not selective to hydrogen sulfate, as $H_2PO_4^-$ induces a similar response. AIE was also used in nitrate sensor, **4.74**, which features an anthracene fluorophore and two pyridinium anion receptor groups (Figure 4.41) [194]. In this case, aggregation (and a resulting fluorescence increase) is induced by the electrostatic interaction between positively charged pyridinium groups and negative nitrate. Fortunately, **4.74** exhibits excellent selectivity for NO_3^- over other anions, including $H_2PO_4^-$ and HSO_4^-, in HEPES buffer, enabling its use for nitrate detection in HeLa cells.

4.4 Conclusions

It is evident that recognition-based fluorescent sensors are useful tools for detecting cations and anions in biology. Their reversible nature allows for real-time analysis of dynamic biological processes, such as the movement of ions across membranes and their localisation in various organelles. Careful design of the receptor group of recognition-based sensors is crucial for achieving a selective response to an analyte of interest. Key considerations for designing receptor groups that have been addressed in this chapter include size, number of binding atoms, Lewis acidity, coordination geometries, and relative binding affinities.

Despite the realisation of these considerations, obtaining a selective receptor group for a single analyte remains challenging. In Section 4.2, successful, selective receptor groups were achieved for commonly occurring group I and II metal ions and well-studied essential transition metals such as Zn(II) and Cu(I). However, selective receptor groups for lesser-studied essential transition metals, such as Co(II) and Ni(II), and toxic metals still remain scarce, warranting further research.

Section 4.3 presented the advantages and limitations of three strategies for recognition-based anion sensing: hydrogen bonding, displacement, and metal coordination. A key challenge of sensing anions in aqueous conditions is that they are often heavily hydrated and are therefore less-likely to interact with recognition motifs. Therefore, most anion sensors rely on the use of organic solvents, limiting their utility in biology.

For halogen ions, selectivity can be achieved through receptor group size and differential solubilities of metal salts with reasonable levels of success. Sensors for phosphate typically experienced interference from sulfates, but sensors for pyrophosphate were generally made more selective through the incorporation of more coordination sites into their receptor groups. Overall, sensors for other biologically relevant anions, such as bicarbonate, hydrogen sulfate, and nitrate, are rare, and further research is needed to generate selective receptor groups for these anions that function in biological environments.

References

1 Kakhlon, O. and Cabantchik, Z.I. (2002). *Free Radic. Biol. Med.* 33 (8): 1037–1046.

2 Cotruvo, J.J.A., Aron, A.T., Ramos-Torres, K.M., and Chang, C.J. (2015). *Chem. Soc. Rev.* 44 (13): 4400–4414.

3 Job, P. (1928). *Ann. Chim.* 9 (10): 113–134.

4 Renny, J.S., Tomasevich, L.L., Tallmadge, E.H., and Collum, D.B. (2013). *Angew. Chem. Int. Edit.* 52 (46): 11998–12013.

5 Benesi, H.A. and Hildebrand, J. (1949). *J. Am. Chem. Soc.* 71 (8): 2703–2707.

6 Wang, R. and Yu, Z. (2007). *Acta Phys. -Chim. Sin.* 23 (9): 1353–1359.

7 Anderegg, G. (1993). *Anal. Chim. Acta* 282 (3): 485–488.

8 Wiseman, T., Williston, S., Brandts, J.F., and Lin, L.-N. (1989). *Anal. Biochem.* 179 (1): 131–137.

9 Sigurskjold, B.W. (2000). *Anal. Biochem.* 277 (2): 260–266.

10 Shannon, R.D. (1976). *Acta Crystallogr. Sect. A: Cryst.* 32 (5): 751–767.

11 Schwarzenbach, G. (1952). *Helv. Chim. Acta* 35 (7): 2344–2359.

12 Greenwood, N.N. and Earnshaw, A. (ed.) (1997). Coordination and organometallic compounds. In: *Chemistry of the Elements*, 2ee, 905–911. Oxford: Butterworth-Heinemann.

13 Hancock, R.D. and Martell, A.E. (1989). *Chem. Rev.* 89 (8): 1875–1914.

14 Aronson, A., Hammond, P., and Strafuss, A. (1968). *Toxicol. Appl. Pharmacol.* 12 (3): 337–349.

15 Catsch, A. and Harmuth-Hoene, A.-E. (1976). *Pharmacol. Ther. A.* 1 (1): 1–118.

16 Schwarzenbach, G. and Ackermann, H. (1949). *Helv. Chim. Acta* 32 (5): 1682–1689.

17 Hancock, R.D., Melton, D.L., Harrington, J.M. et al. (2007). *Coord. Chem. Rev.* 251 (13–14): 1678–1689.

18 Pearson, R.G. (1963). *J. Am. Chem. Soc.* 85 (22): 3533–3539.

19 Pearson, R.G. (1968). *J. Chem. Educ.* 45 (9): 581–587.

20 Pople, J.A. and Beveridge, D.L. (1970). *Approximate Molecular Orbital Theory*. McGraw-Hill Book Company.

21 Tsuchida, R.B. (1938). *Chem. Soc. Jpn.* 13 (5): 388–400.

22 House JE. Ligand fields and molecular orbitals. In: House JE. *Inorganic Chemistry (Second Edition)*: Academic Press; 2013. p. 591–616.

23 Lima-de-Faria, J., Hellner, E., Liebau, F. et al. (1990). *Acta Crystallogr. A.* 46 (1): 1–11.

24 Kuppuraj, G., Dudev, M., and Lim, C. (2009). *J. Phys. Chem. B* 113 (9): 2952–2960.

25 Rulíšek, L. and Vondrášek, J. (1998). *J. Inorg. Biochem.* 71 (3–4): 115–127.

26 Bertini, I. (2007). *Biological Inorganic Chemistry: Structure and Reactivity*. Sausalito, CA: University Science Books.

27 Frausto da Silva, J.J.R. and Williams, R.J.P. (2001). *The Biological Chemistry of the Elements: The Inorganic Chemistry of Life*. OUP Oxford.

28 Baldessarini, R.J., Tondo, L., and Hennen, J. (1999). *J. Clin. Psychiatry* 60: 77–84; discussion 111–116.

29 Brini, M., Ottolini, D., Calì, T., and Carafoli, E. (2013). Calcium in health and disease. In: *Interrelations between Essential Metal Ions and Human Diseases* (ed. A. Sigel, H. Sigel and R.K.O. Sigel), 81–137. Dordrecht, Netherlands: Springer.

30 Romani, A.M.P. (2013). Magnesium in health and disease. In: *Interrelations Between Essential Metal Ions and Human Diseases* (ed. A. Sigel, H. Sigel and R.K.O. Sigel), 49–79. Dordrecht, Netherlands: Springer.

31 Roat-Malone, R.M. (2002). *Bioinorganic Chemistry: A Short Course*. Wiley.

32 Pedersen, C.J. (1967). *J. Am. Chem. Soc.* 89 (26): 7017–7036.

33 Gokel, G.W., Leevy, W.M., and Weber, M.E. (2004). *Chem. Rev.* 104 (5): 2723–2750.

34 Lakowicz, J.R. (2006). *Fluorescence Sensing. Principles of Fluorescence Spectroscopy*, 623–673. Boston, MA: Springer.

35 Kim, M.K., Lim, C.S., Hong, J.T. et al. (2010). *Angew. Chem. Int. Edit.* 49 (2): 364–367.

36 Sarkar, A.R., Heo, C.H., Park, M.Y. et al. (2014). *Chem. Commun.* 50 (11): 1309–1312.

37 Martin, V.V., Rothe, A., Diwu, Z., and Gee, K.R. (2004). *Bioorg. Med. Chem. Lett.* 14 (21): 5313–5316.

38 Gao, G., Cao, Y., Liu, W. et al. (2017). *Anal. Methods* 9 (38): 5570–5579.

39 Smith, G.A., Hesketh, T.R., and Metcalfe, J.C. (1988). *Biochem. J.* 250 (1): 227–232.

40 De Silva, A.P., Gunaratne, H.Q.N., and Sandanayake, K. (1990). *Tetrahedron Lett.* 31 (36): 5193–5196.

41 Yin, J., Hu, Y., and Yoon, J. (2015). *Chem. Soc. Rev.* 44 (14): 4619–4644.

42 Ast, S., Schwarze, T., Müller, H. et al. (2013). *Chem. Eur. J.* 19 (44): 14911–14917.

43 Liu, M., Yu, X., Li, M. et al. (2018). *RSC Adv.* 8 (23): 12573–12587.

44 Trapani, V., Farruggia, G., Marraccini, C. et al. (2010). *Analyst* 135 (8): 1855–1866.

45 Romani, A.M. (2011). *Arch. Biochem. Biophys.* 512 (1): 1–23.

46 Afzal, M.S., Pitteloud, J.-P., and Buccella, D. (2014). *Chem. Commun.* 50 (77): 11358–11361.

47 Gruskos, J.J., Zhang, G., and Buccella, D. (2016). *J. Am. Chem. Soc.* 138 (44): 14639–14649.

48 Suzuki, Y., Komatsu, H., Ikeda, T. et al. (2002). *Anal. Chem.* 74 (6): 1423–1428.

49 Komatsu, H., Iwasawa, N., Citterio, D. et al. (2004). *J. Am. Chem. Soc.* 126 (50): 16353–16360.

50 Zhao, Y., Zheng, B., Du, J. et al. (2011). *Talanta* 85 (4): 2194–2201.

51 Maity, S.B. and Bharadwaj, P.K. (2014). *J. Lumin.* 155: 21–26.

52 Poenie, M. (2006). Fluorescent calcium indicators based on BAPTA. In: *Calcium Signaling*, 2e (ed. J.W. Putney), 1–3. Taylor & Francis.

53 Tsien, R.Y. (1980). *Biochemist* 19 (11): 2396–2404.

54 Kim, H.J., Han, J.H., Kim, M.K. et al. (2010). *Angew. Chem. Int. Edit.* 49 (38): 6786–6789.

55 Crichton, R.R. (2020). An overview of the role of metals in biology. In: *Practical Approaches to Biological Inorganic Chemistry*, Seconde (ed. R.R. Crichton and R.O. Louro), 1–16. Elsevier.

56 Andreini, C., Banci, L., Bertini, I., and Rosato, A. (2006). *J. Proteome Res.* 5 (1): 196–201.

57 Carter, K.P., Young, A.M., and Palmer, A.E. (2014). *Chem. Rev.* 114 (8): 4564–4601.

58 Crichton, R.R. (2017). Metal toxicity – an introduction. In: *Metal Chelation in Medicine* (ed. R.R. Crichton, R.J. Ward and R.C. Hider), 1–23. The Royal Society of Chemistry.

59 Zalewski, P.D., Forbes, I.J., and Betts, W. (1993). *Biochem. J.* 296 (2): 403–408.

60 Frederickson, C.J., Kasarskis, E., Ringo, D., and Frederickson, R. (1987). *J. Neurosci. Methods* 20 (2): 91–103.

61 Jiang, P. and Guo, Z. (2004). *Coord. Chem. Rev.* 248 (1–2): 205–229.

62 Walkup, G.K., Burdette, S.C., Lippard, S.J., and Tsien, R.Y. (2000). *J. Am. Chem. Soc.* 122 (23): 5644–5645.

63 Chang, C.J., Nolan, E.M., Jaworski, J. et al. (2004). *Chem. Biol.* 11 (2): 203–210.

64 Burdette, S.C., Frederickson, C.J., Bu, W., and Lippard, S.J. (2003). *J. Am. Chem. Soc.* 125 (7): 1778–1787.

65 Goldsmith, C.R. and Lippard, S.J. (2006). *Inorg. Chem.* 45 (2): 555–561.

66 Nolan, E.M., Ryu, J.W., Jaworski, J. et al. (2006). *J. Am. Chem. Soc.* 128 (48): 15517–15528.

67 Nolan, E.M. and Lippard, S.J. (2004). *Inorg. Chem.* 43 (26): 8310–8317.

68 Nolan, E.M., Jaworski, J., Okamoto, K.-I. et al. (2005). *J. Am. Chem. Soc.* 127 (48): 16812–16823.

69 Zhang, X.-a., Hayes, D., Smith, S.J. et al. (2008). *J. Am. Chem. Soc.* 130 (47): 15788–15789.

70 Ghosh, S.K., Kim, P., Zhang, X.-a. et al. (2010). *Cancer Res.* 70 (15): 6119.

71 Du, P. and Lippard, S.J. (2010). *Inorg. Chem.* 49 (23): 10753–10755.

72 Tang, B., Huang, H., Xu, K. et al. (2006). *Chem. Commun.* 34: 3609–3611.

73 Woodroofe, C.C. and Lippard, S.J. (2003). *J. Am. Chem. Soc.* 125 (38): 11458–11459.

74 Yang, L., McRae, R., Henary, M.M. et al. (2005). *Proc. Natl. Acad. Sci. U. S. A.* 102 (32): 11179–11184.

75 Morgan, M.T., Bagchi, P., and Fahrni, C.J. (2011). *J. Am. Chem. Soc.* 133 (40): 15906–15909.

76 Zeng, L., Miller, E.W., Pralle, A. et al. (2006). *J. Am. Chem. Soc.* 128 (1): 10–11.

77 Dodani, S.C., Domaille, D.W., Nam, C.I. et al. (2011). *Proc. Natl. Acad. Sci. U. S. A.* 108 (15): 5980–5985.

78 Domaille, D.W., Zeng, L., and Chang, C.J. (2010). *J. Am. Chem. Soc.* 132 (4): 1194–1195.

79 Jeong, Y. and Yoon, J. (2012). *Inorg. Chim. Acta* 381: 2–14.

80 Pathak, R.K., Hinge, V.K., Mondal, P., and Rao, C.P. (2012). *Dalton Trans.* 41 (35): 10652–10660.

81 Breuer, W., Epsztejn, S., Millgram, P., and Cabantchik, I.Z. (1995). *Am. J. Physiol. Cell Physiol.* 268 (6): C1354–C1361.

82 Carney, I.J., Kolanowski, J.L., Lim, Z. et al. (2018). *Metallomics* 10 (4): 553–556.

83 Petrat, F., Rauen, U., and de Groot, H. (1999). *Hepatology* 29 (4): 1171–1179.

84 Li, P., Fang, L., Zhou, H. et al. (2011). *Chem. Eur. J.* 17 (38): 10520–10523.

85 García-Beltrán, O., Mena, N., Yañez, O. et al. (2013). *Eur. J. Med. Chem.* 67: 60–63.

86 Tripod, J. (1964). A pharmacological comparison of the binding of iron and other metals. In: *Iron Metabolism: An International Symposium* (ed. F. Gross), 503–524. Berlin, Heidelberg, Springer.

87 Lytton, S.D., Mester, B., Libman, J. et al. (1992). *Anal. Biochem.* 205 (2): 326–333.

88 Cobessi, D., Celia, H., and Pattus, F. (2005). *J. Mol. Biol.* 352 (4): 893–904.

89 Tseng, C.-F., Burger, A., Mislin, G.L.A. et al. (2006). *J. Biol. Inorg. Chem.* 11 (4): 419–432.

90 Brandel, J., Humbert, N., Elhabiri, M. et al. (2012). *Dalton Trans.* 41 (9): 2820–2834.

91 Biswas, S., Sharma, V., Kumar, P., and Koner, A.L. (2018). *Sens. Actuators, B Chem.* 260: 460–464.

92 Fakih, S., Podinovskaia, M., Kong, X. et al. (2009). *J. Pharm. Sci.* 98 (6): 2212–2226.

93 Chang, J.H., Choi, Y.M., and Shin, Y.K. (2001). *Bull. Kor. Chem. Soc.* 22 (5): 527–530.

94 Liang, J. and Canary, J.W. (2010). *Angew. Chem. Int. Edit.* 49 (42): 7710–7713.

95 Mao, X., Su, H., Tian, D. et al. (2013). *ACS Appl. Mater. Interfaces* 5 (3): 592–597.

96 Dodani, S.C., He, Q., and Chang, C.J. (2009). *J. Am. Chem. Soc.* 131 (50): 18020–18021.

97 Kang, M.Y., Lim, C.S., Kim, H.S. et al. (2012). *Chem. Eur. J.* 18 (7): 1953–1960.

98 Mattison, R.L., Bowyer, A.A., and New, E.J. (2020). *Coord. Chem. Rev.* 425: 213522.

99 Ghazali, S., Wang, J., Fan, J., and Peng, X. (2017). *Sens. Actuators, B Chem.* 239: 1237–1242.

100 Zhang, S., Zhao, M., Zhu, W. et al. (2015). *Dalton Trans.* 44 (21): 9740–9743.

101 Lin, W., Yuan, L., Long, L. et al. (2008). *Adv. Funct. Mater.* 18 (16): 2366–2372.

102 Li, P.-F., Dietz, R., and Harsdorf, R. (1997). *Circulation* 96 (10): 3602–3609.

103 Langford, N. and Ferner, R. (1999). *J. Hum. Hypertens.* 13 (10): 651–656.

104 Rasheed, T., Bilal, M., Nabeel, F. et al. (2018). *Sci. Total Environ.* 615: 476–485.

105 National Health and Medical Research Council (2016). *Managing Individual Exposure to Lead in Australia – A Guide for Health Professionals.* Canberra: National Health and Medical Research Council. https://www.nhmrc.gov.au/about-us/publications/managing-individual-exposure-lead-australia#block-views-block-file-attachments-content-block-1.

106 He, Q., Miller, E.W., Wong, A.P., and Chang, C.J. (2006). *J. Am. Chem. Soc.* 128 (29): 9316–9317.

107 Hayashita, T., Qing, D., Minagawa, M. et al. (2003). *Chem. Commun.* 17: 2160–2161.

108 Hayashita, T., Qing, D., Bartsch, R.A. et al. (2005). *Supramol. Chem.* 17 (1–2): 141–146.

109 Tian, M.Z., Feng, F., Meng, S.M., and Yuan, Y.H. (2009). *Chin. Chem. Lett.* 20 (3): 326–329.

110 Xia, W.-S., Schmehl, R.H., Li, C.-J. et al. (2002). *J. Phys. Chem. B* 106 (4): 833–843.

111 Chen, C.-T. and Huang, W.-P. (2002). *J. Am. Chem. Soc.* 124 (22): 6246–6247.

112 Kwon, J.Y., Jang, Y.J., Lee, Y.J. et al. (2005). *J. Am. Chem. Soc.* 127 (28): 10107–10111.

113 Wu, F.-Y., Bae, S.W., and Hong, J.-I. (2006). *Tetrahedron Lett.* 47 (50): 8851–8854.

114 Bi, J., Fang, M., Wang, J. et al. (2017). *Inorg. Chim. Acta* 468: 140–145.

115 Zhu, J., Yeo, J.H., Bowyer, A.A. et al. (2020). *Metallomics* 12 (5): 644–648.

116 Peng, X., Du, J., Fan, J. et al. (2007). *J. Am. Chem. Soc.* 129 (6): 1500–1501.

117 Liu, Z., Zhang, C., He, W. et al. (2010). *Chem. Commun.* 46 (33): 6138–6140.

118 Xue, L., Li, G., Liu, Q. et al. (2011). *Inorg. Chem.* 50 (8): 3680–3690.

119 Xue, L., Liu, C., and Jiang, H. (2009). *Org. Lett.* 11 (7): 1655–1658.

120 Xue, L., Liu, Q., and Jiang, H. (2009). *Org. Lett.* 11 (15): 3454–3457.

121 Taki, M., Desaki, M., Ojida, A. et al. (2008). *J. Am. Chem. Soc.* 130 (38): 12564–12565.

122 Cheng, T., Xu, Y., Zhang, S. et al. (2008). *J. Am. Chem. Soc.* 130 (48): 16160–16161.

123 Yang, Y., Cheng, T., Zhu, W. et al. (2011). *Org. Lett.* 13 (2): 264–267.

124 Kim, H.N., Ren, W.X., Kim, J.S., and Yoon, J. (2012). *Chem. Soc. Rev.* 41 (8): 3210–3244.

125 Lin, W., Cao, X., Ding, Y. et al. (2010). *Org. Biomol. Chem.* 8 (16): 3618–3620.

126 Yoon, S., Albers, A.E., Wong, A.P., and Chang, C.J. (2005). *J. Am. Chem. Soc.* 127 (46): 16030–16031.

127 Takeda, E., Taketani, Y., Sawada, N. et al. (2004). *Biofactors* 21 (1–4): 345–355.

128 Knochel, J.P. (1977). *Arch. Intern. Med.* 137 (2): 203–220.

129 Greenwood, N.N. and Earnshaw, A. (ed.) (1998). Hydrogen. In: *Chemistry of the Elements*, 2ee, 32–52. Oxford: Butterworth-Heinemann.

130 Gautam, D. and Thomas, S. (2001). *The Weak Hydrogen Bond: In Structural Chemistry and Biology*. Oxford University Press.

131 Bates, G.W. and Gale, P.A. (2008). An introduction to anion receptors based on organic frameworks. In: *Recognition of Anions*. Mingos D.M.P. Structure and Bonding (ed. R. Vilar), 1. Berlin Heidelberg: Springer-Verlag.

132 Wenzel, M., Hiscock, J.R., and Gale, P.A. (2012). *Chem. Soc. Rev.* 41 (1): 480–520.

133 Choi, K. and Hamilton, A.D. (2001). *J. Am. Chem. Soc.* 123: 2456–2457.

134 Antonisse, M.M.G. and Reinhoudt, D.N. (1998). *Chem. Commun.* 443–448.

135 Katayev, E.A., Ustynyuk, Y.A., and Sessler, J.L. (2006). *Coord. Chem. Rev.* 250 (23–24): 3004–3037.

136 Gale, P.A., Busschaert, N., Haynes, C.J. et al. (2014). *Chem. Soc. Rev.* 43 (1): 205–241.

137 Blondeau, P., Segura, M., Perez-Fernandez, R., and Mendoza, J. (2007). *Chem. Soc. Rev.* 36: 198–210.

138 Llinares, J.M., Powell, D., and Bowman-James, K. (2003). *Coord. Chem. Rev.* 240 (1–2): 57–75.

139 Kumar, R., Sandhu, S., Singh, P., and Kumar, S. (2017). *Chem. Rec.* 17 (4): 441–471.

140 Řezanka, M., Langton, M.J., and Beer, P.D. (2015). *Chem. Commun.* 51 (21): 4499–4502.

141 Tobey, S.L. and Anslyn, E.V. (2003). *J. Am. Chem. Soc.* 125: 14807–14815.

142 Gale, P.A. and Caltagirone, C. (2015). *Chem. Soc. Rev.* 44 (13): 4212–4227.

143 Suganya, S., Naha, S., and Velmathi, S. (2018). *ChemistrySelect* 3 (25): 7231–7268.

144 Ramakrishnam Raju, M.V., Harris, S.M., and Pierre, V.C. (2020). *Chem. Soc. Rev.* 49 (4): 1090–1108.

145 Zhang, D., Cochrane, J.R., Martinez, A., and Gao, G. (2014). *RSC Adv.* 4 (56): 29735–29749.

146 Ashton, T.D., Jolliffe, K.A., and Pfeffer, F.M. (2015). *Chem. Soc. Rev.* 44 (14): 4547–4595.

147 Lee, S., Yuen, K.K., Jolliffe, K.A., and Yoon, J. (2015). *Chem. Soc. Rev.* 44 (7): 1749–1762.

148 Horowitz, H.S. (2003). *J. Public Health Dent.* 63 (1): 3–8.

149 Gokel, G.W. and Barkey, N. (2009). *New J. Chem.* 33 (5): 947–963.

150 Dai, G., Levy, O., and Carrasco, N. (1996). *Nature* 379 (6564): 458–460.

151 Pavelka, S., Babický, A., Vobecký, M., and Lener, J. (2001). *Biol. Trace Elem. Res.* 82 (1–3): 125–132.

152 Diao, Q., Ma, P., Lv, L. et al. (2016). *Sensors Actuators B Chem.* 229: 138–144.

153 Mahapatra, A.K., Roy, J., Sahoo, P. et al. (2012). *Org. Biomol. Chem.* 10 (11): 2231–2236.

154 Ko, Y.G., Na, W.S., Mayank et al. (2019). *J. Fluoresc.* 29 (4): 945–952. https://link.springer.com/content/pdf/10.1007/s10895-019-02407-y.pdf.

155 Wang, Y., Wang, X., Meng, Q. et al. (2017). *Tetrahedron* 73 (38): 5700–5705.

156 Khoshniat, S., Bourgine, A., Julien, M. et al. (2011). *Cell. Mol. Life Sci.* 68 (2): 205–218.

157 Crook, M. and Swaminathan, R. (1996). *Ann. Clin. Biochem.* 33 (5): 376–396.

158 Lee, H.N., Swamy, K.M.K., Kim, S.K. et al. (2007). *Org. Lett.* 9 (2): 243–246.

159 Kim, I.B., Han, M.H., Phillips, R.L. et al. (2009). *Chem Eur J* 15 (2): 449–456.

160 Hem, J. (1992). U.S. Geological Survey Water Supply Paper, 2254.

161 Kondo, S. and Takai, R. (2013). *Org. Lett.* 15 (3): 538–541.

162 Kumar, R. and Srivastava, A. (2016). *Chem. Eur. J.* 22 (10): 3224–3229.

163 Yoon, J., Kim, S.K., Singh, N.J. et al. (2004). *J. Organomet. Chem.* 69 (2): 581–583.

164 Zhang, D., Jiang, X., Yang, H. et al. (2013). *Chem. Commun.* 49 (55): 6149–6151.

165 Meng, Q., Wang, Y., Yang, M. et al. (2015). *RSC Adv.* 5 (66): 53189–53197.

166 Chen, Z., Wang, L., Zou, G. et al. (2013). *Spectrochim. Acta A Mol. Biomol. Spectrosc.* 114: 323–329.

167 Shi, B., Zhang, Y., Wei, T. et al. (2014). *Sensors Actuators B Chem.* 190: 555–561.

168 Kumar, A., Kumar, V., and Upadhyay, K.K. (2013). *Analyst* 138 (6): 1891–1897.

169 Gunnlaugsson, T., Davis, A.P., O'Brien, J.E., and Glynn, M. (2002). *Org. Lett.* 5 (15): 2449–2452.

170 Caltagirone, C., Bazzicalupi, C., Isaia, F. et al. (2013). *Org. Biomol. Chem.* 11 (15): 2445–2451.

171 Vance, D.H. and Czarnik, A.W. (2002). *J. Am. Chem. Soc.* 116 (20): 9397–9398.

172 Lohani, C.R., Kim, J.M., Chung, S.Y. et al. (2010). *Analyst* 135 (8): 2079–2084.

173 Zhu, W., Huang, X., Guo, Z. et al. (2012). *Chem. Commun.* 48 (12): 1784–1786.

174 Sokkalingam, P., Kim, D.S., Hwang, H. et al. (2012). *Chem. Sci.* 3 (6): 1819–1824.

175 Mizukami, S., Nagano, T., Urano, Y. et al. (2002). *J. Am. Chem. Soc.* 124 (15): 3920–3925.

176 Chen, W.-H., Xing, Y., and Pang, Y. (2011). *Org. Lett.* 13 (6): 1362–1365.

177 Kim, H.J., Lee, J.H., and Hong, J.-I. (2011). *Tetrahedron Lett.* 52 (38): 4944–4946.

178 Rao, A.S., Singha, S., Choi, W., and Ahn, K.H. (2012). *Org. Biomol. Chem.* 10 (42): 8410–8417.

179 Guo, Z., Zhu, W., and Tian, H. (2010). *Macromolecules* 43 (2): 739–744.

180 Villamil-Ramos, R. and Yatsimirsky, A.K. (2011). *Chem. Commun.* 47 (9): 2694–2696.

181 Bhowmik, S., Ghosh, B.N., Marjomaki, V., and Rissanen, K. (2014). *J. Am. Chem. Soc.* 136 (15): 5543–5546.

182 Roos, A. and Boron, W.F. (1981). *Physiol. Rev.* 61 (2): 296–434.

183 Brown, D. and Wagner, C.A. (2012). *J. Am. Soc. Nephrol.* 23 (5): 774–780.

184 Hsu, H.H. and Abbo, B.G. (2004). *Biochim. Biophys. Acta* 1690 (2): 118–123.

185 Chen, Y., Cann, M.J., Litvin, T.N. et al. (2000). *Science* 289 (5479): 625–628.

186 Hassel, B. (2000). *Mol. Neurobiol.* 22: 21–40.

187 Mulugeta, E., He, Q., Sareen, D. et al. (2017). *Chem.* 3 (6): 1008–1020.

188 Ma, L., Hu, L., Feng, X., and Wang, S. (2018). *Aging Dis.* 9 (5): 938–945.

189 Bohrer, A.S. and Takahashi, H. (2016). *Int. Rev. Cell Mol. Biol.* 326: 1–31.

190 Suzuki, I., Ui, M., and Yamauchi, A. (2006). *J. Am. Chem. Soc.* 128 (14): 4498–4499.

191 Pivovarenko, V.G., Zamotaiev, O.M., Shvadchak, V.V. et al. (2012). *J. Phys. Chem. A* 116 (12): 3103–3109.

192 Gao, M. and Tang, B.Z. (2017). *ACS Sens.* 2 (10): 1382–1399.

193 Huang, G., Zhang, G., and Zhang, D. (2012). *Chem. Commun.* 48 (60): 7504–7506.

194 Chen, S. and Ni, X.-L. (2016). *RSC Adv.* 6 (9): 6997–7001.

5

Activity-based Fluorescent Sensors and Their Applications in Biological Studies

Liam D. Adair[1,2,3], Nian Kee Tan[1,2,3], and Elizabeth J. New[1,2,3]

[1] School of Chemistry, The University of Sydney, NSW, Australia
[2] Australian Research Council Centre of Excellence for Innovations in Peptide and Protein Science, The University of Sydney, NSW, Australia
[3] The University of Sydney Nano Institute (Sydney Nano), The University of Sydney, NSW, Australia

5.1 Introduction

Over the past two decades, fluorescent sensors have proved invaluable in visualising the chemical species and elucidating the chemical reactions that underpin health and disease. Designing fluorescent sensors to achieve selectivity for an analyte of choice in the incredibly complex intracellular environment is immensely challenging. Fluorescent sensors can be divided into two main categories in their strategy to achieve selectivity: recognition-based and activity-based (Figure 5.1). The latter is the focus of this chapter, where we define activity-based sensors as fluorescent sensors that use molecular reactivity, incurring a structural modification, to detect an analyte of choice. Activity-based sensing is often referred to as reaction-based sensing, with the two terms used interchangeably in the literature.

Recognition-based sensors, as discussed in Chapter 4, are based on static molecular recognition or binding. These types of probes rely on host–guest interactions that are typically weak and usually reversible. This reversibility can be useful for sensing fluctuations in the concentration of analytes. Recognition-based sensing is an effective strategy, particularly for sensing cationic species such as metal ions. However, achieving selectivity with these interactions can be difficult in aqueous and complex environments, due to competing interactions, or when analytes possess similar properties to interfering species. Recognition-based sensors are also often unsuitable for the sensing of transient analytes.

In activity-based sensors, the desired analyte can induce covalent bond formation or breakage in a highly specific manner. Exquisite selectivity can be achieved by harnessing chemical reactivity and specificity to overcome competing interactions. Changes induced by the sensing of an analyte, such as bond formation or cleavage, or functional group transformation, can produce a large change in the emission profile of a sensor either by extending or reducing the extent of conjugation, or changing electronic density or distribution.

Molecular Fluorescent Sensors for Cellular Studies, First Edition. Edited by Elizabeth J. New.
© 2022 John Wiley & Sons Ltd. Published 2022 by John Wiley & Sons Ltd.

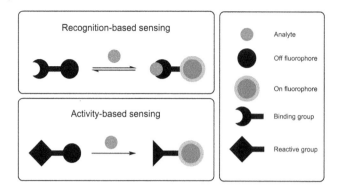

Figure 5.1 Graphical illustration of recognition-based and activity-based sensing.

In most cases, the reaction is irreversible and changes in the concentration of analytes cannot be monitored, but there are notable examples of reversible activity-based sensors. This potential disadvantage of irreversible reactivity can be harnessed for the detection of transient analytes, for instance rapidly metabolised species such as reactive oxygen species (ROS) [1].

5.1.1 Design Principles

The development of activity-based sensors is dependent on the discovery of, or existing knowledge of, chemical reactions specific to the analyte of interest. Selective reactions are pivotal for the design and application of effective activity-based probes.

There are various constraints for reactions that can be used successfully in activity-based sensors for cellular imaging. (i) The reaction must proceed in aqueous conditions. (ii) The reaction should proceed at a fast rate under physiological conditions, as longer reaction times will prevent real-time imaging and will increase the likelihood that metabolic inactivation or clearance from cells could occur before analyte interaction. This is especially problematic for the sensing of short-lived or rapidly metabolised species. (iii) The reaction must be able to operate under biological conditions. As well as tolerance of aqueous solvent, the biological pH range, cellular redox processes, high concentration of various salts, sugars, amino acids, and co-factors, nucleophilic amines, and thiols all need to be tolerated. (iv) Specificity of the reaction for the target analyte must be suitably high. (v) The reaction is also required to be fully bioorthogonal, thereby unable to interfere with cellular metabolism or react with undesired targets. (vi) The probe must also be inert to the normal metabolism of the cell, tissues, or organism until it reacts with the analyte. (vii) Finally, any products generated by the reaction must be non-toxic and preferably inert.

The choice of fluorophore scaffold, as with any fluorescent sensor for cellular imaging, is equally important. The desired fluorophore must be bright, allowing it to be used at low concentrations and be easily observed. A turn-on or ratiometric fluorescence response is also desirable (see Chapter 1). The chosen scaffold must also be readily synthesised, and preferably functionalised in a modular fashion, using synthetic organic chemistry.

This chapter will describe key examples of activity-based fluorescent sensors that have been employed in bioimaging and cellular studies, and detail how the reactions used achieve selectivity. In this chapter, activity-based sensors are classified by their reaction type and analyte.

5.2 Oxidation Reactions for Sensing Oxidative Species

Reactive oxygen species (ROS) and reactive nitrogen species (RNS) are involved in various transduction pathways vital to cellular health [2]. The aberrant production of oxidative species is often linked to oxidative stress, which increases susceptibility of developing chronic diseases in the cardiovascular [3, 4] and nervous systems [5–7] and certain cancers [8]. Monitoring ROS and RNS in biology can provide better insights in their physiological and pathological roles. Given the increasing biomedical interest in these analytes, several fluorescent sensors have been developed, most using an activity-based approach. These sensors tend to operate by utilising the oxidising nature of the analytes to induce an oxidative reaction.

Some of the most widely used sensors for ROS are 2′,7′-dichlorofluorescein diacetate or **DCFH$_2$ (5.1)**, hydroxyphenyl fluorescein or **HPF (5.2)**, and dihydrorhodamine 123 or **DHR 123 (5.3)** (Figure 5.2). These activity-based sensors are oxidised by ROS, giving a turn-on fluorescence response. These sensors have found widespread use for the detection and quantification of specific ROS, despite the fact that they react non-specifically with multiple ROS. For instance, **5.1** was once regarded as a hydrogen peroxide sensor, but it has since been reported that this reactivity only occurs in the presence of protein-bound metal catalyst such as haeme-containing peroxidases and cytochrome c [9]. It is also readily oxidised by nitrogen dioxide [10], the carbonate radical anion [10], and glutathione radical [11]. Additionally, the oxidation of **5.1** generates an intermediate radical that reacts readily with O$_2$ to form superoxide, which produces more hydrogen peroxide. This can lead to self-amplification of the signal [12–14]. Similarly, **5.2** reacts with both the hydroxyl radical and peroxynitrite, while **5.3** can be readily oxidised by peroxynitrite, hypochlorite, and haem-containing peroxidases [15, 16]. There has therefore been much recent research into activity-based strategies that can provide selective sensors for specific ROS. Here, we will discuss approaches to preparing fluorescent sensors for hydrogen peroxide (H$_2$O$_2$), peroxynitrite (ONOO$^-$), hypochlorous acid (HOCl), nitric oxide (NO), and singlet oxygen (^{1}O$_2$).

5.2.1 Fluorescent Sensors for Hydrogen Peroxide

Hydrogen peroxide (H$_2$O$_2$) is one of the main cellular ROS, with roles both in physiological signalling and pathological damage [17]. The selective reaction between aryl boronic acids and H$_2$O$_2$ to produce phenols has been applied extensively to develop activity-based sensors for this analyte [18, 19]. Boronic acid and ester derivatives react selectively with hydrogen peroxide over other ROS at physiological concentrations [20]. In this reaction, the

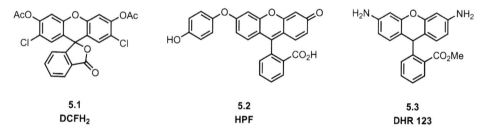

5.1	5.2	5.3
DCFH$_2$	HPF	DHR 123

Figure 5.2 Commonly used commercial activity-based ROS sensors.

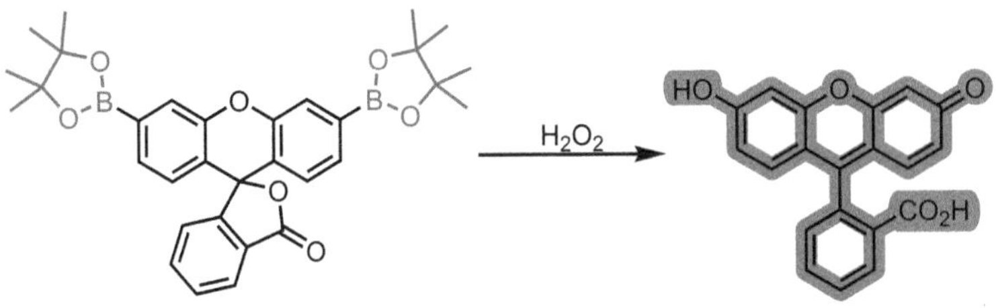

5.4

5.5
Fluorescein
λ_{em} 510 nm

Scheme 5.1 The reaction of **5.4** with H_2O_2 oxidises the two boronate esters, followed by hydrolysis to generate fluorescein **5.5**. Sensing group shown in red.

nucleophilic attack of H_2O_2 at the boron atom forms a negatively charged tetrahedral boronate complex. Subsequent rearrangement and hydrolysis unmask a phenol group. The conversion of aryl boronates to phenols is often used in sensors to generate a turn-on fluorescence response.

5.4 was the first boronate H_2O_2 sensor reported by Chang and co-workers (Scheme 5.1) [21]. It is a fluorescein derivative, where the two hydroxy groups at the 3′ and 6′ positions are replaced by pinacol boronate esters that hold the sensor in the closed non-emissive spiro-lactone confirmation. Upon reaction with H_2O_2, the sensor produced a greater than 1000-fold turn-on in emission, and displayed selectivity over other ROS including superoxide, nitric oxide, *tert*-butyl hydroperoxide and hydroxyl radicals. Proof-of-concept live cell imaging confirmed that **5.4** could image exogenous H_2O_2 in HEK cells. This approach has been used to develop several analogues based on other fluorophores [22, 23].

While sensors bearing two boronate groups were able to report on exogenous H_2O_2 in forced oxidative stress conditions, they were not capable of imaging endogenous H_2O_2. Due to the doubly caged nature, two oxidation events are required to give a fluorescent response, limiting the sensitivity of these sensors towards hydrogen peroxide. This limitation in sensitivity was overcome by using mono-boronate caged sensors (Figure 5.3), as demonstrated by the Chang group's development of **PG1** (**5.6**), based on the fluorescein analogue Tokyo Green [24, 25].

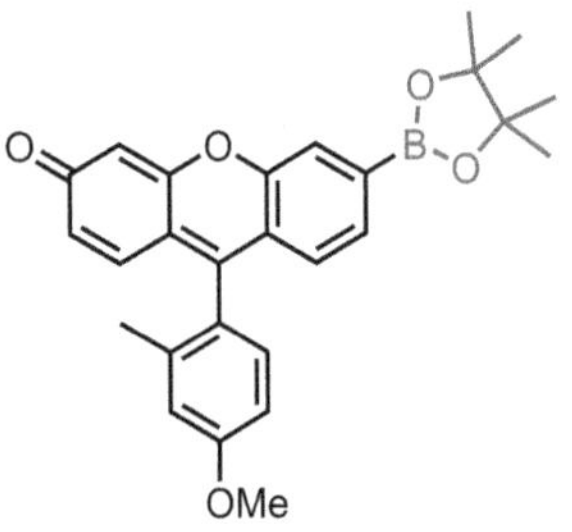

OMe

5.6

Figure 5.3 Mono-boronate-based **5.6** for H_2O_2 sensing (product λ_{em} = 510 nm). Sensing group shown in red.

Chang and colleagues used mono-boronate H_2O_2 sensor **5.6** to investigate the role of H_2O_2 generation in growth factor signalling [25]; **5.6** was able to sense intracellular H_2O_2 produced in A431 cells, which have high epidermal growth factor receptor (EGFR) expression [26], after activation with epidermal growth factor (EGF, 500 ng ml^{-1}) with a bright fluorescence turn-on response; **5.6** was then used successfully to study the pathway of EGF-mediated generation of H_2O_2. Following these results, **5.6** was used in live hippocampal neurons stimulated with EGF to image H_2O_2 produced for brain cell signalling for the first time.

While many sensors of this class involve H_2O_2-induced cleavage of the boronate to a free hydroxy, the boronate group can also be incorporated into probes in such a way that the

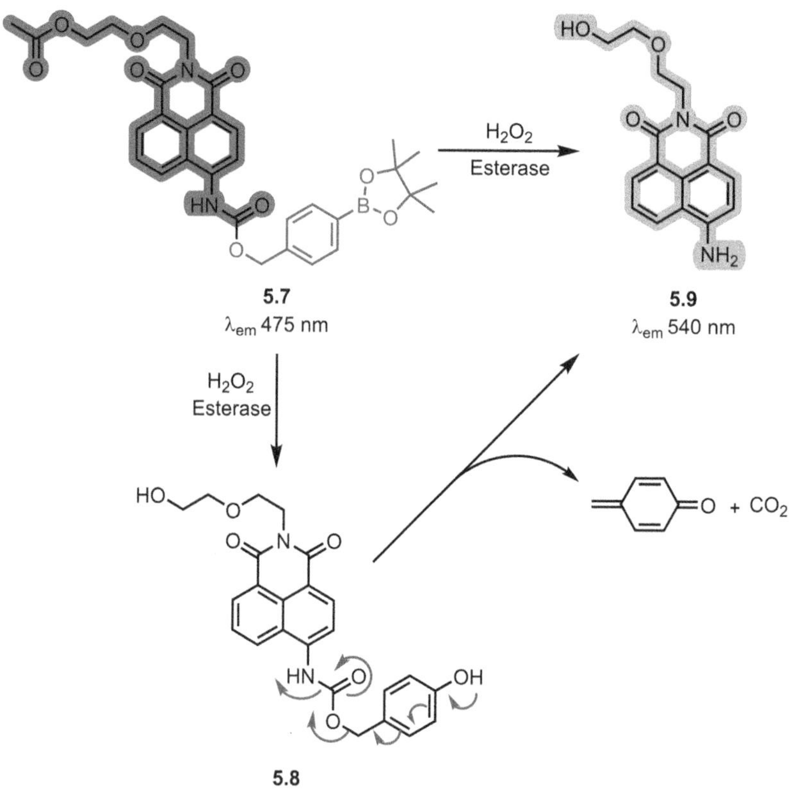

Scheme 5.2 Ratiometric boronate H_2O_2 sensor **5.7**. Oxidation of the boronate to the phenol **5.8** triggers the self-immolative linker, which, after 1,6-elimination and decarboxylation, generates the amine product **5.9** with red-shifted emission. Sensing group shown in red.

H_2O_2 reaction induces a much larger structural change. For example, sensor **5.7** contains a self-immolative linker. The boronate is oxidised to the corresponding hydroxy group, which induces a 1,6-elimination, loss of quinone methide, and decarboxylation to release CO_2, generating the fluorescent 4-aminonaphthalimide product (Scheme 5.2) [27]. This self-immolative linker strategy is commonly used in activity-based fluorescent sensors [28]. **5.7** is a ratiometric sensor, undergoing a bathochromic shift from blue to green fluorescence upon reaction with H_2O_2. **5.7** was shown to be able to report on both exogenous and endogenous H_2O_2 [27].

A key limitation of boronate-based sensors is their reactivity with peroxynitrite ($ONOO^-$), with some sensors showing considerably faster reaction with $ONOO^-$ than H_2O_2 [29]. Boronate-based sensors also tend to show quite slow reaction rates with H_2O_2, often requiring 30 minutes or more to produce a significant fluorescent response. Research to improve the response time and sensitivity of boronate-based H_2O_2 sensors is therefore highly desirable.

Another strategy for the sensing of H_2O_2 is the use of 1,2-diketones, which when decorated with a phenyl group at each end are known as benzils. H_2O_2 induces an oxidative cleavage of benzils to give benzoic acids. Using this strategy, Nagano and co-workers prepared **5.10**, a turn-on sensor for H_2O_2 (Scheme 5.3) [30]. The product of the reaction is a highly

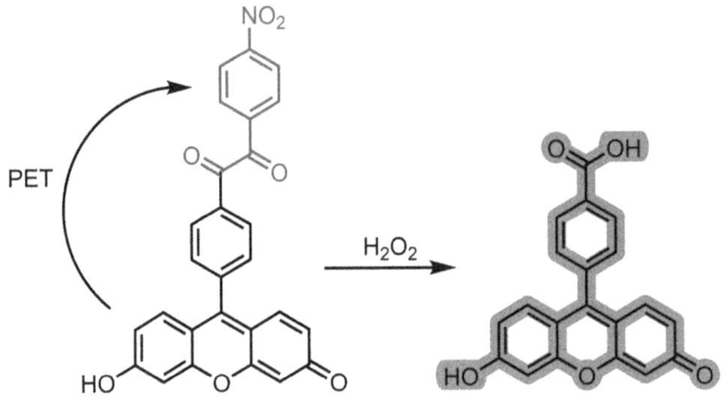

Scheme 5.3 Hydrogen peroxide sensor **NBzF (5.10)** based on benzil group. Red signifies the sensing group.

5.10

5.11
λ_{em} 519 nm

fluorescent 5-carboxyfluorescein derivative **5.11**. **5.10** was shown to have enhanced reactivity towards H_2O_2 when compared to aryl boronate-based sensors, only requiring 10 minutes to generate the fluorescent product. However, the sensor was also found to react with $ONOO^-$ and *tert*-butyl hydroperoxide. The reactivity of 1,2-diketones with H_2O_2 has also been used in other sensors with different fluorophore scaffolds [31, 32]. 1,2-Diketone-based H_2O_2 sensors generally have faster reaction kinetics and lower detection limit than boronate-based sensors. However, they also respond to other ROS and are highly reactive towards a range of nucleophiles.

5.2.2 Fluorescent Sensors for Peroxynitrite

Peroxynitrite ($ONOO^-$) is a highly nucleophilic ROS that is not only an important signalling molecule but also plays a role in stroke [33], Parkinson's disease [34], and other neurodegenerative disorders [35].

ONOO$^-$ reacts with aryl boronates to generate the corresponding phenol in an analogous manner to their reaction with H_2O_2. The rate of reaction with ONOO$^-$ is several orders of magnitude faster [29], and this can be exploited to prepare sensors or experiments that can distinguish ONOO$^-$ from H_2O_2 [36]. For example, the mono-boronate-caged azulene sensor **5.12** (Scheme 5.4) has been reported as an activity-based turn-on ROS sensor that displays high selectivity for ONOO$^-$ and H_2O_2 [37]; **5.12** reacts rapidly with ONOO$^-$, with high selectivity over other ROS, including H_2O_2, at equivalent concentrations (500 nM). At higher

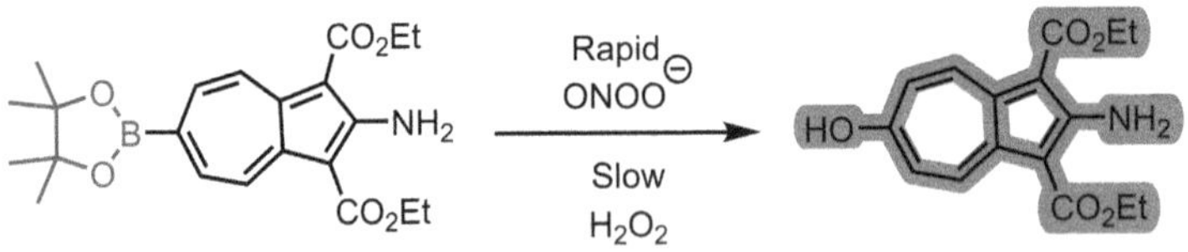

5.12

5.13
λ_{em} = 483 nm

Scheme 5.4 Boronate-based ONOO$^-$ and H_2O_2 sensor **AzuFluor 483-Bpin (5.12)** and fluorescent product **5.13**. Sensing group shown in red.

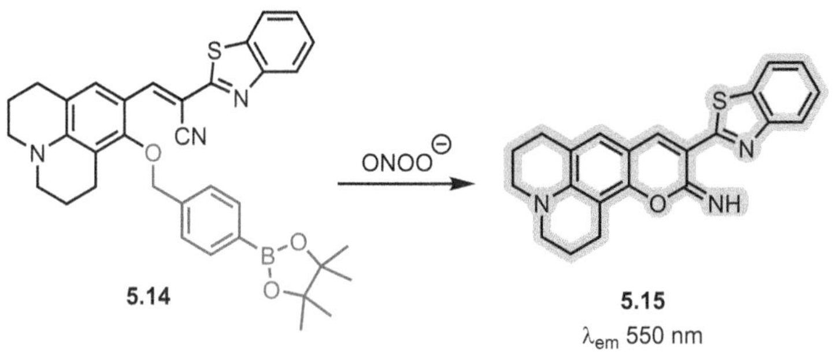

Scheme 5.5 ONOO⁻ sensor **5.14** combining boronate reactivity and a self-immolative linker to generate the highly fluorescent product **5.15**. Sensing group shown in red.

concentrations, the selectivity for ONOO⁻ over H_2O_2 was lost, but high selectivity against other ROS was retained; **5.12** was used to image endogenous ONOO⁻ and H_2O_2 in live cells and rat hippocampal tissue.

Sensor **5.14** also makes use of the rapid reactivity of peroxynitrite towards the boronate group (Scheme 5.5). Initial oxidative hydrolysis of the aryl boronate group by peroxynitrite is followed by self-immolative 1,6-elimination and cyclisation to generate the highly fluorescent benzothiazolyl iminocoumarin **5.15**. **5.14** was successfully used to monitor exogenous and endogenous ONOO⁻ in living cells [38].

An alternate strategy to the use of boronate esters is the use of an electrophilic site that is susceptible to ONOO⁻ attack, as demonstrated in **5.16** (Scheme 5.6) [39]; **5.16** uses a Forster resonance energy transfer (FRET) strategy; **5.16** consists of a coumarin as the FRET donor and a chromenylium fluorophore, which is red emissive, as a FRET acceptor. Excitation at 405 nm leads to FRET from the coumarin donor to the chromenylium acceptor and red emission is observed. In the presence of ONOO⁻, nucleophilic addition, oxidation, elimination, and hydrolysis give fluorescent coumarin **5.17**, and the chromenylium fragment is converted

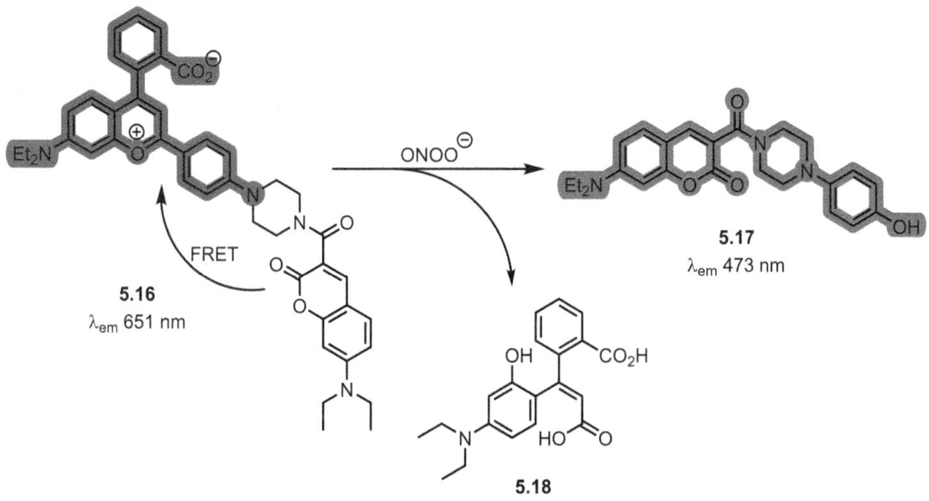

Scheme 5.6 Ratiometric, mitochondria-targeted ONOO⁻ sensor **5.16**.

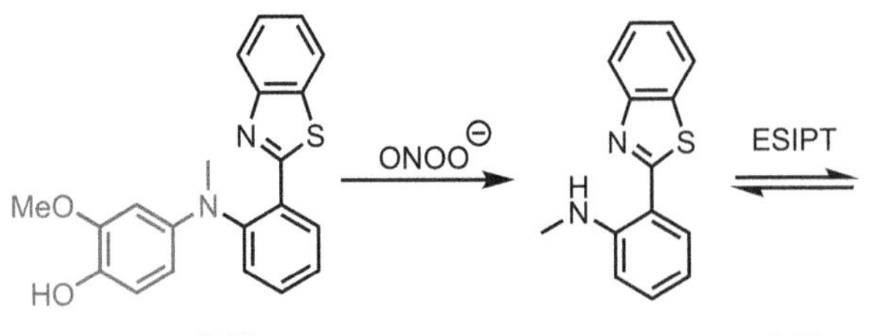

Scheme 5.7 Benzothiazole-based ONOO⁻ sensor **5.19** and reaction product **5.20**, which undergoes ESIPT. Sensing group shown in red.

to a non-fluorescent **5.18**. Excitation at 405 nm then leads to blue emission; **5.16** is therefore a ratiometric sensor for ONOO⁻ and was found to localise in the mitochondria and to be suitable for two-photon excitation microscopy.

A similar strategy for ONOO⁻ detection is used in the benzothiazole-based sensor **5.19** (Scheme 5.7) [40]. Benzothiazoles exist in a non-fluorescent form but can tautomerise when excited giving large fluorescence enhancement by excited-state intramolecular proton transfer (ESIPT) [41]; **5.19** is non-fluorescent due to the addition of a 2-methoxyphenol group on the amine, which blocks the ESIPT process. Oxidation of this electron-rich phenol by peroxynitrite to the corresponding benzoquinone, and subsequent cleavage leads to unmasking of the free secondary methylaniline group, and restoration of the ESIPT process responsible for fluorescence. The sensor is selective to peroxynitrite over H_2O_2, hypochlorite (ClO⁻), and peroxyl radicals (ROO˙); **5.19** has been tested in comprehensive biological experiments for visualising ONOO⁻ in endothelial cells, in an oxygen-glucose deprivation (OGD) treatment as a cerebral ischemia cell model.

5.2.3 Fluorescent Sensors for Hypochlorous Acid

Hypochlorous acid (HOCl) is a neutrophil-derived oxidant produced by the enzyme myeloperoxidase [42]. HOCl has antimicrobial activity, primarily through induction of protein unfolding and aggregation resulting from interactions with sulfur-containing amino acids [43]. Hence, the *N,N*-dimethylthiocarbamate group was first suggested by Tang to be an ideal receptor for HOCl over other biological oxidising species [44]. Since then, a selection of thiocarbamate-based HOCl fluorescent sensors have been reported.

The fluorescent naphthalimide probe **5.21** reported by Tang and co-workers was the first fluorescent probe for detecting HOCl at picomolar level, with a detection limit of 7.6 pM (Scheme 5.8) [44]. The researchers also prepared a lysosome-targeting analogue, replacing the butyl chain at the imide position with a morpholine. The sensors were shown to be suitable for use in two-photon microscopy and were used to image endogenous HOCl in live cells and *in vivo*.

Using the same dimethylthiocarbamate sensing group for HOCl, Guo and co-workers rationally designed a series of probes using coumarin and 2-(2-hydroxyphenyl)benzothiazole-derived fluorescent sensors, that have emission spanning from blue to red, thus providing a wide range of choices for imaging HOCl in cells, with the ability to detect HOCl in the nanomolar range [45].

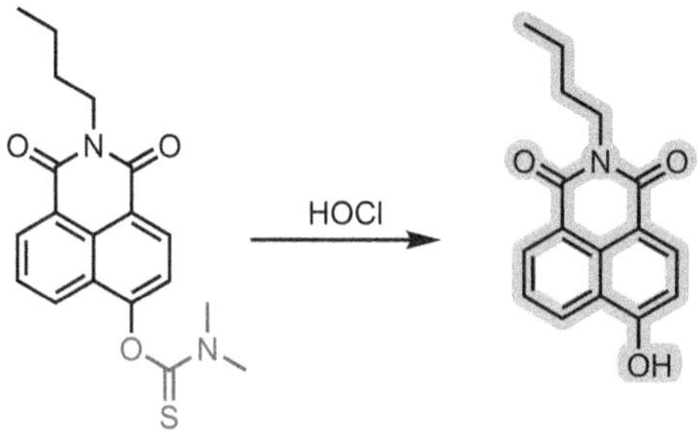

Scheme 5.8 HOCl sensor **5.21** and the fluorescent product **5.22**. Sensing group shown in red.

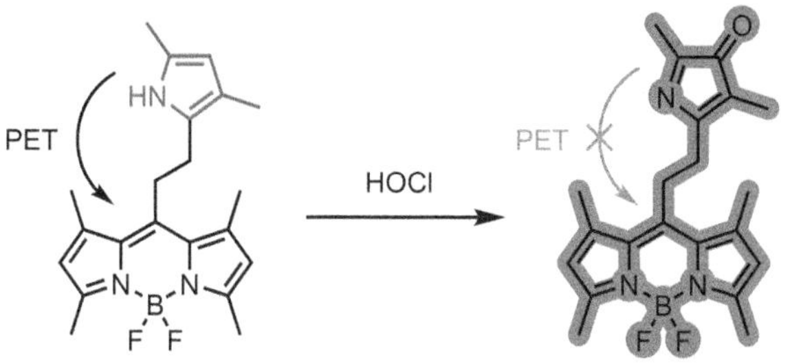

Scheme 5.9 Detection of HClO *via* oxidation of pyrrole in probe **5.23** to give fluorescent product **5.24**. Sensing group shown in red.

Another activity-based HOCl sensor, **5.23**, uses the oxidation of a pyrrole group by HOCl to modulate the fluorescence of the BODIPY core; **5.23** bears a pyrrole group at the meso-position and its high electron density is responsible for the PET quenching of the attached BODIPY fluorophore (Scheme 5.9) [46]. Oxidation of the appended pyrrole by HOCl forms the less electron-dense pyrrolone, **5.24**, which inhibits PET and restores the green BODIPY fluorescence; **5.23** exhibits a fast reaction time with HOCl and a low limit of detection of 0.56 nM.

5.2.4 Fluorescent Sensors for Nitric Oxide

Another important signalling ROS in the body is nitric oxide (NO) [47]. Activity-based sensing of NO can be achieved using aromatic 1,2-diamines, such as in the fluorescein-based **DAF-2-DA (5.25)**. **5.25** contains two diacetate groups that increase membrane permeability, ensuring the sensor can permeate live cells. The cleavage of the acetate groups by intracellular esterases help cellular retention. Subsequent reaction with NO generates the triazole form **DAF-2-T (5.26)**, with greatly increased quantum yield and therefore fluorescence intensity (Scheme 5.10) [48]. Sensor **5.25** has a detection limit of 5 nM and is commercially available. It has therefore been used extensively in biological studies.

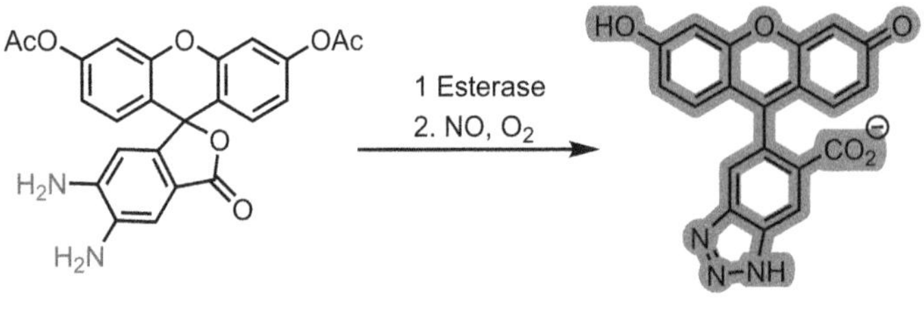

5.25
DAF-2-DA

5.26
DAF-2-T
λ_{em} 513 nm

Scheme 5.10 Commercially available NO probe, **DAF-2-DA** (**5.25**), and the fluorescent reaction product **DAF-2-T** (**5.26**). Sensing group shown in red.

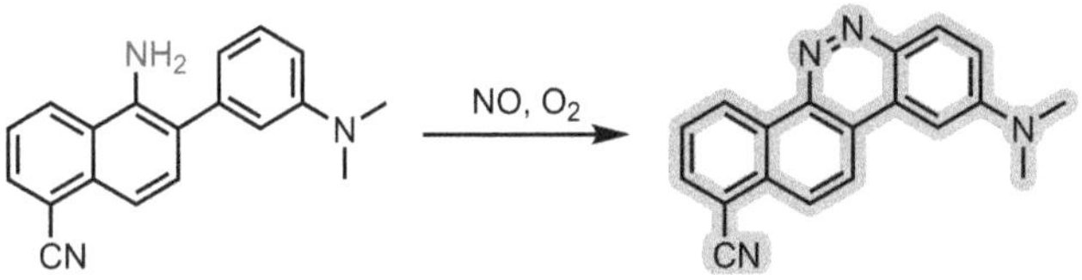

5.27

5.28
λ_{em} 550 nm

Scheme 5.11 Highly specific fluorogenic NO sensor, **5.27** and the fluorescent reaction product **5.28**.

Despite the widespread use of **5.25**, the sensor has some shortcomings. The diamine sensing group can react with ascorbic acid and dehydroascorbic acid, forming conjugates that have emission that is indistinguishable from the NO reaction product **5.26** [49]. As well as condensing with carbonyl compounds, aromatic 1,2-diamines are also prone to oxidation and can react with ROS, giving inaccurate results. These competing reactivities can prove challenging, but there are many examples of NO sensors that successfully utilise this strategy [50].

Anslyn and co-workers used an alternative strategy in their sensor **5.27**, which generates a fluorescent diazo ring system **5.28** *in situ* when reacted with NO (Scheme 5.11) [51]. The fluorophore is constructed by the reaction with NO, meaning the fluorescence turn-on is very high (>1500 fold), there is no background emission, and the interferences associated with aromatic 1,2-diamines are eliminated. Compared to **5.25**, **5.27** also has higher specificity towards NO, lower pH dependence, and higher signal–noise ratio in cells. It has limited solubility in aqueous solutions but was able to detect endogenous NO production in IFN-γ/IL-1β stimulated astrocytes. This strategy has been used to prepare NO sensors based on other fluorophores, such as BODIPY [52].

5.2.5 Fluorescent Sensors for Singlet Oxygen

Singlet oxygen ($^1O_2$) is an excited state of O_2, in an electronic configuration where all electrons are spin paired. It is highly reactive and a very strong oxidant, which can oxidise DNA, lipids, membranes, and proteins [53], and has also been reported as an activator of gene expression [54]. Photodynamic therapy uses oxygen sensitisers to produce $^1O_2$ [55]. Sensing both endogenous $^1O_2$ in cells, and $^1O_2$ produced by photodynamic therapies, is therefore

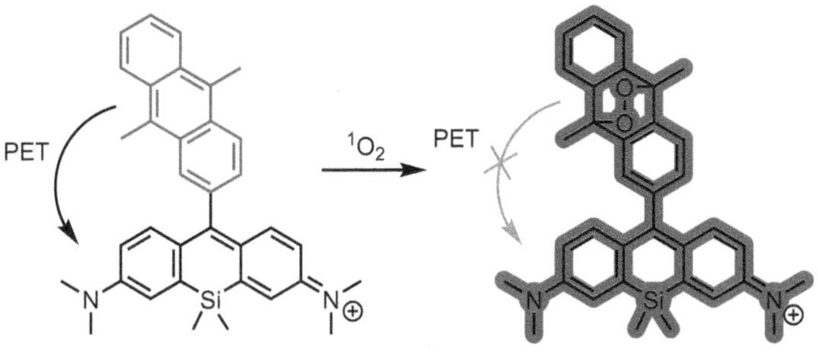

5.29
Si-DMA

5.30
Si-DMEP
λ_{em} 680 nm

Scheme 5.12 Activity-based $^1O_2$ sensor **Si-DMA** (**5.29**) and the fluorescent endoperoxide product **5.30**. Sensing group shown in red.

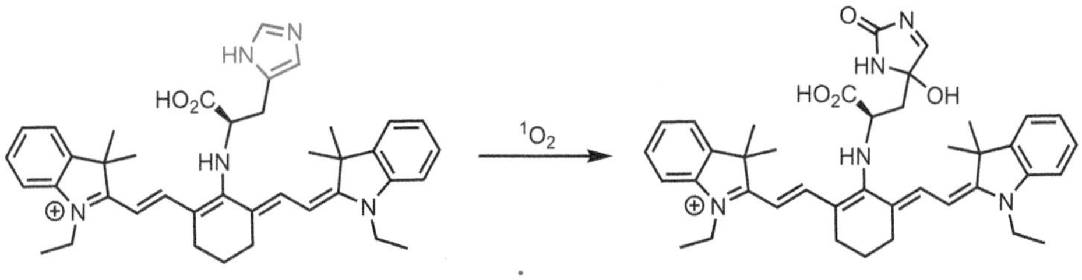

5.31

5.32
λ_{em} 794 nm

Scheme 5.13 Imidazole-based singlet oxygen sensor **5.31**. Sensing group shown in red.

important. A common approach to $^1O_2$ sensing is to harness its ability to participate in cycloaddition reactions [56, 57]. For example, the far-red analogue **Si-DMA** (**5.29**) is an activity-based sensor used to monitor generation of $^1O_2$ during photodynamic therapy [58]. It uses a 9,10-dimethylanthracene (DMA) sensing group that reacts with $^1O_2$ to form an endoperoxide. DMA quenches the fluorescence of the rhodamine *via* PET, which is halted by formation of the endoperoxide, resulting in a large fluorescence turn-on. The sensor was used to visualise the generation of $^1O_2$ within the mitochondria during photodynamic therapy for the first time (Scheme 5.12).

Another selective activity-based sensor for $^1O_2$ is the imidazole-cyanine derivative **5.31** reported by Tang and co-workers. The sensor consists of a tricarbocyanine scaffold with a histidine at the *meso*-position (Scheme 5.13) [59]. $^1O_2$ can react with the imidazole ring of histidine *via* a cycloaddition reaction to give an oxygenated product **5.32** that is NIR fluorescent; **5.31** demonstrates selectivity for $^1O_2$ over other ROS such as ClO^-, H_2O_2, $O_2^{\bullet-}$, and reductants like GSH and vitamin C. The sensor was used to image the induced $^1O_2$ production in PMA-stimulated RAW 264.7 macrophages.

One limitation to sensing $^1O_2$ by fluorescence is the possibility of the fluorophore generating $^1O_2$ from intersystem crossing to the triplet state, then sensitisation of triplet oxygen forming $^1O_2$. This complication has been demonstrated in the commercially available $^1O_2$ sensor Singlet Oxygen Sensor Green (**SOSG**) [60, 61].

5.3 Reduction Reactions for Sensing Reductive Species

Reductive species underpin the antioxidant system and can prevent and remedy the detrimental effects of ROS. Recent reviews have highlighted that reductive stress induced by excessive levels of reductive species is as harmful as oxidative stress and is also implicated in many pathological processes [62, 63]. This section will discuss sensors for reductive species including hydrogen sulfide (H_2S), glutathione (GSH) and selenocysteine (Sec).

5.3.1 Fluorescent Sensors for Hydrogen Sulfides

Hydrogen sulfide is a signalling gas that has many roles in physiology including antioxidant activity [64, 65] and signal transduction [66], but perturbed levels are associated with disease [67, 68]. The reduction of azides to amines by H_2S is a useful reaction that has been applied to a variety of fluorophore scaffolds including rhodamine, coumarin, rhodol, and 1,8-naphthalimide. The azide functional group acts as a fluorescence quencher and the reductive reaction of H_2S with azides gives the corresponding amine, restoring fluorescence. There are numerous H_2S sensors in the literature using this azide strategy, the first of which was **5.33**. This sensor contains an azide group tethered to a rhodamine (Scheme 5.14) [69]. As azides are weakly electron withdrawing [70], they can keep the spirolactone of a rhodamine scaffold in the non-fluorescent closed spiro-lactone form. It is likely that hydrogen sulfide exists predominantly as hydrosulfide ion (HS^-) at physiological conditions [71]. In **5.33**, the reduction of the azide with H_2S gives the corresponding amino group and the ring open form, which is highly emissive, resulting in a fluorogenic sensor.

The sensor **5.35** undergoes azide reduction with H_2S and subsequent self-immolative cleavage in the detection of hydrogen sulfide, leading to a fluorescence enhancement (Scheme 5.15) [72]. The fluorophore has an extended π-conjugated system comprised of phenolic dihydroxanthene and cyanine. Its NIR excitation and fluorescence and low detection limit of 0.26 μM enabled monitoring of H_2S in live cells in a time- and dose-dependent manner.

Another strategy to sense H_2S exploits the nucleophilicity of H_2S in the reduction of disulfide bonds. Xian and co-workers exploited this reactivity in a fluorescein-based sensor **5.37** to release a fluorescent 3-methoxyfluorescein **5.39** and a benzodithiolone upon reaction with H_2S (Scheme 5.16) [73]. Selectivity is gained from the fact that H_2S as a non-substituted thiol forms intermediate **5.38** that can undergo a second nucleophilic reaction releasing benzodithiolone, while substituted thiols like cysteine can only undergo one reaction forming a non-fluorescent adduct.

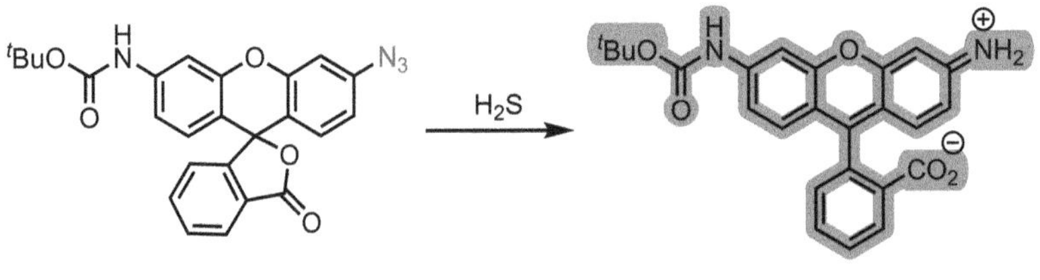

Scheme 5.14 Sulfidefluor-1 (**5.33**) as a hydrogen sulfide sensor *via* azide reduction to amine **5.34**. Sensing group shown in red.

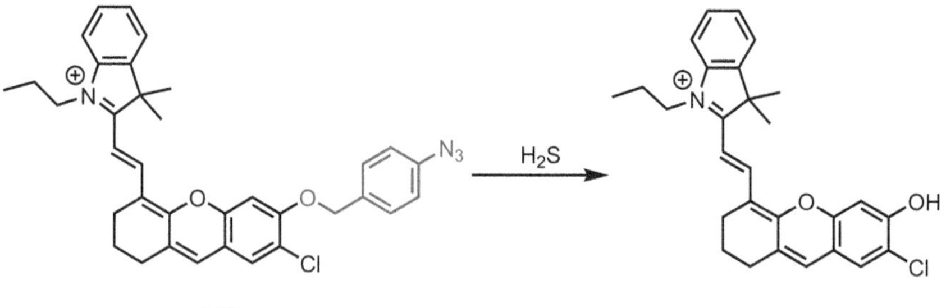

Scheme 5.15 A hydrogen sulfide sensor **5.35** sensing through azide reduction and self-immolative linker to reveal the fluorescent form **5.36**. Sensing group shown in red.

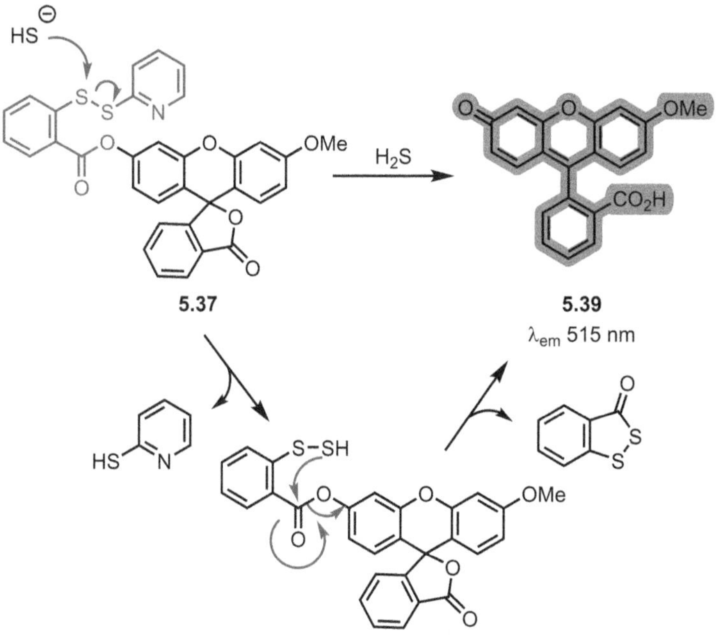

Scheme 5.16 The mechanism of hydrogen sulfide sensing by benzodithiolone formation in sensor **5.37**. Sensing group shown in red.

5.3.2 Fluorescent Sensors for Glutathione, Cysteine, and Homocysteine

Thiol-containing compounds such as glutathione (GSH), cysteine (Cys), and homocysteine (Hcy) underpin cellular redox homeostasis. They are abundant in cells, with glutathione (GSH) present at millimolar concentrations. However, these three analytes have similar reactivities, making it difficult to achieve selective discrimination solely through an electrophilic centre.

As a result of this challenge, a combination of two or more sensing groups is often used. Hence, many sensors that claim to be selective for GSH are in fact dual response sensors with one reactive site selective for glutathione, cysteine, and homocysteine, and another reactive

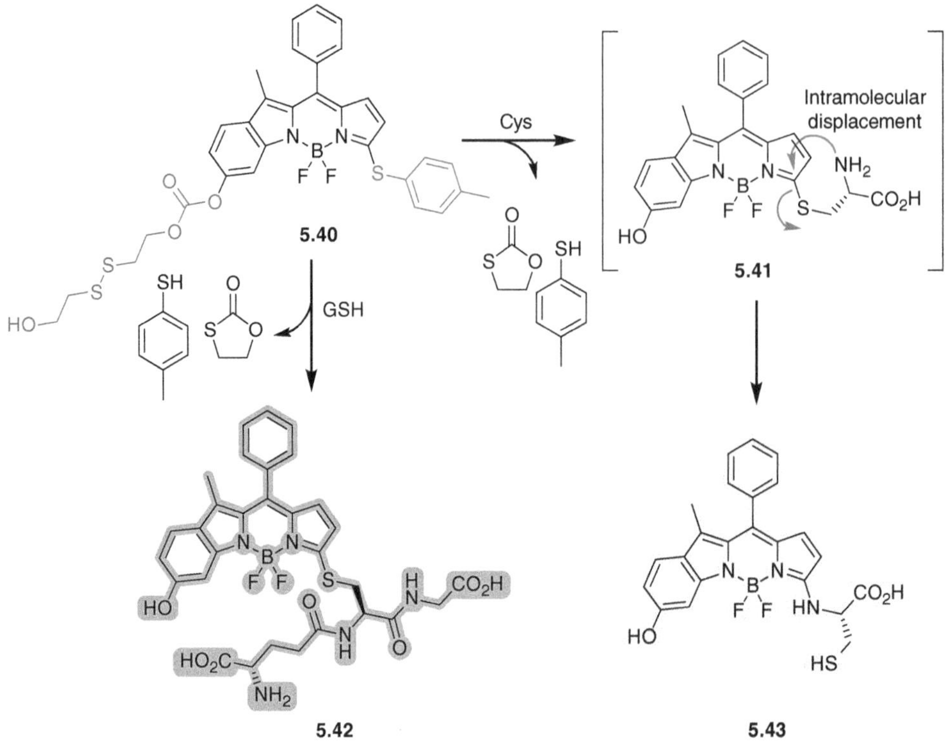

Scheme 5.17 Activity-based sensor **5.40** discriminates between GSH and Cys/Hcy (product of reaction with Hcy not shown here). The reaction with GSH forms **5.42**, while the reaction with Cys forms intermediate **5.41** followed by intramolecular displacement of the thioether to give **5.43**. Sensing groups shown in red.

site selective for cysteine and homocysteine only. For instance, the BODIPY-based fluorescent probe **5.40** reported by Wang *et al.* is an example of a sensor combining two independent reactive sites: a disulfide bond that is cleavable by the thiol groups in GSH, Cys and Hcy, and a thioether reaction site that can undergo nucleophilic aromatic substitution with thiols, then a subsequent intramolecular displacement reaction exclusively with the amino groups of Cys and Hcy (Scheme 5.17) [74].

Initially, cleavage of the disulfide linker by the GSH, Cys, or Hcy, followed by intramolecular cyclisation and cleavage at the carbonate to release the side product 1,3-oxathiolan-2-one, affords the free phenol. Next, a nucleophilic aromatic substitution reaction with the thiol of GSH, Cys, or Hcy displaces 4-methylbenzenethiol. In the case of GSH, this leads to the fluorescent product thiol substituted BODIPY **5.42**. Reaction with Cys or Hcy forms an intermediate **5.41** that can undergo an intramolecular displacement reaction *via* a kinetically favourable 5- or 6-membered cyclic transition state to give the amino-substituted BODIPY **5.43**. The intramolecular rearrangement does not take place after reaction with GSH, due to both the kinetically unfavourable 10-membered transition state that would be required and the sterically encumbered amino group. Thus, **5.40** can distinguish between GSH and Cys or Hcy. The ratio of orange to red channels can be obtained in microscopy experiments and allows the calculation of Cys/HCy to GSH ratios. This was achieved in MKN-45 and HeLa cells.

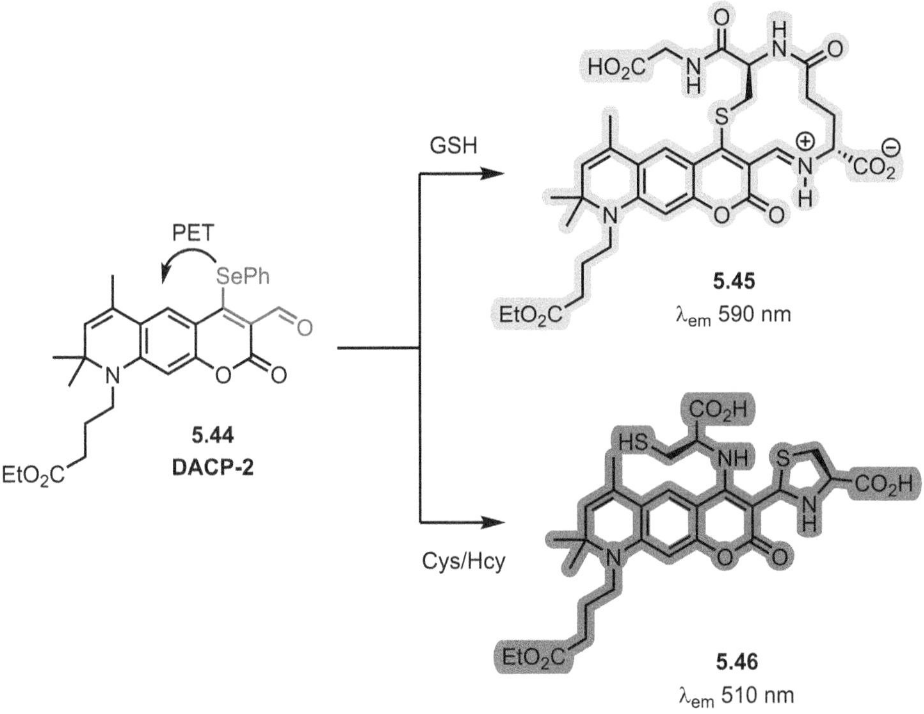

Scheme 5.18 Sensor **DACP-2** was used to discriminate between GSH and Cys/Hcy. The reaction with GSH forms product **5.45**; the reaction with Cys forms product **5.46** (product of reaction with Hcy not shown). Sensing group shown in red.

Another strategy that exploits the strongly nucleophilic sulfhydryl group of GSH, Cys, and Hcy are nucleophilic conjugate addition reactions. Churchill and co-workers reported a coumarin fluorophore in an organoselenium-based system **DACP-2** (**5.44**), which generates product **5.45** upon reaction with GSH and product **5.46** with Cys or Hcy (Scheme 5.18) [75].

Two structural components were important for activity-based sensing in **5.44**. Firstly, the incorporation of the phenylselenium group at the 4-position of the coumarin helped to reduce background fluorescence *via* PET quenching and acts as a good leaving group after nucleophilic addition of the sulfhydryl. Secondly, the adjacent aldehyde at the 3-position enhances the electrophilicity of the 4-position, and the primary amine and sulfhydryl groups can participate in a cyclisation reaction. GSH reacts firstly by conjugate addition of the thiol at the 4-position and elimination of phenylselenide, then secondly the primary amine forms the iminium after reaction with the aldehyde to generate product **5.45**. Cys, or Hcy, also reacts by conjugate addition of the sulfhydryl at the 4-position followed by an intramolecular rearrangement (in the same manner as **5.40** shown in Scheme 5.17) to give the amino substituted product. The aldehyde then reacts with Cys to form the thiazolidine (or thiazinane with Hcy) to generate product **5.46**.

5.44 was used in A549 cells treated with *N*-ethylmaleimide, a GSH alkylating agent that consumes free GSH, and demonstrated a significant decrease in fluorescence emission at 590 nm correlating to inhibited formation of **5.45** and reduced GSH levels. **5.44** demonstrated selective sensing of GSH and Cys/Hcy both in live cell and *in vivo* imaging and is commercially available.

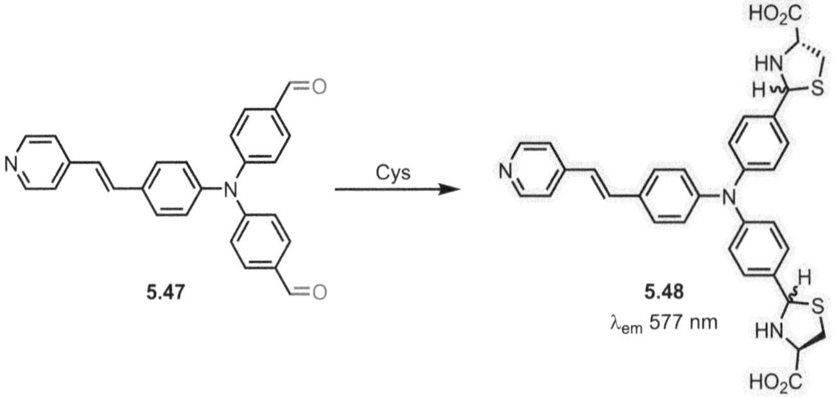

Scheme 5.19 Fluorogenic selective sensor for cysteine **5.47** and fluorescent product **5.48**. Sensing group shown in red.

Sensors that detect cysteine exclusively have been reported, exploiting the reactivity of Cys with aldehydes. For example, sensor **5.47** contains an aldehyde-sensing group and is initially non-fluorescent, attributed to the lack of an electron donor necessary for ICT fluorescence (Scheme 5.19) [76]. The reaction with Cys to form a thiazolidine provides an electron donor and therefore formation of **5.48** gives a fluorogenic response. **5.47** demonstrated suitability for two-photon fluorescence microscopy in HeLa cells.

5.3.3 Fluorescent Sensors for Selenocysteine

Selenocysteine (Sec), an analogue of Cys with a selenol group in place of the thiol group, is the predominant chemical form of selenium in biological systems, with a role in immune function [77].

Compared to biothiols like GSH, Cys and Hcy, Sec has a lower pK_a of 5.8 (cf. 8.3 in other biothiols) and therefore primarily exists in the ionised form (selenolate) at physiological pH [78, 79]. This results in Sec being more nucleophilic than most biothiols, and this difference in reactivity makes it possible to selectively sense Sec over biothiols. The relatively low abundance of Sec means that high sensitivity is also required. These properties make activity-based sensing an ideal tool for imaging of Sec.

2,4-Dinitrobenzene sulfonate has been used as an activity-based sensing group for sensing thiols. Modification to 2,4-dinitrobenzenesulfonyl amides allows for selective sensing of thiophenols which have a lower pK_a (~6.6) than aliphatic thiols. Thiophenols are not found in biological systems, therefore, switching to a 2,4-dinitrobenzenesulfonyl amide group makes the sensing group selective for Sec in biological systems. The 2,4-dinitrobenzenesulfonyl group acts as a quencher of fluorescence, and after nucleophilic aromatic substitution gives the selenoether, sulfur dioxide, and the amine with a turn on in fluorescence. Examples of sensors that use this strategy are the coumarin-containing **Sel-green** (**5.49**) and the NIR sensor **5.50** [79, 80].

5.49 was used to successfully image exogenous and endogenous Sec in live cells. It is commercially available but has a relatively high limit of detection for Sec (0.5 μM) (Figure 5.4) [79]. A NIR sensor **5.50** using the same mechanism has also been reported, and imaging was successful in live MCF-7 cells treated with sodium selenite and exogenous Sec *in vivo* [80].

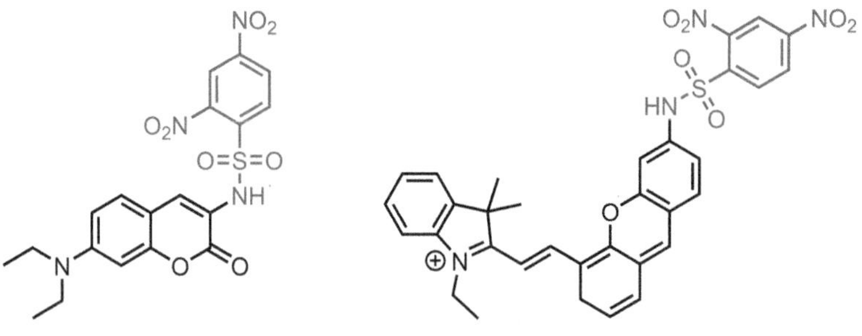

5.49
Sel-green

5.50

Figure 5.4   Selective selenocysteine sensors at physiological pH (pH 7.0–7.4) **Sel-green** (product λ_{em} = 502 nm), and **HD-Sec** (product λ_{em} = 712 nm). Sensing groups shown in red.

5.4   Reactions for Sensing Carbonyl Species

Reactive carbonyl species (RCS) such as formaldehyde and methylglyoxal (MGO) are endogenously produced and are cytotoxic. Formaldehyde and methylglyoxal induce DNA crosslinking and glycate proteins, which can lead to genetic alterations and loss of protein function, respectively [81]. Detecting these species in cells is therefore desirable.

5.4.1   Fluorescent Sensors for Formaldehyde

Formaldehyde (FA) is the simplest aldehyde and a reactive carbonyl species (RCS) known for its carcinogenicity and environmental toxicity [82, 83]. FA is produced endogenously in one carbon metabolism and during demethylation events of *N*-methylated amino acids [84], which supports biomolecule synthesis and epigenetic maintenance [85]. It can cause DNA crosslinking, which in turn can cause cancer [86–88]. Several different activity-based approaches have been reported for the sensing of FA.

Lin and co-workers reported the first example of a formimine FA activity-based sensor, **5.51** (Scheme 5.20) [89]. This comprised a 1,8-naphthalimide scaffold and a hydrazine at the 4-position as the FA reactive group, which reacts with aldehydes forming imines. **5.51** was used to image endogenous FA in cells and tissue [89]. Targeted analogues were also reported;

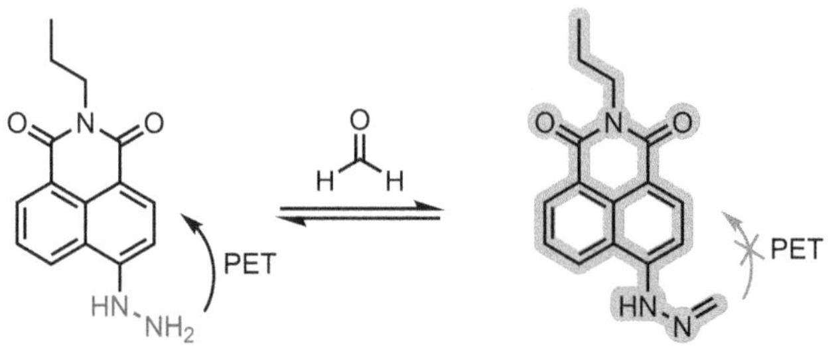

5.51

5.52
λ_{em} 543 nm

Scheme 5.20   Formimine-based FA sensor **5.51** and fluorescent formimine product **5.52**. Sensing group shown in red.

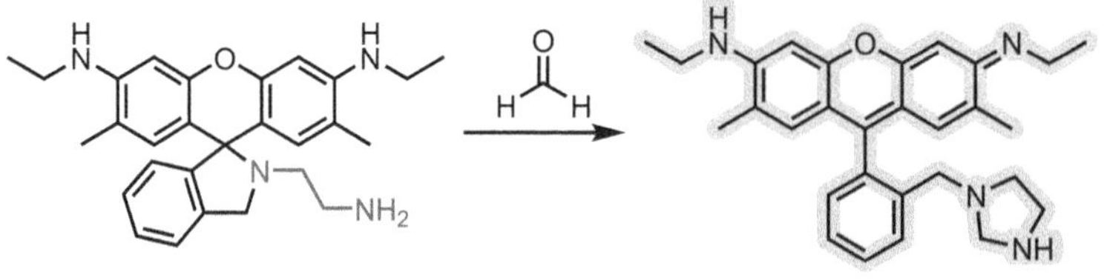

5.53　　　　　　　　　　　　　　**5.54**

λ_{em} 560 nm

Scheme 5.21 Aminal-based FA sensor **5.53** and fluorescent product **5.54**. Sensing group shown in red.

the imide position was substituted with an ester linked biotin group, for cancer cell targeting [90], and a morpholine group, for targeting lysosomes [91]. **5.51** is PET quenched by the hydrazine group, and the reaction with FA generates the formimine **5.52**, which does not PET quench the naphthalimide and fluorescence is restored. **5.51** was used to image endogenous FA in cells and two-photon microscopy for tissue imaging.

Another similar mechanism for sensing FA is aminal formation; **5.53** is the first example of an aminal activity-based sensor for FA (Scheme 5.21) [92]. **5.53** is a rhodamine 6G derivative that bears a primary amine which can react with FA. The resulting iminium is highly electrophilic and induces spirocycle ring opening resulting in a highly fluorescent imidazolidine product, **5.54**. However, the high reactivity of the primary amine means that there are interferences from other carbonyl-containing compounds, including methylglyoxal and acetaldehyde.

The laboratories of Chang and Chan both reported the use of a homoallylic amine as an FA reactive group that undergoes a 2-aza-Cope reaction as a sensing reaction for the detection of FA [93, 94]. The reaction mechanism of this homoallylic amine sensing group is shown in Scheme 5.22. Nucleophilic addition of the amine in **5.55** to FA forms the iminium **5.56**, and subsequent 2-aza-Cope rearrangement generates iminium **5.57**, β-elimination, and then hydrolysis converts the iminium to the aldehyde, and yields a fluorescent product **5.58** (Scheme 5.22). Both **5.55** reported by Chang and co-workers [93] and **5.59** developed by Chan and colleagues [94] are based on silicon rhodamine fluorophores.

Formimine and aminal formation are notably faster reactions than aza-Cope rearrangements, meaning that sensors such as **5.51** and **5.53** have much faster response rates than sensors such as **5.55** and **5.59**. However, while formimine and aminal-based FA probes have faster response rates, they have issues with selectivity over other aldehydes and RCS, such as acetaldehyde and methylglyoxal. Additionally, the formation of an imine or an aminal is a reversible reaction (in equilibrium), and therefore, this could affect fluorescence output. Aza-Cope-based FA sensors display exquisite selectivity for FA over other biologically relevant RCS.

5.4.2 Fluorescent Sensors for Methylglyoxal

Methylglyoxal (MGO) is a highly reactive α-dicarbonyl formed during glucose, amino acid, and fatty acid metabolism. Excess MGO can increase ROS production and cause oxidative stress [95, 96], and MGO is also a major precursor of advanced glycation end products (AGEs) that have been shown to trigger pro-inflammatory signals and protein dysfunction, contributing to the aging process and other age-related diseases [97, 98].

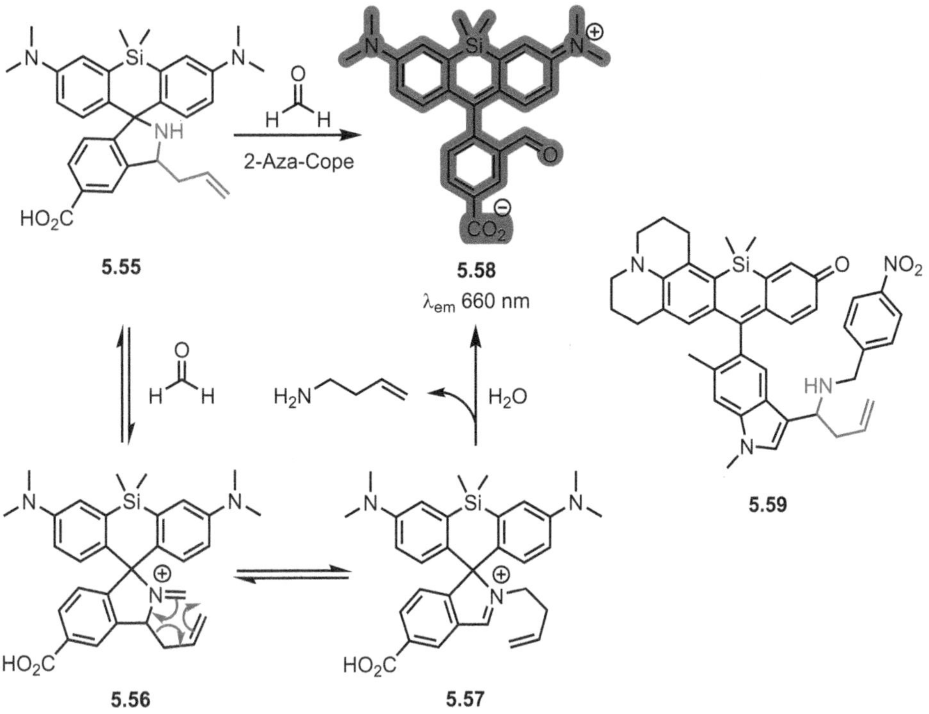

Scheme 5.22 Structure and mechanism of action of **5.55**: FA reacts with the amine to form the iminium **5.56**, that then undergoes a 2-aza-cope rearrangement to generate **5.57**, followed by hydrolysis to give the fluorescent aldehyde product **5.58**. Structure of **5.59** (product λ_{em} = 645 nm). Sensing groups shown in red.

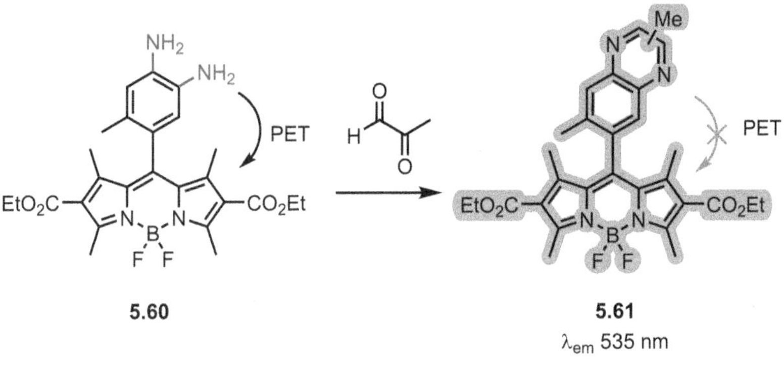

Scheme 5.23 MGO sensor **5.60** and fluorescent reaction product **5.61**. Sensing group shown in red.

Aromatic 1,2-diamine groups have been used as sensing groups for MGO. For example, **5.60** is a BODIPY-based MGO sensor with selectivity for MGO over a variety of α-dicarbonyl species including glyoxal, glucosone, and pyruvate (Scheme 5.23) [99]. **5.60** is an MGO sensor that uses a 1,2-diamine scaffold, originally employed in fluorescent sensors to detect NO, as in **5.25** (Scheme 5.10). **5.60** is quenched by PET and upon reaction with MGO, **5.61** is formed and PET is inhibited resulting in a turn-on fluorescence response. Although cross-reactivity studies found that 50 μM NO caused some fluorescence, this is unlikely to be a concern in cellular studies due to the nanomolar physiological concentration and short half-life of NO [100, 101]. **5.60** was used in a study that showed that MGO-mediated nuclear accumulation of a transcriptional regulator protein, YAP, in human breast cancer [100].

5.5 Metal-mediated Reactions

The reactivity of metals has been capitalised upon for the development of activity-based sensors. Many metals have unique reactivity and therefore selectivity can be achieved using this. This can be a selective reaction specific to a metal or reactivity of an analyte with a metal. Biological systems or processes involving metals, particularly metalloenzymes, can inspire the design of effective sensors.

Activity-based fluorescent sensors with metal-mediated reactions can be broadly categorised into sensors that contain a metal atom as a sensing group, and sensors where the metal is the analyte and triggers a reaction.

Over half of all proteins require a metal to function, and a third of all enzymes contain a metal at their active site [102]. In these systems, the metal cofactor often plays a role in binding to a small molecule and exemplifies the exquisite analyte selectivity that can be achieved by using a metal reactive site.

Carbon monoxide (CO) is a reactive gas that is formed during incomplete combustion of carbon-based fuels. It is toxic and can stop respiration by reacting with haemoglobin to form carboxyhaemoglobin [103]. However, it is also produced endogenously from the breakdown of haem, and a recent appreciation of its physiological functions has led to the development of CO releasing molecules (CORMs) with potential therapeutic uses [104]. CO reactivity and coordination with organometallic complexes has been exploited for synthetic chemistry and in particular palladium-mediated catalysis [105, 106].

Inspired by this reactivity, Chang and colleagues designed a BODIPY sensor, **5.62**, containing a palladacycle for the selective activity-based imaging of CO in living cells [107]. **5.62** is quenched by the palladium and upon coordination of CO, a carbonylation reaction takes place, releasing Pd(0) and the fluorescent carboxylic acid product, **5.63**. **5.62** was used to image CO released by a water-soluble ruthenium CORM in HEK293T cells (Scheme 5.24).

An interesting example of a fluorescent sensor incorporating a metal-based reactive group was reported by Michel and colleagues [108]. The sensor, **5.64**, consists of a BODIPY fluorophore with a ruthenium recognition group based on Hoveyda-Grubbs second-generation

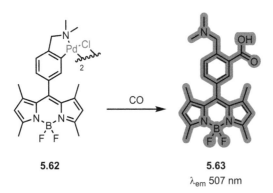

Scheme 5.24 Activity-based sensor for carbon monoxide **5.62** used in live cell imaging, and fluorescent product **5.63**. Sensing group shown in red.

5.62

5.63
λ_{em} 507 nm

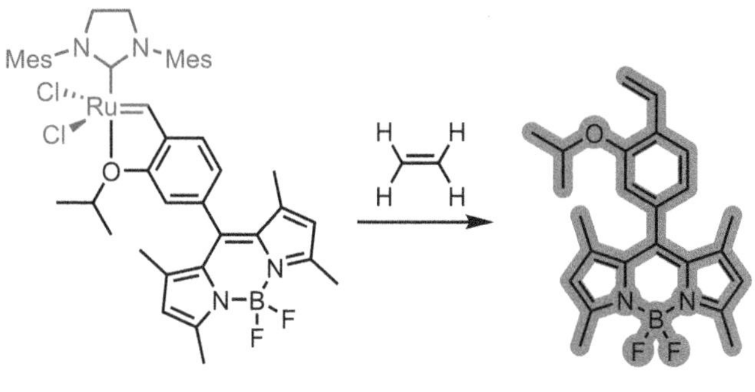

Scheme 5.25 Ethylene sensor
5.64 uses a metathesis
reaction to generate
fluorescent product **5.65** for
live cell detection and imaging.
Sensing group shown in red.

Scheme 5.25 Ethylene sensor **5.64** uses a metathesis reaction to generate fluorescent product **5.65** for live cell detection and imaging. Sensing group shown in red.

5.64

5.65

λ_{em} 512 nm

catalyst for alkene-metathesis [109]. Using this approach, they obtained a selective fluorescence turn-on for ethylene, with good specificity over other small olefins, *via* a bioorthogonal metathesis reaction. This is a notable achievement as ethylene is typically challenging to sense in a complex biological environment due to its structural simplicity; **5.64** was successfully used for imaging ethylene in *Chlamydomonas reinhardtii* algae, as well as HEK293-T and PC12 cells (Scheme 5.25).

Displacement of a metal upon reaction with an analyte can be used to generate a fluorescence response. **HSip-1 (5.66)** contains a Cu(II) cyclen complex that, upon reaction with H_2S and subsequent formation of insoluble copper sulfide, displays a turn-on fluorescence response [110]. This metal-displacement strategy overcomes issues with the slow reduction of azides by H_2S (Section 5.3.1) at physiological conditions and avoids interferences from other ROS. In the parent sensor, the paramagnetic Cu(II) quenches fluorescence, and so this sensing group can theoretically be applied to a wide range of fluorophores.

Metals can also be used to mimic the active site of a metalloenzyme and undertake redox chemistry to cleave a covalent bond and induce a change in fluorescence [111]. A notable example of this strategy is used in sensor **5.67** [112]. This sensor uses a resorufin fluorophore that is *O*-alkylated with a linker to a peroxidase active site mimetic iron(III) tetradentate monoamido complex, Fe(III)mpaq (mpaq = 2-[*N*-methyl-*N*-pyridin-2-ylmethyl]-amino-*N'*-q uinolin-8-yl-acetamido) [113]. Due to this alkylation, the sensor is non-fluorescent, and upon reaction with H_2O_2, the iron centre is oxidised from Fe(III) to Fe(V). This Fe(V) species can then oxidise the linker, leading to elimination of the resorufin resulting in a turn-on in fluorescence. **5.67** was used to image both exogenous and endogenous H_2O_2 in cultured cells (Figure 5.5).

5.6 Metal-sensing Reactions

Metal ions are of considerable interest in biological study: perturbations of essential metal homeostasis can lead to disease; while heavy metals can be toxic to biological systems, as outlined in Chapter 4. While recognition-based metal sensors predominate the literature, activity-based sensing offers an alternative strategy and can be particularly beneficial in cases where the analyte concentration is low. Unlike reversible recognition-based sensors, which equilibrate over time, irreversible reaction-based sensors will accumulate over time, leading to a signal enhancement.

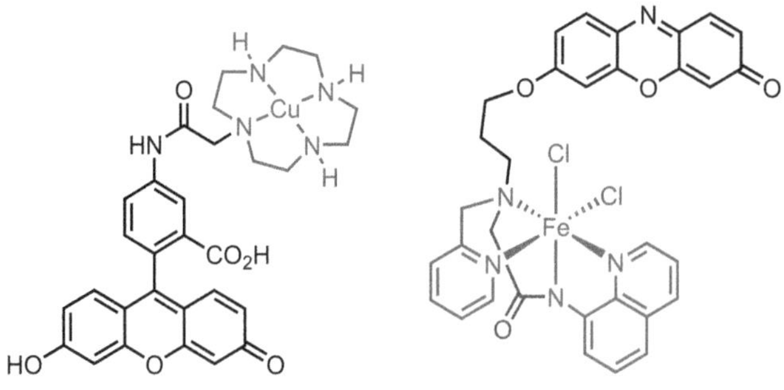

5.66　　　　　　　　　　**5.67**

Figure 5.5 Activity-based sensors employing a metal: **HSip-1 (5.66)** H_2S sensor with displacement of Cu (product λ_{em} = 516 nm); **5.67** H_2O_2 sensor with oxidation of ligand and β-elimination of resorufin (product λ_{em} = 590 nm). Sensing groups shown in red.

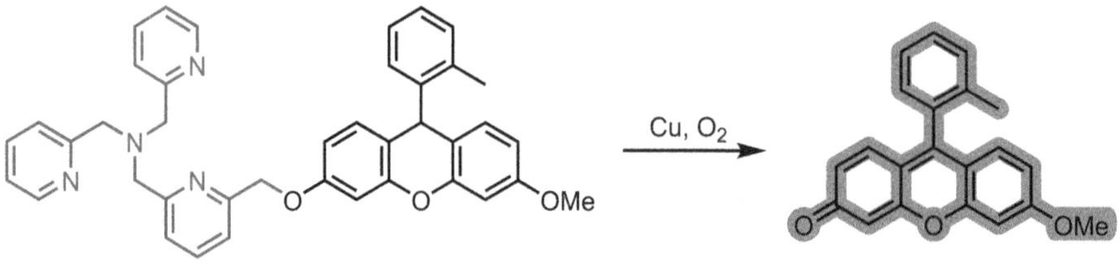

5.68　　　　　　　　　　**5.69**
λ_{em} 513 nm

Scheme 5.26 Activity-based Cu probe **5.68**, used to image exogeneous Cu in Hela cells. Sensing group shown in red.

A common strategy for the reaction-based sensing of metals is to design a metal receptor group that is cleaved upon binding to the metal of interest. This is a particularly useful strategy for the sensing of metals with low physiological concentration, such as Cu(I), or metals that tend to quench appended fluorophores, such as Cu(II), Fe(II), and Co(II).

There are multiple recognition-based probes for the fluorescent sensing of copper in biological systems [114]. The first activity-based sensors for Cu(I) were reported by Taki and colleagues [115]. They used fluorescein analogues where the phenolic oxygen was appended with a tetradentate tris[(2-pyridyl)-methyl]amine (TPA) ligand through a benzyl ether linker, as shown in probe **5.68**. Upon binding to Cu(I) the C–O bond is cleaved, followed by an oxidation, which gives the fluorescent fluorescein **5.69**, with a turn-on response of more than 1500-fold. The reaction takes place selectively under aerobic physiological conditions. Furthermore, the probe displayed minimal reactivity towards Fe(II), Zn(II), Ni(II), Co(II), or Mn(II) and biologically relevant oxidants H_2O_2, ClO^-, and $\cdot OH$ (Scheme 5.26).

Inspired by this approach, subsequent modifications of the TPA ligand led to the development of activity-based sensors for other metals. For example, **5.70** is a fluorescein-based sensor for Fe(II) [116], while **5.71** is based on the Tokyo Green fluorophore and can selectively sense Co(II) [117] (Figure 5.6).

Hydrazine *n*-butyl thiourea groups have been shown to be effective reaction-based sensing groups for Cu(II). For example, incorporation of this group of the lactam of a rhodamine B

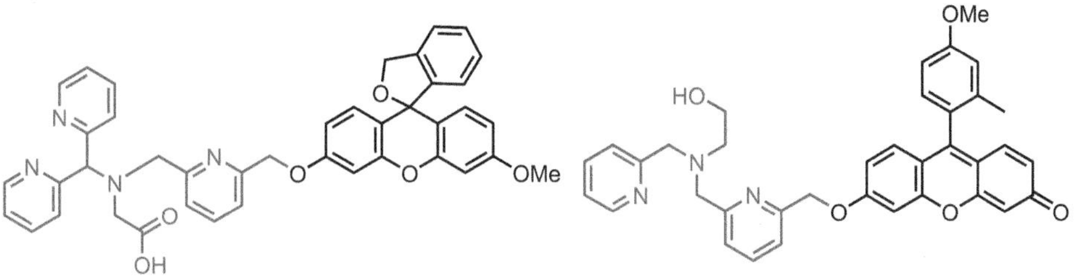

Figure 5.6 Fe(II) activity-based sensor **5.70** (product λ_{em} = 508 nm); Co(II) activity-based sensor **5.71** (product λ_{em} = 508 nm). Sensing groups shown in red.

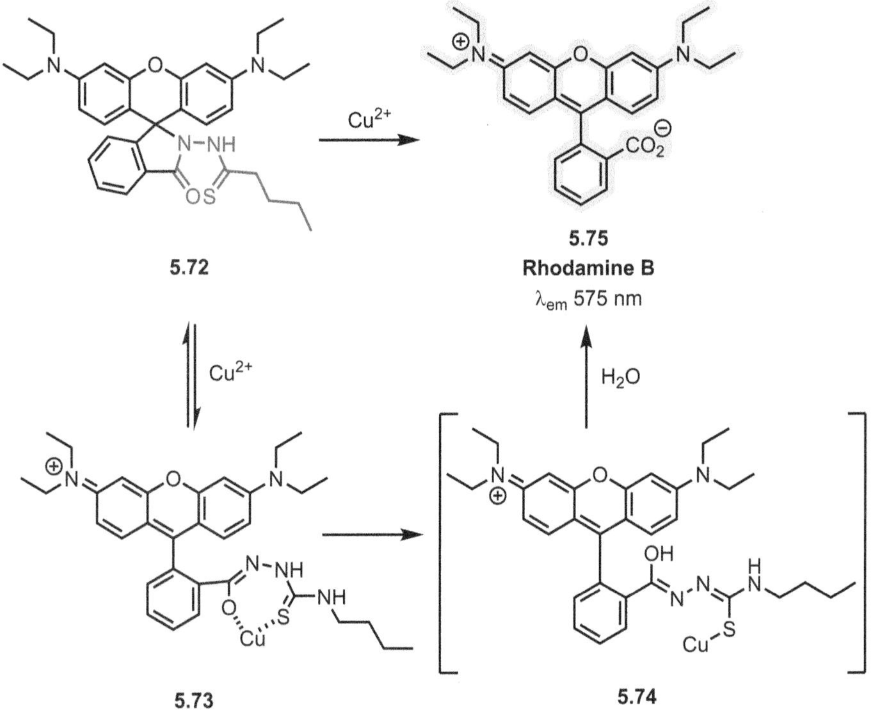

Scheme 5.27 Cu(II) activity-based sensor **5.72** and postulated sensing mechanism. Sensing group shown in red.

scaffold with sensing group introduced at the lactam gives a turn-on sensor **5.72** for Cu(II) [118]. When **5.72** is treated with Cu(II), the metal coordinates to the electron-rich sensing group and a redox reaction reduces Cu(II) to Cu(I). The Cu(I) is then coordinated to the S-atom (**5.73**) and the sensing group tautomerises to the hydrazonic acid to give intermediate **5.74**, which is then hydrolysed to give rhodamine B (**5.75**), resulting in a fluorescence turn-on (Scheme 5.27). The limit of detection for the probe was less than 10 ppb, with excellent selectivity over other metal ions, particularly biologically relevant ions. The sensor was used to sense exogenously-added Cu(II) in live HeLa cells using both confocal and two-photon microscopy.

An alternative, more universally applicable, method for sensing Cu(II) is the selective hydrolysis of picolinic esters in the presence of Cu(II) [119]. This sensing group can be used to mask a phenolic oxygen and the Cu(II) reactivity has been harnessed to give a turn-on activity-based fluorescent response. **5.76** is a Cu(II) sensor that uses this sensing strategy [120]. The sensor has an asymmetrical BODIPY core with a phenol that has been masked by a picolinic acid group that is cleaved in the presence of Cu(II) (Scheme 5.28). The sensor displayed excellent selectivity for Cu(II) over a range of metal ions. Cu(I) was not included in the selectivity screen and no oxidative species were tested for reactivity, which in theory could react with the ester functionality and cause a fluorescence response.

As discussed in Chapter 4, there are many recognition-based probes for Zn(II). Activity-based sensing can result in improved fluorescence turn-on responses. Radford and colleagues designed a mitochondrially targeted activity-based probe for Zn(II) **5.78** (Scheme 5.29) [121]. The sensor has a dichlorofluorescein scaffold, where the phenolic oxygens are protected as the diacetate, locking the fluorescein in the non-fluorescent spirolactone form. Two dipicolylamine (DPA) groups, a widely used Zn(II) binding group, were appended to the scaffold and a triphenylphosphonium cation was added as a mitochondria-targeting group. Upon complexation of Zn(II) to the DPA groups, the Lewis acidity of the

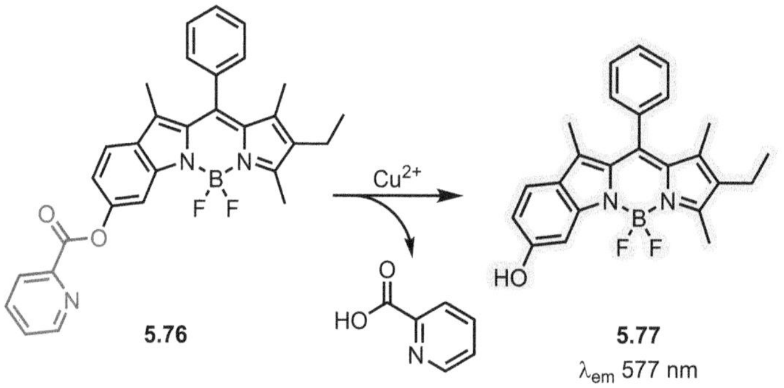

Scheme 5.28 Fluorogenic activity-based sensor for Cu(II) **5.76** and fluorescent product **5.77**. Sensing group shown in red.

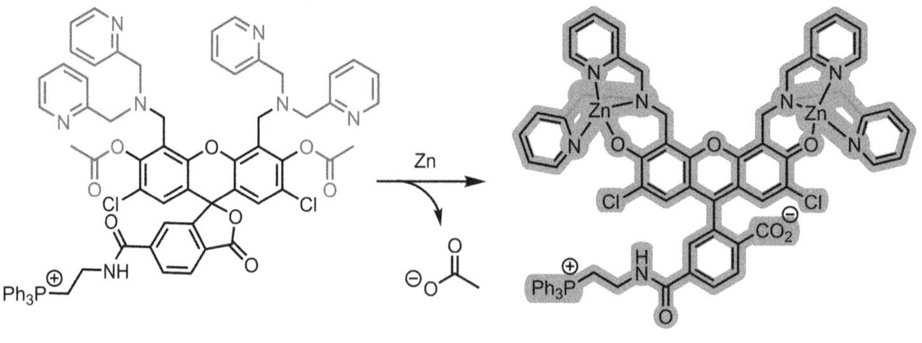

Scheme 5.29 Mitochondrially targeted activity-based Zn sensor **5.78** and fluorescent product **5.79**. Sensing group shown in red.

metal results in the hydrolysis of the acetate groups and coordination of the oxygen to the metal. This results in the fluorescein ring-opening to the quinoid form **5.79**, giving a large fluorescence turn-on. The acetate groups were shown to be stable to intracellular esterases and stopped the endosomal sequestration that resulted in non-responsiveness to Zn(II), which had been observed in the non-acetylated analogue. **5.78** was used in three different prostate cell lines and demonstrated that healthy epithelial prostate cells accumulated labile Zn(II) in their mitochondria, whereas tumorigenic prostate cells did not.

Chang and colleagues developed a ratiometric probe **5.80** using a FRET strategy for imaging Fe(II) labile pools in live cells [122]. The activity-based sensing group is an endoperoxide group inspired by the antimalarial artemisinin [123]. The probe uses a fluorescein coupled to a Cy3 cyanine with a linker containing the endoperoxide sensing group. When **5.80** reacts with Fe(II), the endoperoxide is cleaved, the fluorophores are separated giving products **5.81** and **5.82**, and FRET no longer occurs between the fluorophores. This results in emission at 556 nm, in the absence of Fe(II) and upon treatment with Fe(II), after cleavage of the linker, there is a large increase in fluorescence emission at 515 nm. The authors demonstrated that **5.80** had very high selectivity for Fe(II) over biologically relevant metal ions, GSH, and myoglobin but did show a response to 10 equivalents of Cu(I). **5.80** was able to image changes in labile Fe(II) during ferroptosis (Scheme 5.30).

Iron probe **RhoNox-1** (**5.83**) uses a different metal-mediated reaction, a reduction by iron, and was successfully designed to sense labile Fe(II) in live cells [124]. **5.83** capitalises on the fact that the tertiary amine N-oxide can be reduced using relatively mild chemistry. Reduction of N-oxide as an activity-based strategy has since been used in a variety of Fe(II) fluorescent sensors [125, 126]. Tertiary amines can be easily oxidised to the N-oxide [127], and so **5.83** consists of rhodamine B (**5.75**) where one amine has been oxidised to the corresponding

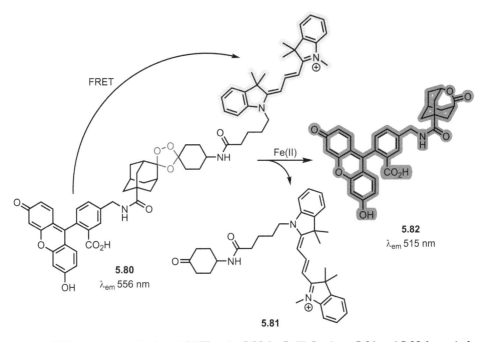

Scheme 5.30 Endoperoxide-based FRET probe **5.80** for Fe(II). Products **5.81** and **5.82** formed after reaction with Fe(II). Sensing group shown in red.

N-oxide. This means the lone pair of electrons on the nitrogen are bonded to oxygen making the N-oxide less electron-donating than the corresponding amine, leading to reduced fluorescence. Reduction by Fe (II) restores fluorescence. **5.83** demonstrated good selectivity for Fe(II) over several metal ions and biological reductants, and was able to sense both exogenously added and endogenous Fe(II) in live HepG2 cells (Scheme 5.31).

5.84 is another Fe(II) sensor in which an N-nitroxide radical is reduced by Fe(II) to generate a turn-on in fluorescence [128]. The stable TEMPO ((2,2,6,6-tetramethylpiperidin-1-yl) oxyl) radical quenches the fluorescence of attached fluorophores, a property that has been used to prepare probes for oxidative and free radical reactions [129]. TEMPO is used as the reactive sensing group in **5.84** and the N-nitroxide is reduced to the corresponding hydroxylamine selectively by Fe(II) generating **5.85**. TEMPO is paramagnetic, and as a result the fluorescence of the rhodamine in **5.84** is effectively quenched. Upon reduction, there is a strong fluorescence emission as the reduced hydroxylamine **5.85** is diamagnetic. This response was confirmed using electron paramagnetic resonance (EPR) spectroscopy in solution and in cells. **5.84** successfully sensed both endogenous and exogeneous Fe(II) and was used to image a labile Fe(II) pool in the mitochondria of human primary fibroblast cells (Scheme 5.32).

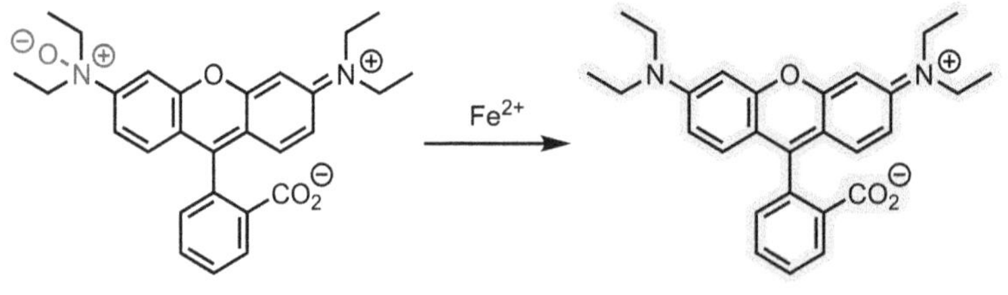

5.83
RhoNox-1

5.75
Rhodamine B
λ_{em} 575 nm

Scheme 5.31   N-oxide activity-based sensor for Fe (II) **RhoNox-1 (5.83)**. Sensing group shown in red.

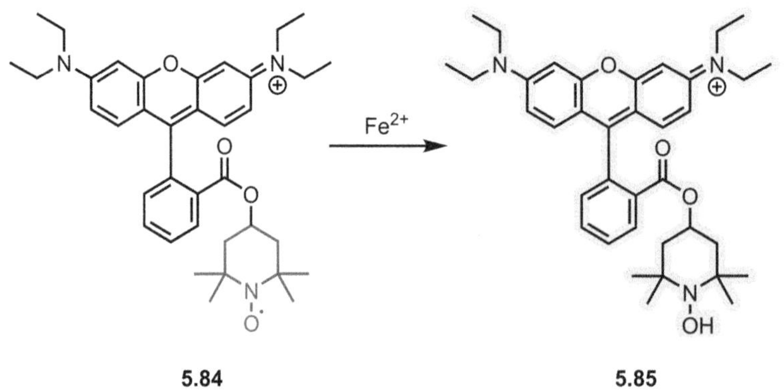

5.84

5.85
λ_{em} 580 nm

Scheme 5.32   N-Nitroxide radical activity-based Fe(II) sensor **5.84** and fluorescent product **5.85**. Sensing group shown in red.

5.7 Enzymatic Reactions

The majority of enzymes act by forming or breaking covalent bonds: indeed, six of the seven major enzyme classes (oxidoreductases, transferases, hydrolases, lyases, isomerases, and ligases) catalyse such reactions. A common strategy to study such enzymes is therefore to develop an activity-based sensor that makes use of the enzyme's specific reaction.

While there are numerous activity-based fluorescent probes for enzymes, their design principles vary little: the sensor is designed to mimic the enzyme's natural substrates in such a way that enzyme activity will cause an analogous bond cleavage or formation. The other aspect of sensor design is to ensure that this reaction will lead to a change in fluorescence. Here, we exemplify different design principles by describing notable examples of activity-based enzyme sensors.

Tyrosinase is an enzyme that controls the synthesis of melanin in the melanosome, with abnormal expression or activation associated with multiple conditions and diseases including melanoma [130], albinism [131], vitiligo [132], and Parkinson's disease [133]. In the presence of oxygen, tyrosinase causes ortho-hydroxylation of tyrosine, which can then be further oxidised to the *ortho*-quinone. Other phenol-containing groups can also be hydroxylated by the enzyme and several activity-based sensors for tyrosinase have been developed using this reactivity [134].

Ma and co-workers designed an activity-based sensor for tyrosinase, **5.86**, with excellent selectivity [135]. **5.86** is based on a xanthene-cyanine fluorophore with a 3-hydroxybenzyloxy reactive group, which inhibits ICT and therefore quenches fluorescence. Tyrosinase catalyses the hydroxylation of the 3-hydroxybenzyloxy group at the 4-position, which is followed by 1,6-elimination. This gives the emissive phenolic compound **5.87**, with restored ICT, accompanied by a turn-on in emission with maximum intensity at 708 nm (Scheme 5.33). **5.86** was used to selectively image tyrosinase activity in live cells and zebrafish. The use of a 3-hydroxybenzyl rather than 4-hydroxybenzyl reactive group ensured that the probe did not suffer from interference from ROS.

TRFS-green (**5.88**) is a fluorescent sensor for mammalian thioredoxin reductase (TrxR) reported by Fang and colleagues [136]. The thioredoxin system is ubiquitous and highly conserved, and is involved in maintaining redox homeostasis and various redox signalling processes [137]. In the thioredoxin system, TrxR reduces thioredoxin (Trx) in the presence of NADPH [138].

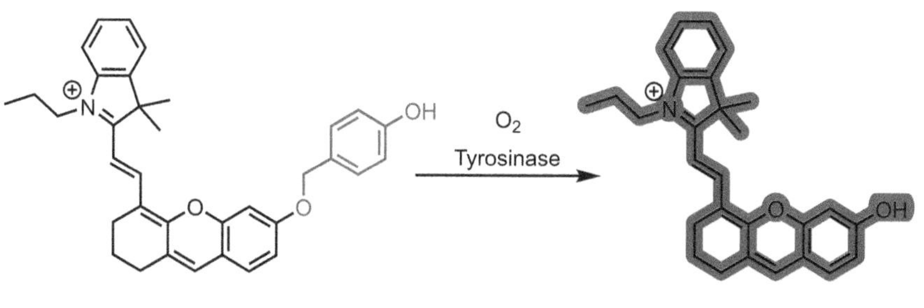

5.86

5.87
λ_{em} 708 nm

Scheme 5.33 NIR sensor **5.86** for tyrosinase activity used in live cells and zebrafish. Sensing group shown in red.

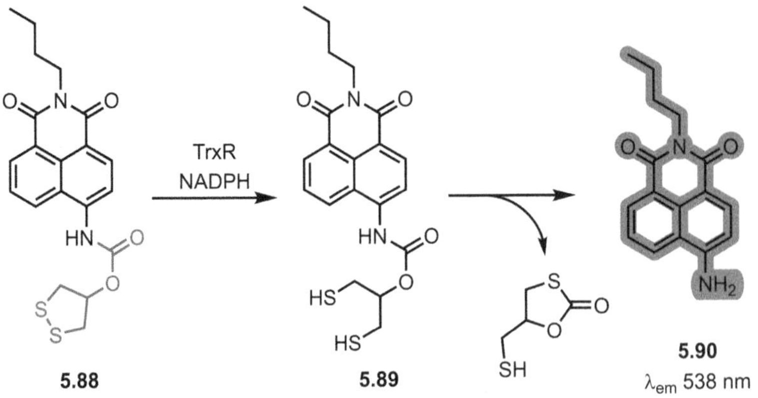

Scheme 5.34 **TRFS-green** (**5.88**) sensor used for imaging of thioredoxin reductase (TrxR) in live cells. Sensing group shown in red.

5.88 consists of a 1,2-dithiolane sensing group coupled to a 4-amino-1,8-naphthalimide fluorophore *via* a carbamate linker. The disulfide bond of the dithiolane is reduced by TrxR in the presence of NADPH, giving the free dithiol/dithiolate **5.89**, which then undergoes an intramolecular cyclisation to give a 5-membered *S*-thiocarbamate ring, with release of the fluorescent 4-amino-1,8-naphthalimide **5.90** (Scheme 5.34). **5.88** demonstrated selectivity for TrxR over other reductase enzymes and reducing agents. The probe was used to image TrxR activity in live cells and is commercially available.

Activity-based sensors, with their potential for exquisite selectivity through chemical reactivity, can be used to design isoform-specific enzyme sensors. **Green-AlDeSense** (**5.91**) is a turn-on fluorescent probe for aldehyde dehydrogenase 1A1 (ALDH1A1) [139]. Aldehyde dehydrogenases (ALDHs) are a family of non-P450 oxidising enzymes that convert aldehydes to carboxylic acids [140]. The 1A1 isoform has been identified as a particularly useful biomarker of cancer stem cells (CSCs) across multiple cancer types.

5.91 is based on the fluorescein derivative, Pennsylvania Green [141]. The acetoxymethyl ether group provides cell permeability, and this group is hydrolysed by intracellular esterases, giving a negatively charged product at physiological pH, which is retained in cells. The activity-based sensing group is a benzaldehyde, which gave the probe selectivity for the ALDH1A1 isoform [142]. The benzaldehyde is electron deficient and PET quenches the fluorescence. Upon oxidation to the acid **5.92**, the probe's emission turn-on was 20-fold. The reaction product **5.92** is dianionic, leading to even better retention in the CSCs; **5.91** was able to detect ALDH1A1 in K562 cells (human chronic myeloid leukaemia) and was used to image elevated ALDH1A1 activity in *ex vivo* and *in vivo* mice models and showed correlation with elevated ALDH1A1 activity and higher tumorigenicity (Scheme 5.35).

(*S*)-**Sulfox-1** is an activity-based ratiometric probe for imaging the activity of methionine sulfoxide reductase [143]. Methionine sulfoxide reductase A (MsrA) is a ubiquitous enzyme, which reduces free and protein-bound methionine sulfoxide back to methionine [144], with an important role in responding to oxidative stress. MsrA selectively reduces the (*S*)-epimer of the methionine sulfoxide, while its isoform, MsrB, reduces the (*R*)-epimer *in vivo* [145]. (*S*)-**Sulfox-1** contains a BODIPY fluorophore with an appended phenyl sulfoxide group, with an (*S*) configuration. The reduction of the sulfoxide to the sulfide gave a bathochromic shift in emission maximum from 541 to 566 nm (Scheme 5.36). In contrast, the

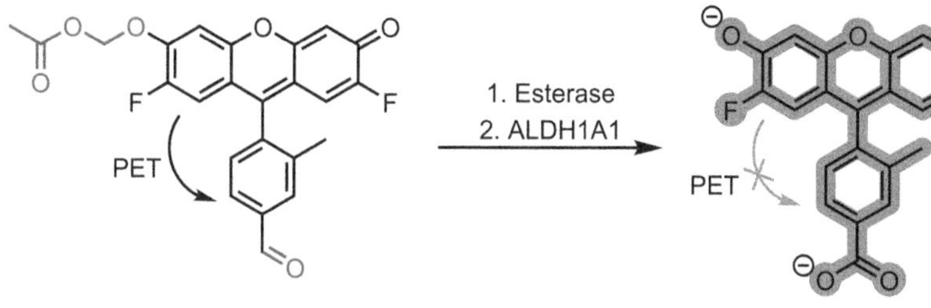

5.91
Green-AlDeSense

5.92
λ_{em} 515 nm

Scheme 5.35 **Green-AlDeSense** activity-based sensor for ALDH1A1. Sensing groups shown in red.

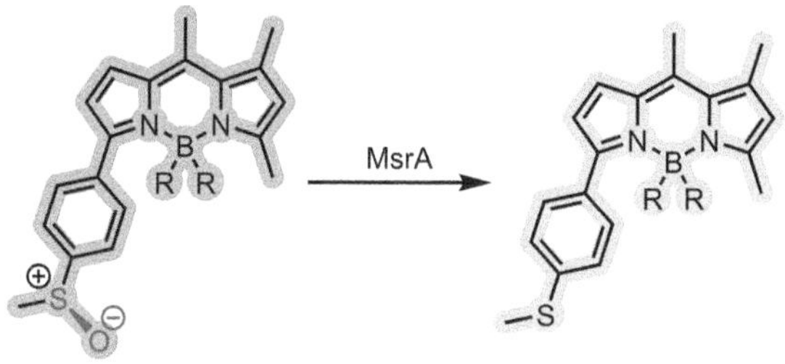

5.93
λ_{em} 541 nm

5.94
λ_{em} 566 nm

$R = $

Scheme 5.36 Ratiometric activity-based sensor for MsrA, **5.93** and fluorescent product, **5.94**. Sensing group shown in red.

(R)-enantiomer shows no activity towards MsrA: the chirality of the sulfoxide provides selectivity for the MsrA isoform. Replacement of the fluorine atoms on the BODIPY core with triethyleneglycol arms improved the water solubility. The emission wavelength shift was large enough for the probe to be used in real-time ratiometric (F_{535}/F_{570}) imaging of *E. coli* to visualise natural variations in MsrA activity.

5.95 is an activity-based fluorogenic probe for nucleotide pyrophosphatase/phosphodiesterase 6 (NPP6) [146]. NPPs are ubiquitous enzymes that are involved in the regulation of extracellular nucleotide and phospholipid levels [147]. **5.95** is based on the fluorescein derivative, Tokyo Green [148], with a phosphorylcholine group sensing group, which quenches the fluorescence by PET. Hydrolysis of the phosphorylcholine by NPP6 leads to formation of **5.96** and restoration of the green fluorescence. **5.95** shows high selectivity for NPP6 over six different NPP isoforms, including NPP2, which acts on the same substrates as NPP6 but with differing regioselectivity. **5.95** reacts with NPP2 giving a non-fluorescent reaction product **5.97** (Scheme 5.37). The probe was used in a high-throughput screening assay for developing NPP6 inhibitors.

CoxFluor (**5.98**) is a small-molecule activity-based sensor that allows for monitoring of cyclooxygenase-2 (COX-2) [149], which is overexpressed in inflammatory diseases and neurodegenerative disorders [150, 151]; COX-2 initiates the synthesis of prostaglandins,

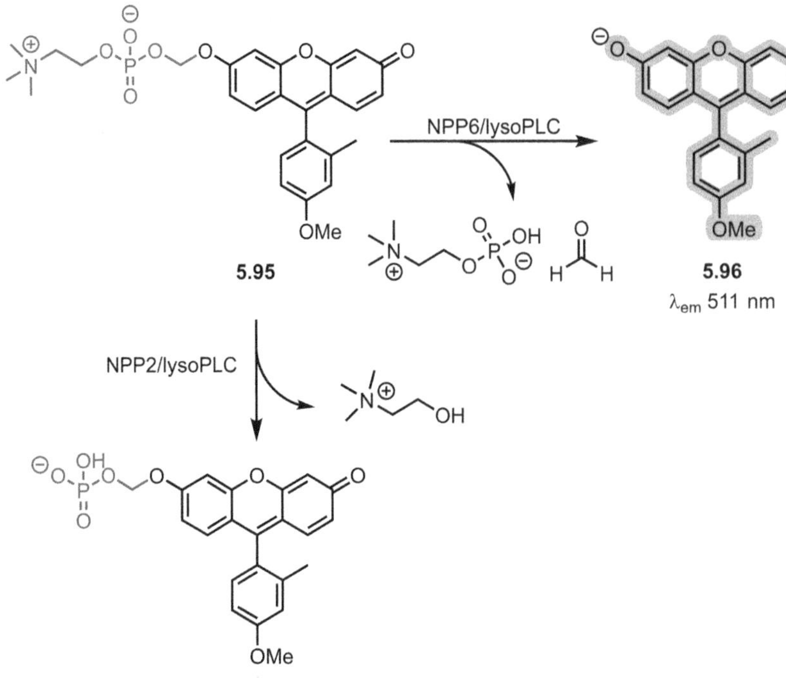

Scheme 5.37 The regioselective reaction of **5.95** with NPP6 generates the fluorescent product **5.96**, and reaction with NPP2 leads to the non-fluorescent product **5.97**. Sensing group shown in red.

pro-inflammatory compounds, from the precursor arachidonic acid, *via* production of prostaglandin H2 (PGH$_2$).

5.98 is based on the non-fluorescent, reduced form of the resorufin fluorophore, coupled to the COX-2 substrate arachidonic acid through an amide bond. The large dye blocks the arachidonic acid from fitting in the COX-1 substrate pocket, giving the probe selectivity for the COX-2 isoform, which has a larger substrate pocket and can tolerate amide derivatives of arachidonic acid [152]. Upon reaction with COX-2 endoperoxide **5.99** is generated, and in a subsequent peroxidase-mediated reaction, the amide bond is cleaved and PGH$_2$ is released along with the intensely fluorescent resorufin (**5.100**). The probe was able to detect oxygen-dependent changes in the activity of COX-2 in macrophages (Scheme 5.38).

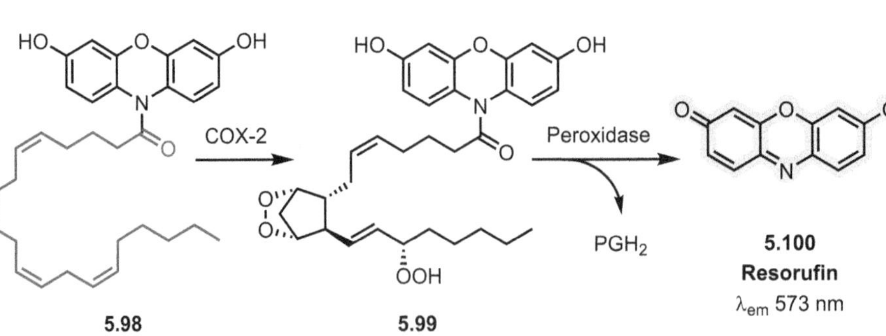

Scheme 5.38 CoxFluor (**5.98**) a selective activity-based sensor for COX-2 activity. Sensing group shown in red.

5.8   Reversible Reactions

The examples discussed in previous sections have involved irreversible covalent bond formation or cleavage. Such systems are useful for measuring low analyte concentrations but cannot be used for temporal imaging. Reversible activity-based fluorescent sensors can be advantageous for imaging fluctuations of transient analytes in real-time. Using a reversible reaction, it can be possible to detect and distinguish between chronic and temporary changes in levels of biologically active species, something that irreversible activity-based sensors cannot.

The development of selective, reversible reactions that can detect analytes in biological systems is challenging. For example, there are many examples of reversible reaction-based redox probes, but these sensors usually react to global changes in redox status, as opposed to individual analytes. Redox sensors are discussed in Chapter 6.

A reversible reaction for an activity-based sensor must have an appropriate equilibrium constant, with both the forward and reverse reactions having suitably fast kinetics. The sensor should be chemically robust to allow several cycles of sensing reaction. This section details a selection of reversible reactions that can be used to prepare activity-based sensors to image a specific analyte in cellular studies.

5.8.1   Nucleophilic Conjugate Additions

Nucleophilic conjugate addition, or 1,4-addition, involves reaction of a nucleophile with an α,β-unsaturated electrophile. Depending on the choice of nucleophile and electrophile, this reaction can be a reversible process, and therefore there has been in interest in developing reversible activity-based sensors using this strategy. The ability of GSH to act as a nucleophile presents an opportunity to use reversible conjugate additions as the basis for GSH sensors. There are many reported examples of sensors using this strategy. **5.102** is a reversible activity-based ratiometric sensor for GSH and is the first fluorescent sensor reported for quantitative imaging of GSH in live cells [153]. The sensor is based on a 7-diethylaminocoumarin fluorophore tethered to the α,β-unsaturated electrophilic Michael acceptor. The reversibility of the reaction and the time taken to reach an equilibrium in the reaction with GSH was altered by changing the substitution on the double bond of the Michael acceptor and the substitution on the phenyl ring. An unsubstituted electrophilic acceptor and 2-bromo-4-carboxylate were found to be optimal. Reaction of glutathione gives **5.103** and leads to a decrease in conjugation of the system, with a corresponding blue-shift in fluorescence emission (Scheme 5.39); **5.102** is administered as the acetoxymethyl ester **5.101** to allow the probe to efficiently cross cell membranes and was successfully used in fluorescent activated cell sorting (FACS) experiments to measure changes in GSH.

The limitations of **5.102** are the low quantum yield in aqueous solvents (<0.01 in PBS), and the slow reaction kinetics of the forward reaction with GSH ($k = 0.144\,M^{-1}\,s^{-1}$), as well as the sluggish reverse reaction.

RealThiol (5.104) [154] and the mitochondrially targeted **Mito-RealThiol (5.106)** [155] improved upon the GSH sensing ability of **5.102** using a similar scaffold. A cyano group was incorporated at the α-position of the α,β-unsaturated carbonyl electrophilic acceptor of **5.104** and **5.106** to enhance the nucleophilicity, increasing the rate of the reaction with GSH (**5.104** $k = 7.5\,M^{-1}\,s^{-1}$) that forms adducts **5.105** and **5.107**. To improve the quantum yield in aqueous solvents, the diethylamino group was replaced with an azetidine (>0.97 in PBS for product). **5.106** was the first example of a sensor that could monitor *in vivo* mitochondrial GSH fluctuations and was used in confocal microscopy, super-resolution fluorescence imaging, and flow cytometry. [156]. Despite the vastly improved photophysical properties, **5.104** and **5.106** still possess slow reaction rates with GSH (Scheme 5.40).

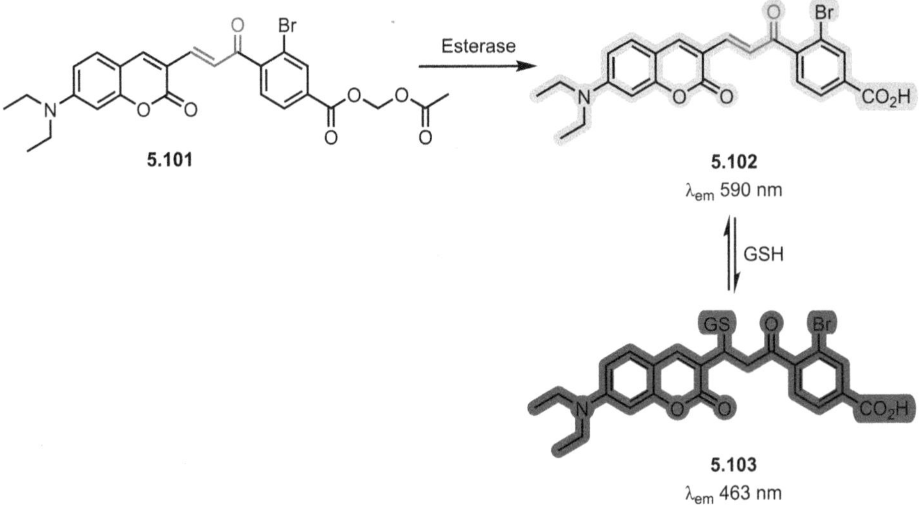

Scheme 5.39 Reversible activity-based GSH sensor **5.101** (AM ester cell-permeable analogue) and **5.102**. Sensing group shown in red.

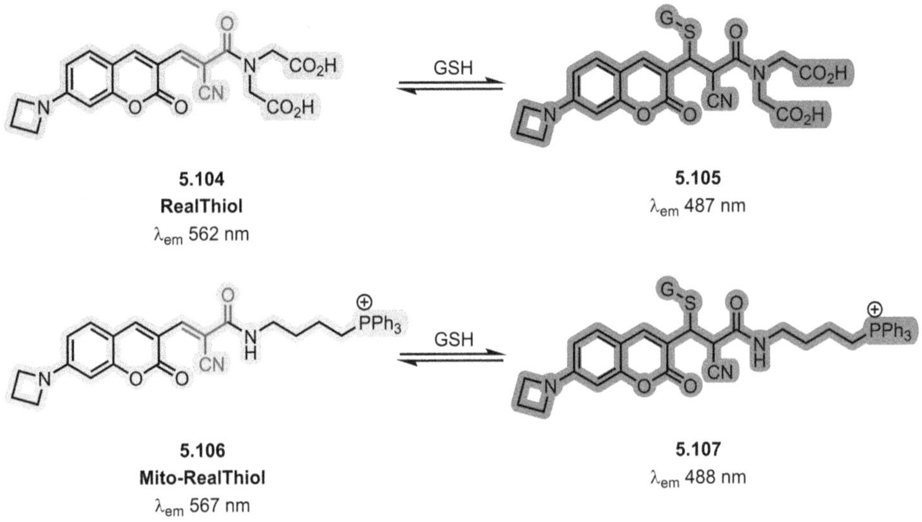

Scheme 5.40 Reversible activity-based ratiometric sensors for GSH **RealThiol (5.104)** and mitochondrially targeted **Mito-RealThiol (5.106)**. Sensing group shown in red.

Urano and co-workers reported a rationally designed sensor **QuicGSH3.0 (5.108)** [157]. They improved the rate of reaction with GSH ($k = 560\,M^{-1}\,s^{-1}$) by using a positively charged silicon rhodamine as the electrophile, in place of the neutral α,β-unsaturated carbonyl electrophile used in the previously-mentioned sensors. In the absence of GSH, excitation of the rhodamine results in FRET to the silicon-rhodamine and emission at 632 nm is observed. After nucleophilic addition of GSH (**5.109**), the conjugation in the silicon-rhodamine is disrupted, FRET is inhibited, and emission at 593 nm is observed giving a ratiometric response (Scheme 5.41). This rapid and reversible reaction with GSH, and ratiometric response, enabled real-time imaging of GSH dynamics and quantification of GSH concentrations in live cells with temporal resolution of seconds.

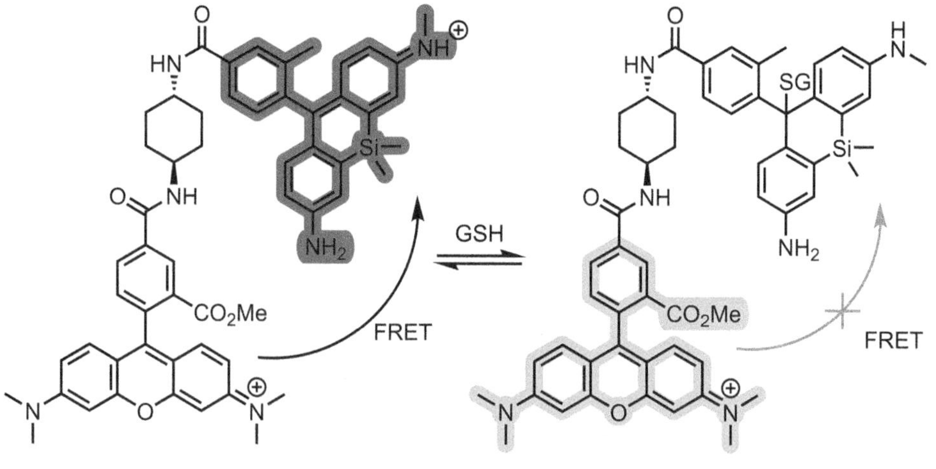

5.108
QuicGSH3.0
λ_{em} 632 nm

5.109
λ_{em} 593 nm

Scheme 5.41 FRET-based reversible activity-based sensor for GSH **QuicGSH3.0** (**5.108**).

Sulfur dioxide (SO_2) and its reduced products sulfite (SO_3^-) and bisulfite (HSO_3^-) are produced endogenously by the oxidation of sulfur-containing species [158], and at high concentrations are associated with respiratory diseases and cancer [159, 160]. SO_3^- and HSO_3^- are known to add rapidly to α,β-unsaturated electrophiles [161], and harnessing this reactivity is therefore a valid strategy for the development of reversible activity-based sensors.

5.110 was reported as a reversible fluorescent probe for HSO_3^- [162]. The structure consists of a hemicyanine conjugated to a benzothiazole. **5.110** displays dual emission at 450 and 590 nm upon excitation at 390 nm. After reaction with bisulfite, the emission at 590 nm is decreased and the 450 nm emission increases, giving a ratiometric output (Scheme 5.42). Excellent selectivity for bisulfite over a range of biologically relevant anions and nucleophiles was demonstrated, with negligible responses for all other species tested. Cyanide (CN^-) and bisulfide (HS^-) show a slight turn-on in the absence of bisulfite, but do not affect the fluorescence output in the presence of bisulfite. The reaction could be reversed by oxidative species, with the most effective being H_2O_2. **5.110** showed good reversibility over four cycles of the addition-elimination reaction and was used to sense both exogenous and endogenous bisulfite in MCF-7 cells.

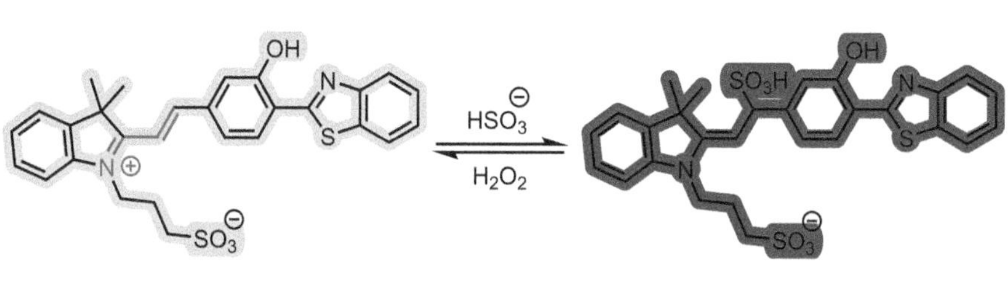

5.110
λ_{em} 590 nm

5.111
λ_{em} 450 nm

Scheme 5.42 Bisulfite activity-based sensor **5.110** and product of reaction with HSO_3^- **5.111**. Sensing group shown in red.

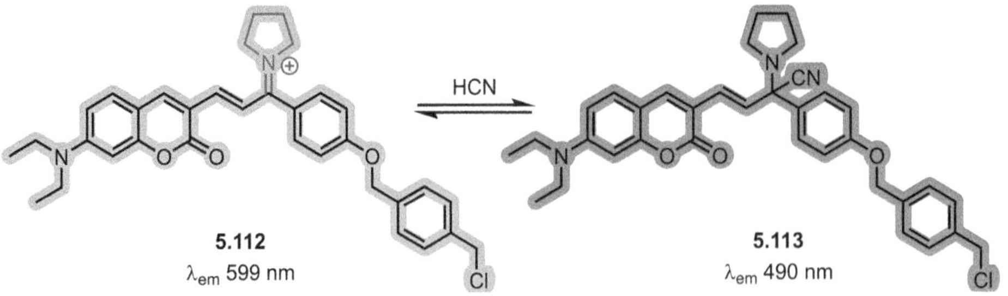

Scheme 5.43 Mitochondrially targeted reversible cyanide sensor **5.112** and reaction product **5.113**. Sensing group shown in red.

5.8.2 Nucleophilic Addition

Many cellular analytes are good nucleophiles, and this reactivity can be harnessed by developing probes with suitable electrophilic sensing groups.

Sessler and colleagues used this strategy to develop a mitochondrially targeted fluorescent sensor **5.112** for cyanide [163], which is a well-documented toxin [164] but also has an important role in some physiological processes [165]. A positively charged iminium group was used as a sensing group, tethered to a coumarin fluorophore. Reaction with cyanide leads to a decrease in conjugation, and a concomitant decrease in emission maximum from 599 to 490 nm, giving the sensor a ratiometric response (Scheme 5.43). The positive charge on the sensor also dictates its mitochondrial localisation, with a benzyl chloride group installed for mitochondrial immobilisation, ensuring that the sensor is retained in the mitochondria [166].

5.112 was shown to display excellent selectivity for cyanide over a series of biologically relevant anions and cations, and glucose, glycine, and cysteine. However, there is a notable omission of screening against ROS and GSH. The sensor was used for imaging of HepG2 cells treated with potassium cyanide. **5.112** was then used in rat neurons and was used to visualise endogenous cyanide produced after stimulation with hydromorphine. The reversibility of the probe was demonstrated by treatment with methaemoglobin, a scavenger of cyanide.

5.8.3 Imine Formation

The term imine describes compounds containing a carbon-nitrogen double bond. The process of imine formation is reversible and begins with the nucleophilic addition of an amine to a carbonyl compound. This reaction can therefore theoretically by applied to the reversible sensing of either an amino- or carbonyl-containing analyte.

One biologically important carbonyl-containing analyte is formaldehyde. As discussed in Section 5.4.1, there are a number of selective reactions employed for studying this analyte, one of which is the formation of an imine, specifically a formimine. While the examples in Section 5.4.1 make no reference to the reaction being reversible, other sensors have been reported as reversible FA sensors.

The BODIPY-based sensor **5.114** uses this activity-based sensing strategy for the reversible sensing of FA in the cytoplasm, both in live cells and *in vivo* [167]. A primary amine was chosen as a FA-reactive group. **5.114** is highly fluorescent, and upon formation of the formimine **5.115**, non-radiative decay from rotation of the C=N bond results in the fluorescence being quenched (Scheme 5.44).

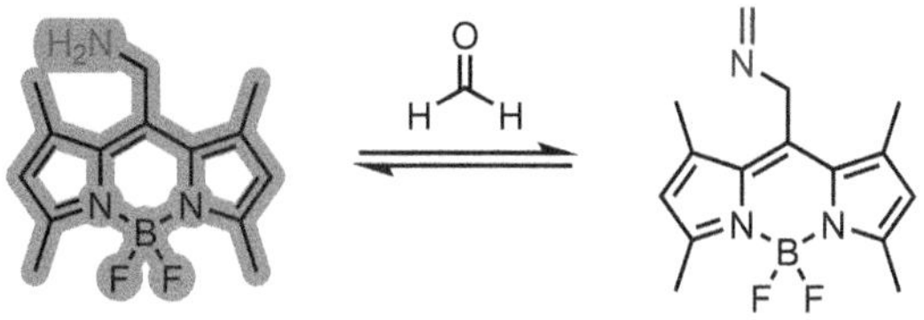

Scheme 5.44 **5.114** a BODIPY reversible activity-based turn-off sensor for FA; **5.116** a coumarin reversible activity-based turn-on sensor for FA. Sensing groups shown in red.

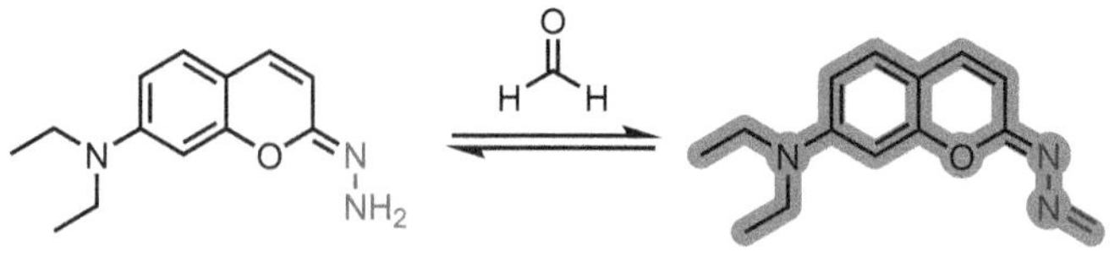

5.114 was used initially to image exogenous and endogenous fluctuations of formaldehyde in several cultured cell lines. The reversibility of the sensor was demonstrated both in cuvette and in live cells by sequential dosing of FA and NaHSO$_3$ (sodium bisulfite); **5.114** was then used to assess FA in a mouse model in specific organs *ex vivo*, for imaging FA levels in hippocampal tissue sections, and *in vivo*.

Such a reactive primary amine giving selectivity for FA over other reactive carbonyl species is surprising, but the sensor was screened against acetaldehyde, methylglyoxal, glyoxal, benzaldehyde, and pyridoxal without significant effect, albeit at 10 times lower concentration than FA. The obvious disadvantage of this probe is the turn-off response to FA and the increase in fluorescence of the turned-off reaction product **5.115** with increasing viscosity.

Another example of a reversible formaldehyde sensor, **5.116**, was reported using a hydrazonate reactive group on a coumarin scaffold [168]. **5.116** is almost non-fluorescent and upon reaction with formaldehyde forms product **5.117** which exhibits a strong green fluorescence emission (Scheme 5.44). It also demonstrated good selectivity to formaldehyde against biologically relevant species and other reactive carbonyl species. As with **5.114**, the reversibility of **5.116** was shown by addition of NaHSO$_3$, and the sensor was used to image formaldehyde levels in cells with induced exoplasmic reticulum stress.

5.8.4 Oxidation–Reduction Reactions

Redox processes are inherently reversible and therefore present a promising strategy for reversible reaction-based sensing. This is particularly relevant for biological systems, which contain numerous oxidising and reducing species of interest, as outlined in Sections 5.2 and 5.3. For the study of specific analytes, the challenge is to tune selectivity.

A notable example of a selective reversible sensor for a specific ROS is the reversible ONOO$^-$ sensor, **5.118**, reported by Han and colleagues [169]. The probe uses a heptamethine carbocyanine scaffold with 4-(phenylselenyl)aniline as a selective reversible sensing group

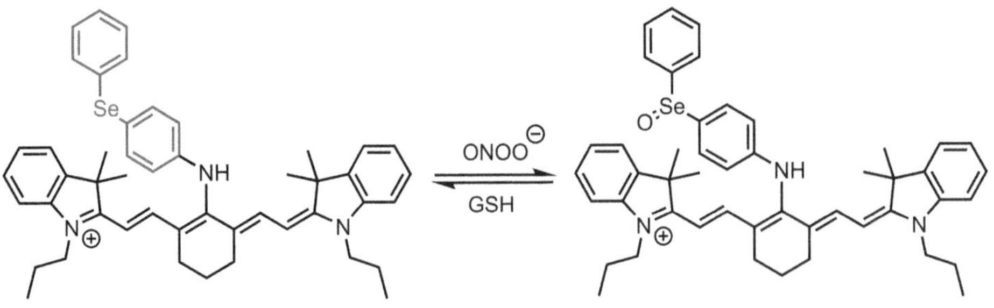

5.118

5.119
λ_{em} 775 nm

Scheme 5.45 NIR reversible fluorescent sensor **5.118** for ONOO⁻. Sensing group shown in red.

for peroxynitrite. This selenium-containing sensing group quenches the fluorescence of the cyanine by PET; **5.118** is almost non-fluorescent, and upon reaction with ONOO⁻ generates **5.119**, with a 23-fold increase emission at 775 nm (Scheme 5.45). Selectivity and reversibility of the fluorescence response was thoroughly demonstrated in cuvette studies. The sensor was used in various live cell imaging experiments demonstrating the selective response to ONOO⁻ and showing cytoplasmic localisation in RAW264.7 cells. The reversibility of response was also demonstrated by using a ONOO⁻ donor and ROS scavenger.

5.9 Analyte Regeneration

A challenge facing activity-based fluorescent probes is that they generally consume the analyte to produce fluorescence. This could affect local environment and perturb any biological processes that are being interrogated. One solution to negate this potential problem is the development of probes that regenerate the analyte. There are limited examples of analyte regeneration sensors that have been employed in live cell imaging, but this is a promising domain for future work.

Wang and co-workers reported a particularly interesting example of an activity-based sensor for FA, with the analyte regenerated as a result of the sensing reaction [170]. They reported a lysosomally targeted sensor, **5.120**, and a non-targeted analogue, both based on a naphthalimide scaffold. The activity-based sensing mechanism uses an induced intramolecular cyclisation strategy as the sensing mechanism. The fluorescence is PET quenched and additionally the succinimide group lowers the electron-donating ability of the nitrogen at the 4-position, inhibiting ICT. The secondary amine reacts with FA to give a hemiaminal, **5.121**, which then results in a 5-*exo*-trig cyclisation to give **5.122**. Hydrolysis of **5.122** generates intermediate **5.123**, which then releases FA resulting in formation of the product **5.124** with a turn-on fluorescence response and regeneration of the analyte (Scheme 5.46); **5.120** was successfully used to detect exogenous FA in lysosomes in HeLa cells. Using **5.120** and the non-targeted analogue in combination allowed comparison of the cytosolic and lysosomal concentrations of FA.

Another example of an analyte regeneration sensor was reported by Berreau and colleagues [171]. This analyte replacement sensor, **5.125**, senses CO, and then subsequently releases CO; **5.125** forms **5.126** after the sensing reaction with CO, providing a ratiometric response. However, after release of CO, the product of the reaction, **5.127**, is non-fluorescent (Scheme 5.47). The product of the sensing reaction with CO, **5.126**, is a known CO donating

Scheme 5.46 Activity-based analyte regeneration probe **5.120**.

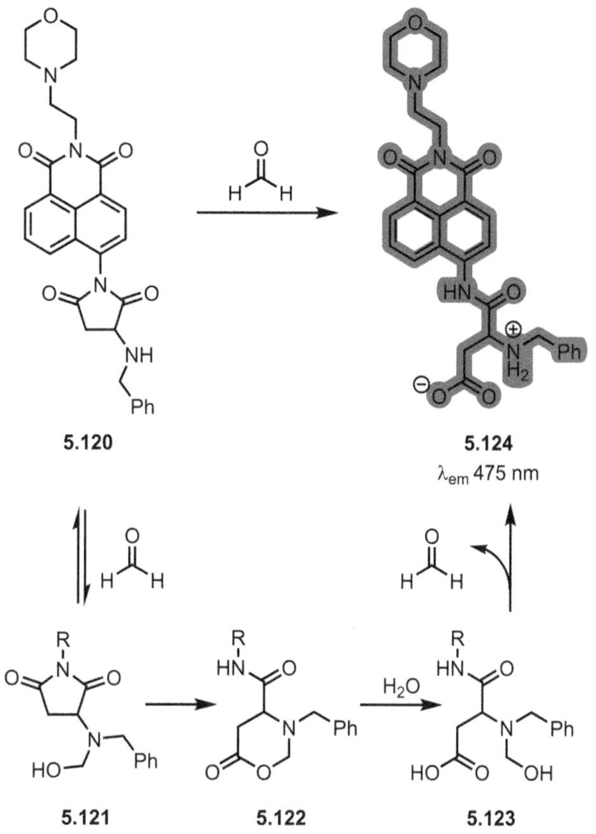

5.120
5.124
λem 475 nm
H2O
5.121
5.122
5.123

Scheme 5.47 CO analyte regeneration activity-based sensor **5.125.**

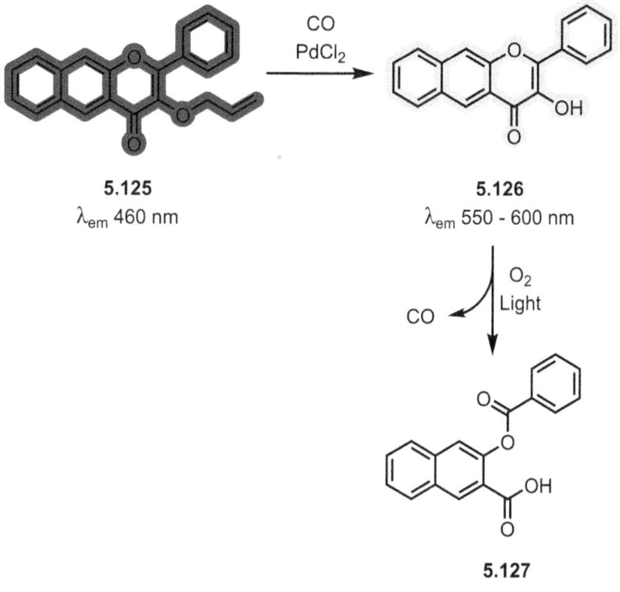

CO
PdCl2
5.125
λem 460 nm
5.126
λem 550 - 600 nm
O2
Light
CO
5.127

fluorophore; **5.126** is a visible light-induced CO releasing molecule (photoCORM) [172]. This fluorescent flavone derivative regenerates CO after illumination with 488 nm light. For the sensing mechanism to occur, the probe requires the addition of $PdCl_2$. Palladium causes an issue in cuvette studies as **5.126** can coordinate to the metal, meaning the CO release is minimised. Fortunately, this did not translate to cellular studies. The sensor was used successfully in cells to image CO produced by incubation with CORM-2 (tricarbonyldichlororuthenium (II) dimer).

5.10 Summary

Activity-based sensors provide a promising avenue to study biological analytes in a specific, selective manner by harnessing the diverse range of intrinsic chemical reactivity. The concentration of sensor used in imaging experiments must be kept to a minimum to ensure it does not perturb the biological processes it is being used to observe. This means that the concentration of sensor is almost always much lower than that of the analyte. Because of this, irreversible activity-based sensors sometimes cannot provide quantitative analyses. Furthermore, the analyte of interest is usually consumed by the sensing reaction in activity-based sensing. This has the potential to perturb the system being studied.

Reversible sensing reactions can provide information on fluctuations in analyte concentration and provide quantitative analyses. Suitably designed reversible activity-based sensors can be used for real-time monitoring and represent a powerful tool for understanding the temporal dynamics of cellular analytes. Fluorescent probes that regenerate the analyte of interest after sensing could provide a sensing platform that minimises perturbation of the system being investigated. However, this is an emerging area of research and examples of this type of activity-based sensor are currently limited.

The development and discovery of new reactivity in synthetic chemistry, enzymatic reactions, and bio-orthogonal reactions will drive the advances in activity-based sensing. Further advances will increase the impact of activity-based sensors in the progress of imaging improvements and elucidation of biology.

References

1 Wu, L., Sedgwick, A.C., Sun, X. et al. (2019). *Accounts of Chemical Research* 52 (9): 2582–2597.

2 Finkel, T. (2011). *The Journal of Cell Biology* 194 (1): 7–15.

3 Ballinger, S.W. (2005). *Free Radical Biology and Medicine* 38 (10): 1278–1295.

4 Chen, Y.-R. and Zweier, J.L. (2014). *Circulation Research* 114 (3): 524–537.

5 Markesbery, W.R. (1997). *Free Radical Biology and Medicine* 23 (1): 134–147.

6 Nunomura, A., Perry, G., Aliev, G. et al. (2001). *Journal of Neuropathology & Experimental Neurology* 60 (8): 759–767.

7 Sultana, R., Perluigi, M., and Butterfield, D.A. (2013). *Free Radical Biology & Medicine* 62: 157–169.

8 Li, X., Fang, P., Mai, J. et al. (2013). *Journal of Hematology & Oncology* 6: 19.

9 Burkitt, M.J. and Wardman, P. (2001). *Biochemical and Biophysical Research Communications* 282 (1): 329–333.

10 Wrona, M., Patel, K., and Wardman, P. (2005). *Free Radical Biology and Medicine* 38 (2): 262–270.

11 Wrona, M., Patel, K.B., and Wardman, P. (2008). *Free Radical Biology and Medicine* 44 (1): 56–62.

12 Yazdani, M. (2015). *Toxicology in vitro* 30 (1, Part B): 578–582.

13 Andina, D., Leroux, J.-C., and Luciani, P. (2017). *Chemistry - A European Journal* 23 (55): 13549–13573.

14 Kalyanaraman, B., Darley-Usmar, V., Davies, K.J.A. et al. (2012). *Free Radical Biology & Medicine* 52 (1): 1–6.

15 Tarpey, M.M. and Fridovich, I. (2001). *Circulation Research* 89 (3): 224–236.

16 Scientific, T. (2010). *The Molecular Probes Handbook.* https://www.thermofisher.com/au/en/home/references/molecular-probes-the-handbook.html.

17 Gough, D.R. and Cotter, T.G. (2011). *Cell Death & Disease* 2 (10): e213.

18 Ainley, A.D. and Challenger, F. (1930). *Journal of the Chemical Society* 2171–2180.

19 Kuivila, H.G. (1954). *Journal of the American Chemical Society* 76 (3): 870–874.

20 Rhee, S.G., Chang, T.-S., Jeong, W., and Kang, D. (2010). *Molecules and Cells* 29 (6): 539–549.

21 Chang, M.C.Y., Pralle, A., Isacoff, E.Y., and Chang, C.J. (2004). *Journal of the American Chemical Society* 126 (47): 15392–15393.

22 Miller, E.W., Albers, A.E., Pralle, A. et al. (2005). *Journal of the American Chemical Society* 127 (47): 16652–16659.

23 Albers, A.E., Dickinson, B.C., Miller, E.W., and Chang, C.J. (2008). *Bioorganic & Medicinal Chemistry Letters* 18 (22): 5948–5950.

24 Dickinson, B.C., Huynh, C., and Chang, C.J. (2010). *Journal of the American Chemical Society* 132 (16): 5906–5915.

25 Miller, E.W., Tulyathan, O., Isacoff, E.Y., and Chang, C.J. (2007). *Nature Chemical Biology* 3 (5): 263–267.

26 Ullrich, A., Coussens, L., Hayflick, J.S. et al. (1984). *Nature* 309 (5967): 418–425.

27 Srikun, D., Miller, E.W., Domaille, D.W., and Chang, C.J. (2008). *Journal of the American Chemical Society* 130 (14): 4596–4597.

28 Yan, J., Lee, S., Zhang, A., and Yoon, J. (2018). *Chemical Society Reviews* 47 (18): 6900–6916.

29 Sikora, A., Zielonka, J., Lopez, M. et al. (2009). *Free Radical Biology and Medicine* 47 (10): 1401–1407.

30 Abo, M., Urano, Y., Hanaoka, K. et al. (2011). *Journal of the American Chemical Society* 133 (27): 10629–10637.

31 Gao, C., Tian, Y., Zhang, R. et al. (2017). *Analytical Chemistry* 89 (23): 12945–12950.

32 Xie, X., Xe, Y., Wu, T. et al. (2016). *Analytical Chemistry* 88 (16): 8019–8025.

33 Chen, H., Chen, X., Luo, Y., and Shen, J. (2018). *Free Radical Research* 52 (11–12): 1220–1239.

34 Kouti, L., Noroozian, M., Akhondzadeh, S. et al. (2013). *European Review for Medical and Pharmacological Sciences* 17 (7): 964–970.

35 Torreilles, F., Salman-Tabcheh, S.D., Guérin, M.-C., and Torreilles, J. (1999). *Brain Research Reviews* 30 (2): 153–163.

36 Yik-Sham Chung, C., Timblin, G.A., Saijo, K., and Chang, C.J. (2018). *Journal of the American Chemical Society* 140 (19): 6109–6121.

37 Murfin, L.C., Weber, M., Park, S.J. et al. (2019). *Journal of the American Chemical Society* 141 (49): 19389–19396.

38 Kim, J., Park, J., Lee, H. et al. (2014). *Chemical Communications* 50 (66): 9353–9356.

39 Cheng, D., Pan, Y., Wang, L. et al. (2017). *Journal of the American Chemical Society* 139 (1): 285–292.

40 Li, X., Tao, R.-R., Hong, L.-J. et al. (2015). *Journal of the American Chemical Society* 137 (38): 12296–12303.

41 Sedgwick, A.C., Wu, L., Han, H.-H. et al. (2018). *Chemical Society Reviews* 47 (23): 8842–8880.

42 Ulfig, A. and Leichert, L.I. (2021). *Cellular and Molecular Life Sciences* 78 (2): 385–414.

43 Armesto, X.L., Canle, L.M., Fernández, M.I. et al. (2000). *Tetrahedron* 56 (8): 1103–1109.

44 Zhu, B., Li, P., Shu, W. et al. (2016). *Analytical Chemistry* 88 (24): 12532–12538.

45 Shi, D., Chen, S., Dong, B. et al. (2019). *Chemical Science* 10 (13): 3715–3722.

46 Zhu, H., Fan, J., Wang, J. et al. (2014). *Journal of the American Chemical Society* 136 (37): 12820–12823.

47 Luiking, Y.C., Engelen, M.P., and Deutz, N.E. (2010). *Current Opinion in Clinical Nutrition and Metabolic Care* 13 (1): 97–104.

48 Kojima, H., Nakatsubo, N., Kikuchi, K. et al. (1998). *Analytical Chemistry* 70 (13): 2446–2453.

49 Zhang, X., Kim, W.-S., Hatcher, N. et al. (2002). *Journal of Biological Chemistry* 277 (50): 48472–48478.

50 Chen, Y. (2020). *Nitric Oxide* 98: 1–19.

51 Yang, Y., Seidlits, S.K., Adams, M.M. et al. (2010). *Journal of the American Chemical Society* 132 (38): 13114–13116.

52 Lv, X., Wang, Y., Zhang, S. et al. (2014). *Chemical Communications* 50 (56): 7499–7502.

53 Cadet, J. (2003). *Mutation Research: Fundamental and Molecular Mechanisms of Mutagenesis* 531 (1–2): 5–23.

54 Grether-Beck, S., Olaizola-Horn, S., Schmitt, H. et al. (1996). *Proceedings of the National Academy of Sciences of the United States of America* 93 (25): 14586–14591.

55 Dolmans, D.E.J.G.J., Fukumura, D., and Jain, R.K. (2003). *Nature Reviews Cancer* 3 (5): 380–387.

56 Umezawa, N., Tanaka, K., Urano, Y. et al. (1999). *Angewandte Chemie International Edition* 38 (19): 2899–2901.

57 Tanaka, K., Miura, T., Umezawa, N. et al. (2001). *Journal of the American Chemical Society* 123 (11): 2530–2536.

58 Kim, S., Tachikawa, T., Fujitsuka, M., and Majima, T. (2014). *Journal of the American Chemical Society* 136 (33): 11707–11715.

59 Xu, K., Wang, L., Qiang, M. et al. (2011). *Chemical Communications* 47 (26): 7386–7388.

60 Kim, S., Fujitsuka, M., and Majima, T. (2013). *The Journal of Physical Chemistry B* 117 (45): 13985–13992.

61 Ragàs, X., Jiménez-Banzo, A., Sánchez-García, D. et al. (2009). *Chemical Communications* 20: 2920.

62 Xiao, W. and Loscalzo, J. (2020). *Antioxidants & Redox Signaling* 32 (18): 1330–1347.

63 Schafer, F.Q. and Buettner, G.R. (2001). *Free Radical Biology and Medicine* 30 (11): 1191–1212.

64 Kimura, Y. and Kimura, H. (2004). *The FASEB Journal* 18 (10): 1165–1167.

65 Kimura, Y., Goto, Y.-I., and Kimura, H. (2010). *Antioxidants & Redox Signaling* 12 (1): 1–13.

66 Li, L., Rose, P., and Moore, P.K. (2011). *Annual Review of Pharmacology and Toxicology* 51 (1): 169–187.

67 Paul, B.D., Sbodio, J.I., Xu, R. et al. (2014). *Nature* 509 (7498): 96–100.

68 Szabo, C. (2016). *Nature Reviews Drug Discovery* 15 (3): 185–203.

69 Lippert, A.R., New, E.J., and Chang, C.J. (2011). *Journal of the American Chemical Society* 133 (26): 10078–10080.

70 Hansch, C., Leo, A., and Taft, R.W. (1991). *Chemical Reviews* 91 (2): 165–195.

71 Henthorn, H.A. and Pluth, M.D. (2015). *Journal of the American Chemical Society* 137 (48): 15330–15336.

72 Park, C.S., Ha, T.H., Choi, S.-A. et al. (2017). *Biosensors and Bioelectronics* 89: 919–926.

73 Liu, C., Pan, J., Li, S. et al. (2011). *Angewandte Chemie International Edition* 50 (44): 10327–10329.

74 Wang, F., Zhou, L., Zhao, C. et al. (2015). *Chemical Science* 6 (4): 2584–2589.

75 Mulay, S.V., Kim, Y., Choi, M. et al. (2018). *Analytical Chemistry* 90 (4): 2648–2654.

76 Yang, Z., Zhao, N., Sun, Y. et al. (2012). *Chemical Communications* 48 (28): 3442–3444.

77 Hoffmann, P.R. and Berry, M.J. (2008). *Molecular Nutrition & Food Research* 52 (11): 1273–1280.

78 Huber, R.E. and Criddle, R.S. (1967). *Archives of Biochemistry and Biophysics* 122 (1): 164–173.

79 Zhang, B., Ge, C., Yao, J. et al. (2015). *Journal of the American Chemical Society* 137 (2): 757–769.

80 Chen, H., Dong, B., Tang, Y., and Lin, W. (2015). *Chemistry – A European Journal* 21 (33): 11696–11700.

81 Baig, M.H., Jan, A.T., Rabbani, G. et al. (2017). *Scientific Reports* 7 (1): 5916.

82 Heck, D.H.A., Casanova, M., and Starr, T.B. (1990). *Critical Reviews in Toxicology* 20 (6): 397–426.

83 Salthammer, T., Mentese, S., and Marutzky, R. (2010). *Chemical Reviews* 110 (4): 2536–2572.

84 Ducker, G.S. and Rabinowitz, J.D. (2017). *Cell Metabolism* 25 (1): 27–42.

85 Burgos-Barragan, G., Wit, N., Meiser, J. et al. (2017). *Nature* 548 (7669): 549–554.

86 Moore, L.D., Le, T., and Fan, G. (2013). *Neuropsychopharmacology* 38 (1): 23–38.

87 Hoopes, L. (2008). *Nature Education* 1 (1): 160.

88 Barbosa, E., dos Santos, A.L.A., Peteffi, G.P. et al. (2019). *Environmental Science and Pollution Research* 26 (2): 1304–1314.

89 Tang, Y., Kong, X., Xu, A. et al. (2016). *Angewandte Chemie International Edition* 55 (10): 3356–3359.

90 Lee, Y.H., Tang, Y., Verwilst, P. et al. (2016). *Chemical Communications* 52 (75): 11247–11250.

91 Tang, Y., Kong, X., Liu, Z.-R. et al. (2016). *Analytical Chemistry* 88 (19): 9359–9363.

92 He, L., Yang, X., Ren, M. et al. (2016). *Chemical Communications* 52 (61): 9582–9585.

93 Brewer, T.F. and Chang, C.J. (2015). *Journal of the American Chemical Society* 137 (34): 10886–10889.

94 Roth, A., Li, H., Anorma, C., and Chan, J. (2015). *Journal of the American Chemical Society* 137 (34): 10890–10893.

95 Kim, J., Son, J.-W., Lee, J.-A. et al. (2004). *Journal of Korean Medical Science* 19 (1): 95.

96 Rabbani, N. and Thornalley, P.J. (2011). *Seminars in Cell & Developmental Biology* 22 (3): 309–317.

97 Thornalley, P.J. (1996). *General Pharmacology: The Vascular System* 27 (4): 565–573.

98 Desai, K.M., Chang, T., Wang, H. et al. (2010). *Canadian Journal of Physiology and Pharmacology* 88 (3): 273–284.

99 Wang, T., Douglass, E.F., Fitzgerald, K.J., and Spiegel, D.A. (2013). *Journal of the American Chemical Society* 135 (33): 12429–12433.

100 Nokin, M.-J., Durieux, F., Peixoto, P. et al. (2016). *eLife* 5: e19375.

101 Pacher, P., Beckman, J.S., and Liaudet, L. (2007). *Physiological Reviews* 87 (1): 315–424.

102 Finkelstein, J. (2009). *Nature* 460 (7257): 813.

103 Peterson, J.E. and Stewart, R.D. (1975). *Journal of Applied Physiology* 39 (4): 633–638.

104 Wu, L. and Wang, R. (2005). *Pharmacological Reviews* 57 (4): 585–630.

105 Brennführer, A., Neumann, H., and Beller, M. (2009). *Angewandte Chemie International Edition* 48 (23): 4114–4133.

106 Wu, X.-F., Neumann, H., and Beller, M. (2013). *Chemical Reviews* 113 (1): 1–35.

107 Michel, B.W., Lippert, A.R., and Chang, C.J. (2012). *Journal of the American Chemical Society* 134 (38): 15668–15671.

108 Toussaint, S.N.W., Calkins, R.T., Lee, S., and Michel, B.W. (2018). *Journal of the American Chemical Society* 140 (41): 13151–13155.

109 Garber, S.B., Kingsbury, J.S., Gray, B.L., and Hoveyda, A.H. (2000). *Journal of the American Chemical Society* 122 (34): 8168–8179.

110 Sasakura, K., Hanaoka, K., Shibuya, N. et al. (2011). *Journal of the American Chemical Society* 133 (45): 18003–18005.

111 Joshi, T., Graham, B., and Spiccia, L. (2015). *Accounts of Chemical Research* 48 (8): 2366–2379.

112 Hitomi, Y., Takeyasu, T., and Kodera, M. (2013). *Chemical Communications* 49 (85): 9929.

113 Hitomi, Y., Hiramatsu, K., Arakawa, K. et al. (2013). *Dalton Transactions* 42 (36): 12878.

114 Cotruvo, J.J.A., Aron, A.T., Ramos-Torres, K.M., and Chang, C.J. (2015). *Chemical Society Reviews* 44 (13): 4400–4414.

115 Taki, M., Iyoshi, S., Ojida, A. et al. (2010). *Journal of the American Chemical Society* 132 (17): 5938–5939.

116 Au-Yeung, H.Y., Chan, J., Chantarojsiri, T., and Chang, C.J. (2013). *Journal of the American Chemical Society* 135 (40): 15165–15173.

117 Au-Yeung, H.Y., New, E.J., and Chang, C.J. (2012). *Chemical Communications* 48 (43): 5268.

118 Yu, M., Shi, M., Chen, Z. et al. (2008). *Chemistry - A European Journal* 14 (23): 6892–6900.

119 Kovács, J. and Mokhir, A. (2008). *Inorganic Chemistry* 47 (6): 1880–1882.

120 Zhao, C., Feng, P., Cao, J. et al. (2012). *Organic & Biomolecular Chemistry* 10 (15): 3104.

121 Chyan, W., Zhang, D.Y., Lippard, S.J., and Radford, R.J. (2014). *Proceedings of the National Academy of Sciences* 111 (1): 143–148.

122 Aron, A.T., Loehr, M.O., Bogena, J., and Chang, C.J. (2016). *Journal of the American Chemical Society* 138 (43): 14338–14346.

123 Klayman, D. (1985). *Science* 228 (4703): 1049–1055.

124 Hirayama, T., Okuda, K., and Nagasawa, H. (2013). *Chemical Science* 4 (3): 1250.

125 Hirayama, T., Tsuboi, H., Niwa, M. et al. (2017). *Chemical Science* 8 (7): 4858–4866.

126 Niwa, M., Hirayama, T., Okuda, K., and Nagasawa, H. (2014). *Organic & Biomolecular Chemistry* 12 (34): 6590–6597.

127 Cymerman Craig, J. and Purushothaman, K.K. (1970). *The Journal of Organic Chemistry* 35 (5): 1721–1722.

128 Maiti, S., Aydin, Z., Zhang, Y., and Guo, M. (2015). *Dalton Transactions* 44 (19): 8942–8949.

129 Blinco, J.P., Fairfull-Smith, K.E., Morrow, B.J., and Bottle, S.E. (2011). *Australian Journal of Chemistry* 64 (4): 373.

130 Brichard, V., Van Pel, A., Wölfel, T. et al. (1993). *Journal of Experimental Medicine* 178 (2): 489–495.

131 Oetting, W.S. and King, R.A. (1999). *Human Mutation* 13 (2): 99–115.

132 Jin, Y., Birlea, S.A., Fain, P.R. et al. (2010). *New England Journal of Medicine* 362 (18): 1686–1697.

133 Asanuma, M., Miyazaki, I., and Ogawa, N. (2003). *Neurotoxicity Research* 5 (3): 165–176.

134 Qu, Y., Zhan, Q., Du, S. et al. (2020). *Journal of Pharmaceutical Analysis* 10 (5): 414–425.

135 Wu, X., Li, L., Shi, W. et al. (2016). *Angewandte Chemie International Edition* 55 (47): 14728–14732.

136 Zhang, L., Duan, D., Liu, Y. et al. (2014). *Journal of the American Chemical Society* 136 (1): 226–233.

137 Lu, J. and Holmgren, A. (2014). *Free Radical Biology and Medicine* 66: 75–87.

138 Arnér, E.S.J. and Holmgren, A. (2000). *European Journal of Biochemistry* 267 (20): 6102–6109.

139 Anorma, C., Hedhli, J., Bearrood, T.E. et al. (2018). *ACS Central Science* 4 (8): 1045–1055.

140 Marchitti, S.A., Brocker, C., Stagos, D., and Vasiliou, V. (2008). *Expert Opinion on Drug Metabolism & Toxicology* 4 (6): 697–720.

141 Mottram, L.F., Boonyarattanakalin, S., Kovel, R.E., and Peterson, B.R. (2006). *Organic Letters* 8 (4): 581–584.

142 Wang, M.-F., Han, C.-L., and Yin, S.-J. (2009). *Chemico-Biological Interactions* 178 (1–3): 36–39.

143 Makukhin, N., Tretyachenko, V., Moskovitz, J., and Míšek, J. (2016). *Angewandte Chemie International Edition* 55 (41): 12727–12730.

144 Drazic, A. and Winter, J. (2014). *Biochimica et Biophysica Acta (BBA) - Proteins and Proteomics* 1844 (8): 1367–1382.

145 Ranaivoson, F.M., Neiers, F., Kauffmann, B. et al. (2009). *Journal of Molecular Biology* 394 (1): 83–93.

146 Kawaguchi, M., Okabe, T., Okudaira, S. et al. (2011). *Journal of the American Chemical Society* 133 (31): 12021–12030.

147 Bollen, M., Gijsbers, R., Ceulemans, H. et al. (2000). *Critical Reviews in Biochemistry and Molecular Biology* 35 (6): 393–432.

148 Urano, Y., Kamiya, M., Kanda, K. et al. (2005). *Journal of the American Chemical Society* 127 (13): 4888–4894.

149 Yadav, A.K., Reinhardt, C.J., Arango, A.S. et al. (2020). *Angewandte Chemie International Edition* 59 (8): 3307–3314.

150 Samad, T.A., Moore, K.A., Sapirstein, A. et al. (2001). *Nature* 410 (6827): 471–475.

151 Teismann, P., Tieu, K., Choi, D.K. et al. (2003). *Proceedings of the National Academy of Sciences* 100 (9): 5473–5478.

152 Yu, M., Ives, D., and Ramesha, C.S. (1997). *Journal of Biological Chemistry* 272 (34): 21181–21186.

153 Jiang, X., Yu, Y., Chen, J. et al. (2015). *ACS Chemical Biology* 10 (3): 864–874.

154 Jiang, X., Chen, J., Bajić, A. et al. (2017). *Nature Communications* 8 (1): 16807.

155 Chen, J., Jiang, X., Zhang, C. et al. (2017). *ACS Sensors* 2 (9): 1257–1261.

156 Grimm, J.B., English, B.P., Chen, J. et al. (2015). *Nature Methods* 12 (3): 244–250.

157 Umezawa, K., Yoshida, M., Kamiya, M. et al. (2017). *Nature Chemistry* 9 (3): 279–286.

158 Gunnison, A.F. (1981). *Food and Cosmetics Toxicology* 19: 667–682.

159 Shapiro, R. (1977). *Mutation Research/Reviews in Genetic Toxicology* 39 (2): 149–175.

160 Leung, K.-H., Post, G.B., and Menzel, D.B. (1985). *Toxicology and Applied Pharmacology* 77 (3): 388–394.

161 Morton, M. and Landfield, H. (1952). *Journal of the American Chemical Society* 74 (14): 3523–3526.

162 Zhang, Y., Guan, L., Yu, H. et al. (2016). *Analytical Chemistry* 88 (8): 4426–4431.

163 Long, L., Huang, M., Wang, N. et al. (2018). *Journal of the American Chemical Society* 140 (5): 1870–1875.

164 Cooper, C.E. and Brown, G.C. (2008). *Journal of Bioenergetics and Biomembranes* 40 (5): 533–539.

165 Gunasekar, P.G., Borowitz, J.L., Turek, J.J. et al. (2000). *Journal of Neuroscience Research* 61 (5): 570–575.

166 Presley, A.D., Fuller, K.M., and Arriaga, E.A. (2003). *Journal of Chromatography B* 793 (1): 141–150.

167 Song, X., Han, X., Yu, F. et al. (2018). *The Analyst* 143 (2): 429–439.

168 Liang, X.-G., Chen, B., Shao, L.-X. et al. (2017). *Theranostics* 7 (8): 2305–2313.

169 Yu, F., Li, P., Li, G. et al. (2011). *Journal of the American Chemical Society* 133 (29): 11030–11033.

170 Xu, H., Xu, H., Ma, S. et al. (2018). *Journal of the American Chemical Society* 140 (48): 16408–16412.

171 Popova, M., Lazarus, L.S., Benninghoff, A.D., and Berreau, L.M. (2020). *ACS Omega* 5 (17): 10021–10033.

172 Anderson, S.N., Richards, J.M., Esquer, H.J. et al. (2015). *ChemistryOpen* 4 (5): 590–594.

6

Fluorescent Sensors of the Cellular Environment

Nian Kee Tan[1,2,3], Jianping Zhu[1,2], and Elizabeth J. New[1,2,3]

[1] *School of Chemistry, The University of Sydney, NSW, Australia*
[2] *Australian Research Council Centre of Excellence for Innovations in Peptide and Protein Science, The University of Sydney, NSW, Australia*
[3] *The University of Sydney Nano Institute (Sydney Nano), The University of Sydney, NSW, Australia*

The cellular environment is heterogeneous, varying between different organelles, membranes, and the cytosol. Each environment is tightly regulated to ensure physiological processes essential for survival to take place. In this chapter, fluorescent sensors for cellular environment will be discussed, with a focus on sensors for cellular polarity, viscosity, pH, and redox status, which are the main physicochemical properties of cellular environment. We will introduce well-studied mechanisms for sensing polarity, viscosity, pH, and redox status, and present typical examples of sensors that have been used in cellular studies.

6.1 Fluorescent Sensors for Polarity and Viscosity

6.1.1 The Biological Significance of Polarity and Viscosity

Cellular polarity, which is the asymmetric organisation of sub-cellular components, establishes and maintains functionally specialised domains within biomembranes and the cytoplasm [1, 2]. Protein compositions and spatial arrangements of these domains facilitate the operation of cellular processes, including localised membrane growth, directional cell migration, activation of immune response, and vectorial transport of molecules across cell layers [3]. Cellular polarity varies with activity, making polarity an important marker for diagnosis of disorders and diseases. Fluorescent sensors have great advantages in the detection and monitoring of dynamic changes in cellular polarity, as they are sensitive to small environmental changes. While cellular polarity is a complex parameter that is influenced by many factors, sensors to date tend to focus on the measurement of solvent polarity as a proxy for cellular polarity.

Cellular viscosity is another key parameter in determining cellular health and function. Large changes in intracellular viscosity accompany a number of diffusion-mediated cellular processes, such as the transportation of biomolecules and cellular organelles, signal transductions, protein–protein interactions and cellular metabolisms [4–6]. Viscosity varies in different cellular organelles and different cell lines, and it has been reported that drug treatments can change the cellular viscosity [6–8]. Abnormal fluctuations in intracellular viscosity are also important indicators for diseases such as diabetes [9], atherosclerosis [10], hypercholesterolemia [11], and Alzheimer's [12]. Developing fluorescent sensors to monitor dynamic changes in cellular viscosity is therefore crucial for the detection of abnormal cellular status and diagnosis of diseases.

There have been several effective mechanisms for developing polarity-sensitive fluorescent sensors, including twisted intramolecular charge transfer (TICT), intramolecular charge transfer (ICT), excited-state intramolecular proton transfer (ESIPT), and photoinduced electron transfer (PET). TICT mechanisms dominate the literature on viscosity sensing. The following section therefore describes how TICT can be tuned to favour each application.

6.1.2 Twisted Intramolecular Charge Transfer as a Mechanism for Polarity and Viscosity Sensing

Twisted intramolecular charge transfer (TICT) is a process observed for molecules that comprise donor and acceptor groups linked by a single covalent bond [13, 14]. Photoexcitation of the molecule is followed by charge transfer from the donor orbital (HOMO) to the acceptor orbital (LUMO), accompanied by intramolecular donor-acceptor twisting around the single bond, and formation of a perpendicular (twisted) conformation to minimise the TICT excited state energy (Figure 6.1) [13–15]. The equilibrium between the coplanar conformation and the perpendicular conformation results in two emission bands: a high-energy band that corresponds to locally excited state (LE); and a low-energy band that corresponds to the TICT state [13, 14].

The equilibrium between the LE state (coplanar conformation) and the TICT state (perpendicular conformation) is highly influenced by surrounding polarity. As polar solvents reduce the energy level of excited states, a high solvent polarity increases the population of the TICT state [16]. The generation of the TICT state is also influenced by the surrounding steric hindrance such as viscosity. This means that TICT systems can be sensitive to polarity and/or viscosity: in order to develop selective sensors, it is therefore important to tune the system to favour one mechanism over the other.

For TICT-based molecules, pyrrole, carbazole, and dialkylaniline usually act as donors, while cyano groups, bispyrazolopyridine, pyrazoloquinoline, anthracene, acridine, naphthalene, and pyrimidine can operate as acceptors [13, 17]. Representative TICT molecules are shown in Figure 6.2 [13]. Such molecules are also referred to as molecular rotors.

The TICT state releases energy either through fluorescence emission or non-radiative decay. This is dependent on the energy gap between the TICT state and the ground state (GS). A large energy gap will allow for photon emission, while a small energy gap leads to non-radiative decay [18–20]. The TICT state tends to be more sensitive to polarity than the LE state, so fluorophores that exhibit fluorescence emission from the TICT state can be used as polarity sensors [13, 15, 19]. In contrast, TICT systems which emit only from the LE state will be favourable for viscosity sensing [5, 19].

Representative examples of such systems are the polarity sensor 4-(N,N-dimethylamino)-benzonitrile (**6.1**) and the viscosity sensor 9-(2,2-dicyanovinyl)julolidine (**6.2**) (Figure 6.3). In nonpolar solvents, **6.1** exhibits one emission band generated from the LE state at 350 nm [13, 22]. In increasingly polar solvents, the emission band at 350 nm decreases in

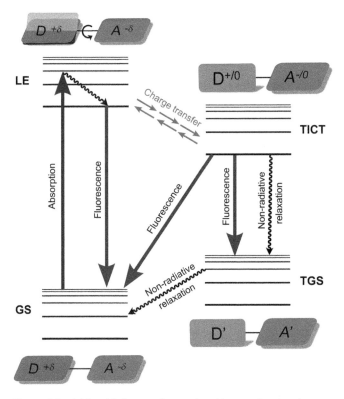

Figure 6.1 Jablonski diagram for a twisted intramolecular charge transfer (TICT) system. The TICT system comprises a donor (D) and acceptor (A). Excitation from the ground state (GS) to the locally excited state (LE) is followed by charge transfer and conformational change to the lower energy TICT state, which can undergo fluorescence decay to the twisted ground state (TGS).

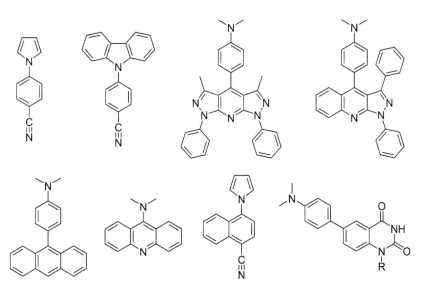

Figure 6.2 Representative TICT molecules [13].

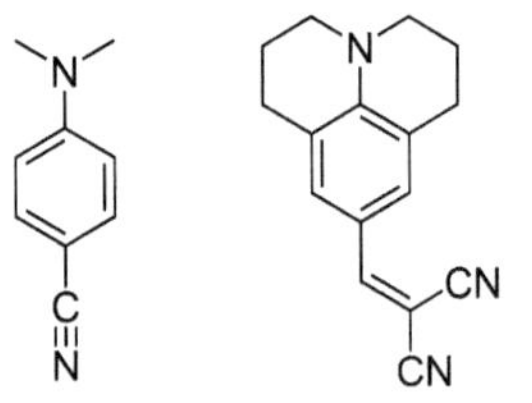

6.1   **6.2**

Figure 6.3   Structure of TICT
based sensors **6.1** [13],
6.2 [21].

intensity and another red-shifted emission band arises from
the TICT state. The ratio of the two emission intensities
(TICT/LE) can be used for quantification of the polarity [13,
22]. In comparison, **6.2**, whose TICT energy gap is too small to
allow photon emission, exhibits only a single emission band
from the LE state [5, 18, 19]; **6.2** is therefore sensitive to viscosity, not polarity.

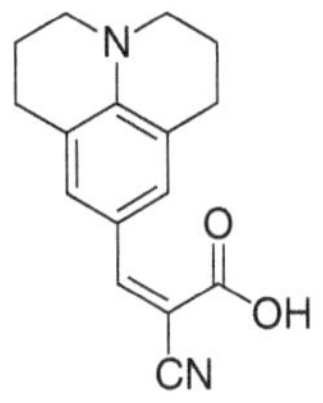

6.1.2.1   TICT-based Viscosity Sensors

There are three main classes of molecular rotors that have
been reported for use as viscosity sensors: vinyljulolidines,
BODIPY derivatives, and aldehyde-based systems. The
vinyljulolidine, **6.2**, is a well-studied molecular rotor for sensing viscosity, because its fluorescence properties are not
affected by solvent polarity (Figure 6.3) [21, 23]. The fluorescence emission of **6.2** increases with environmental viscosity
because the rotational restriction caused by the viscous
medium deactivates the non-radiative energy release pathway. However, its applications in studying viscosity in the
cellular microenvironment are limited due to its poor binding
capacity to biomolecules. 9-(2-Carboxy-2-cyanovinyl)juloli-
dine (**6.3**) was therefore developed to overcome this limitation
by replacing the dicyanovinyl group in **6.2** with
(2-carboxy-2-cyano)vinyl group as the electron acceptor, in
which carboxylic acid group serves as a reaction site for intro-

6.3

Figure 6.4   Structure of
TICT-based viscosity
sensor **6.3** [21].

ducing different targeting moieties (Figure 6.4) [21]; **6.3** derivatives have been applied to
detect viscosity in cell membranes [20, 21].

BODIPY-derived compounds substituted with a phenyl group at the 8-position are another
class of molecular rotor commonly employed for the measurement of viscosity
(Figure 6.5a) [8, 25–27]. Rotation of the *meso* phenyl ring around the single bond that
connects to the dipyrromethene core is dependent on the surrounding environment and

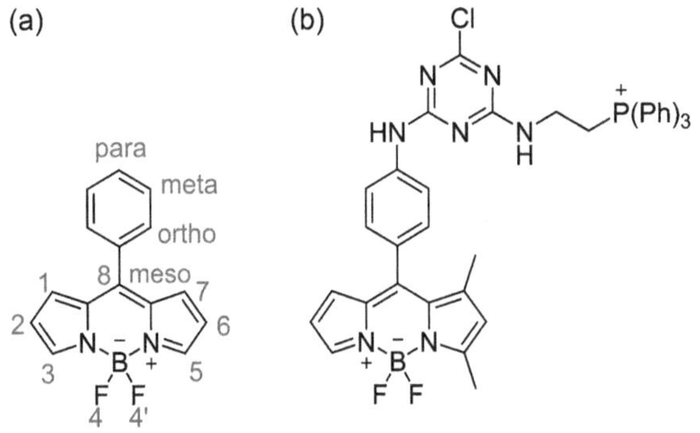

(a)   (b)

BODIPY   **6.4**

Figure 6.5   (a) Generic structure of 8-phenyl BODIPY, showing conventional substituent numbering
and naming. (b) Sensor **6.4**, a viscosity sensor bearing a mitochondrial targeting group tethered
through the *para*-position of the *meso* phenyl ring [24].

therefore affects the fluorescent output [28]. The *meso* phenyl ring rotates freely in low-viscosity environments, resulting in non-radiative decay upon excitation and resulting in weak fluorescence. On the contrary, when the rotation of the phenyl ring is inhibited in high-viscosity environments, fluorescence emission increases. The sensitivity of BODIPY rotors towards viscosity depends on the energy barrier between planar and twisted conformations [28]. A low energy barrier makes it easier to convert from planar conformation to twisted conformation, increasing the sensitivity to viscosity. The energy barrier is mainly affected by the substituents on the BODIPY core and the *meso* phenyl ring. Substituents at the 1- and 7-position of the BODIPY core or at the *ortho*-position of the phenyl ring sterically hinders and restricts the free rotation of the *meso* phenyl ring, making these BODIPY derivatives exhibit low sensitivity towards viscosity. On the other hand, substituents at the *meta*- and *para*-position of the phenyl ring do not induce restrictive effects on the rotation of the phenyl ring, therefore leading to high sensitivity to viscosity. Substitution at the *para*-position of the *meso* phenyl ring is a particularly favourable site for introducing targeting motifs or a second fluorophore, as this will not reduce viscosity sensitivity, no matter how large the targeting motif is [8, 25, 26]. For example, BODIPY-based sensor **6.4** bears a triphenylphosphonium mitochondrial targeting group at the *para*-position (Figure 6.5b) [24]; **6.4** exhibits a greater than 100-fold fluorescence enhancement moving from methanol to glycerol solution and was successfully used to report on mitochondrial swelling in HeLa cells induced by treatment with monensin.

Peng *et al.* reported a novel molecular rotor, **6.5**, where the rotation of an aldehyde group is key to its sensitivity to viscosity (Figure 6.6) [29]. The compound contains an aldehyde group at the *meso* position of the pentamethine chain. The presence of this electron-withdrawing group results in **6.5** exhibiting two emission peaks at 456 and 650 nm in ethanol. The emission at 650 nm is enhanced with increasing viscosity, due to restriction of aldehyde rotation. Meanwhile, the emission at 456 nm is insensitive to viscosity changes, therefore making **6.5** a ratiometric sensor for viscosity. It was used to imaging cytoplasmic viscosity in a number of cell lines using both confocal microscopy and fluorescence lifetime imaging microscopy (FLIM).

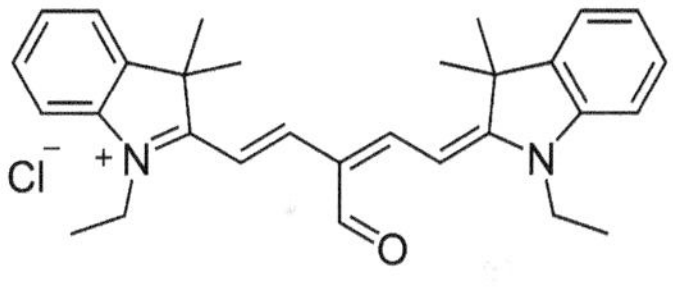

Figure 6.6 Ratiometric CHO-based molecular rotor **6.5** for viscosity sensing [29].

A significant challenge in developing TICT-based viscosity sensors is their tendency to also respond to solvent polarity changes. For example, sensor **6.6** is a push-pull dye based on dioxaborine (Figure 6.7) [30]. Sensor **6.6** shows red-shifted emissions with decreasing intensity as solvent polarity increases. In polar solvents, its emission intensity is highly dependent on the solvent viscosity due to the rotations of diethylamino phenyl ring. Responses towards both polarity and viscosity limit the use of **6.6** for measurement of cellular viscosity.

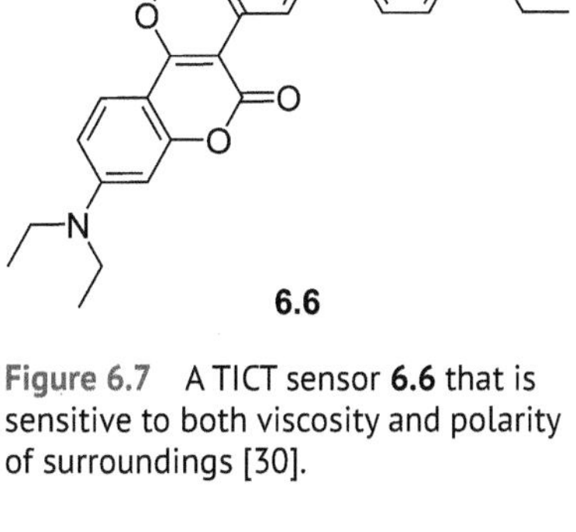

Another challenge in viscosity sensing is that the fluorescence of molecular rotors is also dependent on surrounding temperature, with molecular rotations

Figure 6.7 A TICT sensor **6.6** that is sensitive to both viscosity and polarity of surroundings [30].

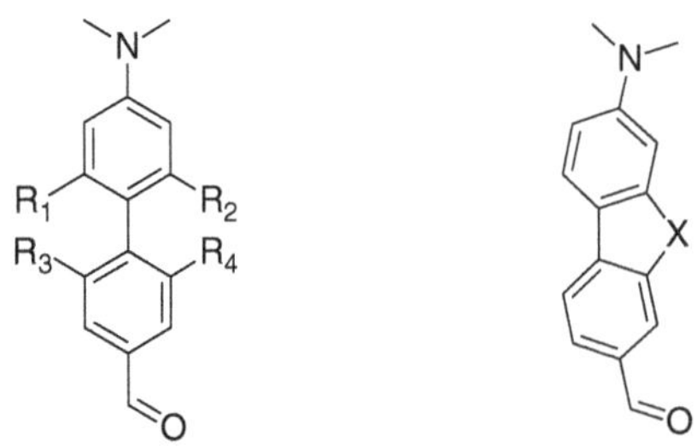

Figure 6.8 Structures of biphenyl analogues **6.7** and **6.8** with sensitivity to polarity [34].

and vibrations increasing at higher temperatures, therefore leading to higher non-radiative decay rates and a decrease in fluorescence intensity [7, 31]. While some molecular rotors, including **6.2** and **6.3**, are inherently insensitive to temperature changes [31–33], other systems do exhibit temperature dependence and therefore require that viscosity measurements should be taken in an environment with stable temperature [4, 33].

6.1.2.2 TICT-based Polarity Sensors

While the effect of polarity on the emission of molecular rotors is well established [18], the potential of such systems to sense polarity in cellular studies is yet to be exploited. A useful strategy for improving the sensitivity of TICT fluorophores to polarity is to pre-twist the geometry of the fluorophore by introducing steric hindrance around the twisting position [34]. Sasaki *et al.* developed a library of push-pull biphenyl analogues with different torsional restrictions by introducing methyl groups or a bridged structure at the *ortho*-position of the biphenyl junction, which reveals the effects of torsional restrictions on solvatochromic properties (Figure 6.8) [34]. The sensitivity of biphenyl derivatives to solvent polarity gradually increases as the biphenyl is restricted to a more twisted conformation [14, 34].

6.1.3 Polarity Sensors Based on Other Mechanisms

While the literature on viscosity sensing is predominated by TICT mechanisms, there are a number of other mechanisms adopted for the preparation of polarity sensors. The following sections explain these approaches, providing key examples of sensors used in biological studies.

6.1.3.1 Polarity Sensors Based on Intramolecular Charge Transfer Mechanism

Solvatochromic dyes are sensitive to the environment, in particular to solvent polarity [35]. Such dyes consist of an electron donor and an electron acceptor that are linked through π-conjugation or aromatic bridges, also known as push-pull systems [2, 35]. Upon photoexcitation, the charge is transferred from the electron donor to the electron acceptor, resulting in an increase in charge separation, which is accompanied by large changes in the dipole moments of the excited state [35]. This process is known as intramolecular charge transfer (ICT). Solvent dipoles can reorient around the dipole moments of the excited state, which reduces the energy level of the excited state and stabilises the molecule [16]. This solvent relaxation effect becomes stronger with increasing solvent polarity, creating a larger bathochromic shift in the excitation and emission of ICT fluorophores [36].

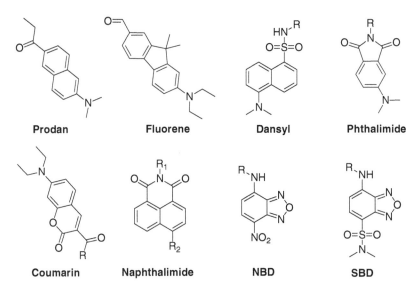

Figure 6.9 Representative classes of solvatochromic dyes.

N,N-Dialkylamines are common electron donors found in solvatochromic dyes. Carbonyl, nitro, sulfonyl amino, and imide are common electron acceptors, and a broad range of fluorophores exhibit solvatochromic properties and can therefore respond to environmental polarity (Figure 6.9) [1, 37–44]. Extension of the π-conjugation in push-pull molecules not only allows for red-shift of emission but also improves the sensitivity to polarity [16, 39].

Red-shifted emission are preferred in bioimaging, encouraging the development of near-infrared fluorescent sensors for detections of polarity. Sensor **6.9**, developed by Lu *et al.*, is constructed with an amine donor and a dicyanomethylenedihydrofuran acceptor, which is linked through the phenyl and thiophene rings (Figure 6.10) [16]. They showed that introduction of a phenyl linker led to red-shift fluorescence properties, with maximum emission at 664 nm. The commercially available polarity-sensitive **Nile Red** is another example of a sensor with an extended conjugated push-pull system [45].

Xiao *et al.* developed an ICT-based near-infrared fluorescent sensor **6.10** for detecting mitochondrial polarity, which is composed of a merocyanine scaffold, a difluoroboronate moiety and a lipophilic side chain (Figure 6.11) [2]. The tertiary amine serves as the electron donor, the difluoroboronate moiety serves as the electron acceptor, and the lipophilic side chain targets the mitochondria through anchoring cardiolipin that exclusively exists in the inner mitochondrial membrane. The strong push-pull effect through the long π-conjugated

Figure 6.10 Examples of ICT-based solvatochromic sensors **6.9** [16] and **Nile red** [45].

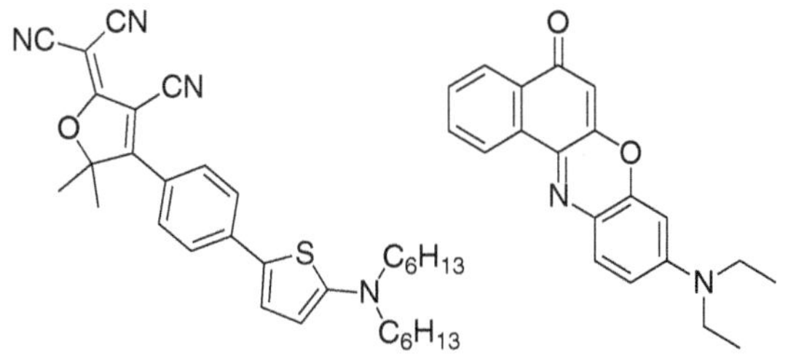

6.9

Nile red

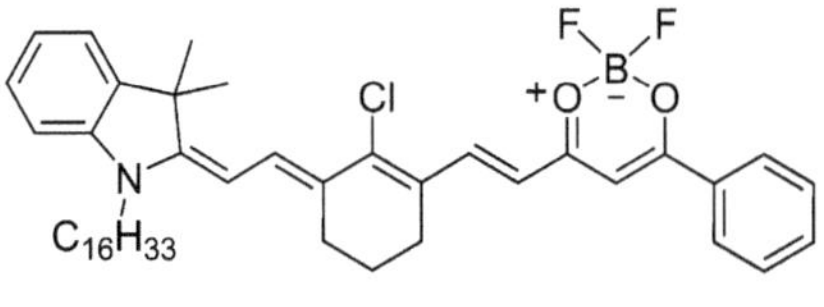

Figure 6.11 Structure of **6.10**, an ICT-based near-infrared fluorescent sensor for mitochondrial polarity [2].

system makes **6.10** highly sensitive to polarity, with a red-shift and decreased intensity in emission intensity as solvent polarity increases. Xiao *et al.* used **6.10** to reveal that the polarity in cancer cells is lower than that in normal cells. Sensor **6.10** was also used to show that the polarity of whole *Caenorhabditis elegans* at the embryonic development stage is lower than at the young adult stage.

6.1.3.2 Polarity Sensors Based on Excited-state Intramolecular Proton Transfer Mechanism

Excited-state intramolecular proton transfer (ESIPT) occurs in aromatic molecules that possess the ability to form intramolecular hydrogen bonds. Hydroxyl or amino groups usually act as the hydrogen bond donors, and carbonyl or aza nitrogen as the hydrogen bond acceptors [46, 47]. Representative ESIPT fluorophores are shown in Figure 6.12.

Fluorophores that undergo excited-state proton transfer exhibit normal (N*) and tautomeric (T*) bands that correspond to the enol and keto forms of molecules [53, 54]. ESIPT is highly sensitive to the surrounding environment, especially to polar and hydrogen bond donating solvents. These solvents form intermolecular hydrogen bonds with ESIPT molecule, resulting in the increase of the proton transfer barrier in the molecule and inhibiting ESIPT [46, 55]. Therefore, the ratio of emission intensities (N*/T*) increases as the solvent polarity increases, leading to a ratiometric response towards polarity.

The sensitivity of ESIPT fluorophores to solvent polarity depends on the substituents and the length of the π-conjugation system in the molecule [51, 56]. Introduction of electron donors into ESIPT fluorophores at appropriate positions and extension of the π-conjugation system not only shifts the emission to longer wavelength but also modulates their sensitivity to solvent polarity.

6.1.3.3 Polarity Sensors Based on Photoinduced Electron Transfer Mechanism

While photoinduced electron transfer (PET) is a useful strategy for constructing fluorescent sensors for environmental polarity [57], few polarity sensors have been reported based on this mechanism [58–61]. Sunahara *et al.* developed a series of BODIPY-based sensors

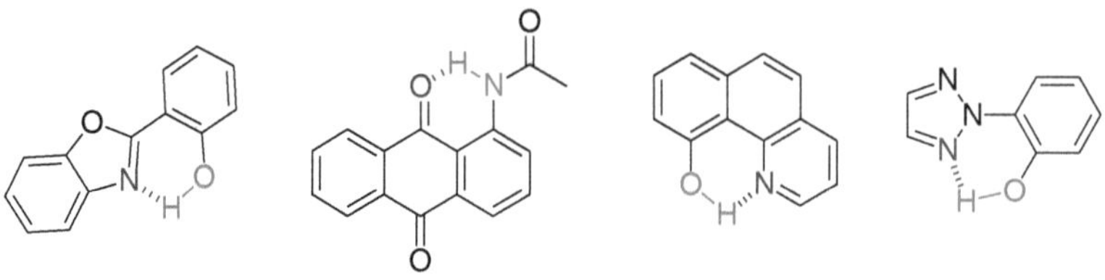

Figure 6.12 Representative ESIPT fluorophores [48–52]. Hydrogen bond donors are shown in red and hydrogen acceptors are shown in blue.

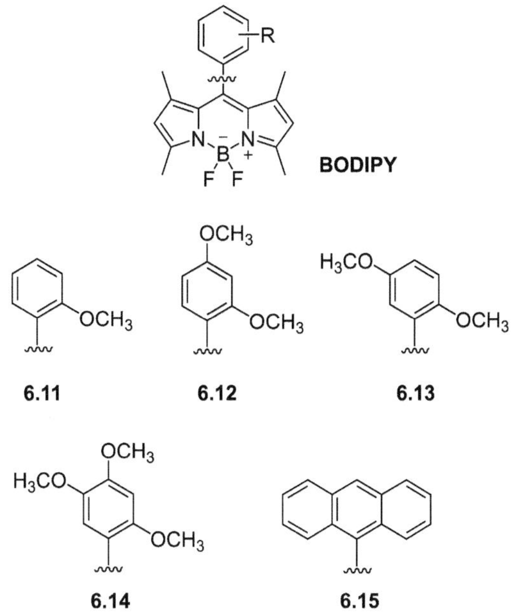

Figure 6.13 A library of BODIPY-based probes **6.11–6.15** for polarity, showing derivatives that were used to determine the polarity of BSA and of the internal cell membranes of HeLa cells [61].

(**6.11–6.15**), which exhibit emission switches in response to solvent polarity (Figure 6.13) [61]. Amongst their derivatives, the most sensitive five compounds were used in more detailed biological studies. Using this set of compounds, the authors concluded that the polarity at the surface of albumin is similar that of acetone. They also determined that the polarity of the internal membranes of HeLa cells was closest to the polarity of dichloromethane.

6.2 Fluorescent Sensors for pH

6.2.1 The Regulation of pH in Health and Disease

Protonation status dictates the charge and structure of macromolecules [62] and hence underpins biological activity. Cellular pH is key to determining this status. In the cytoplasm, the pH is maintained at a narrow range (pH 7.3–7.4) to carry out processes such as cell growth, proliferation, vesicle trafficking, apoptosis, and receptor-mediated signal transduction [63]. pH varies across different organelles: under physiological conditions, the pH in lysosomes is between 4.5 and 6.0, in the cytosol is 7.4, and in the mitochondrial matrix is approximately 7.5. The acidic environment of the lysosome is crucial in maintaining the activity of digestive enzymes to degrade endogenous and exogenous molecules [64]. The slightly basic environment in the mitochondria is a result of the proton gradient established by the electron transport chain [65].

Perturbation of cellular pH, usually in the form of acidification, is associated with impaired cellular functions, growth, and replication processes [66]. These are key characteristics of chronic diseases such as Alzheimer's disease and diabetes [66–68]. Renal failure is another condition implicated with pH imbalance, which can increase the risk of systemic acidosis that may be fatal in severe cases [69]. Additionally, tumours with high metastatic activity have been associated with an acidic tumour extracellular matrix

(TEM) [70]. Under moderate hypoxia, where tumour cells still have limited access to O_2, the acidic TEM is mainly facilitated by hypoxia-induced carbonic anhydrase IX expression [70]. Carbonic anhydrase IX catalyses the hydration of CO_2 and coordinates with acid exporters and bicarbonate importers to establish an alkaline intracellular pH and an acidic TEM [71].

pH can therefore be a diagnostic indicator for disease. Many fluorescent sensors capable of sensing intracellular pH have been developed to study pH-related disease pathology. This section introduces key concepts related to pH sensing and discusses design strategies for fluorescent pH sensors. Selected examples of sensors that have been successfully applied to cellular studies are also provided.

6.2.2 Considerations and Design Strategies for the Preparation of pH Sensors

Almost all pH sensors are designed by incorporating pH-sensitive groups into fluorophores in such a way that protonation influences the conjugation system of the fluorophores, causing a change in emission wavelength or intensity. In a sense, all fluorescent pH sensors are inherently weak acids or bases that have a varying degree of dissociation depending on the pH of the environment. The degree of dissociation is quantifiable by the logarithm of the acid dissociation constant, denoted by the pK_a. The pK_a of a fluorescent pH sensor determines its dynamic range and should be matched to the pH of the experimental system [72]. It is important to note that all ionisable protons will have their own pK_a values: by convention, the pK_a of a pH sensor tends to refer to the group for which protonation and deprotonation generates a fluorescence response.

Weak proton donors and acceptors having pK_a values in the relevant biological range are ideal pH-recognition moieties to be incorporated into fluorescent pH sensors. A number of suitable functional groups are summarised in Figure 6.14 [73]. Notably, functional groups such as carboxylic acids ($-CO_2H$), hydroxy ($-OH$), amines ($-NH_2$), and *N*-heterocycles are widely used in pH sensors [74]. A combination of these acidic and basic functional groups may be used to tune the sensing range of pH sensors.

The lone electron pairs on these pH sensing receptor groups often have high affinities for metal ions. Sensors that comprise multiple electronegative atoms are prone to complexation to metal ions, leading to interference or poor pH sensitivity (as discussed in Chapter 4).

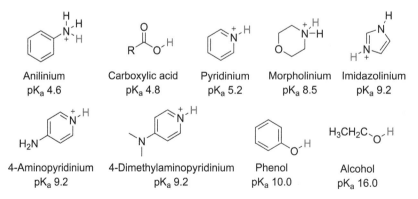

Figure 6.14 A summary of functional groups with their respective pK_a values.

Metal coordination is favoured by the formation of five- or six-membered rings, so pH sensors that avoid this arrangement of donor atoms will minimise selectivity issues [74].

6.2.2.1 Methods to Control pK$_a$

Fluorogenic pH sensors with a low pK$_a$ have capacities to sense acidic environments in stressed or diseased cells. Some examples of studies that have successfully decreased the pK$_a$ of fluorescent sensors are summarised below.

The substitution of halogens has been an effective method to decrease the pK$_a$ of fluorophores. The electron-withdrawing halogens can draw electron density away from the fluorophore. When substituted in positions that are responsible as an electron-donor in an ICT-based sensor, halogens act to decrease the electron density of the ICT donor and consequently increasing the fluorophore's acidity. For instance, in fluorescein-based scaffolds where a spirolactam ring-opening leads to subsequent fluorescent response, introducing fluorine into the 2′ and 7′ positions of the xanthene ring of **6.16** lowered the pK$_a$ from 6.3 to 4.8 (Figure 6.15a) [75]. A similar effect was observed in **6.17**, in which chlorine was substituted in the same position, lowering the pK$_a$ to 4.5 (Figure 6.15a) [76]. The effect of halogen substitution in lowering pK$_a$ is also observed in other fluorophores including coumarin **6.18**–**6.20** (Figure 6.15b) and BODIPY **6.21** and **6.22** (Figure 6.15c) [77, 78].

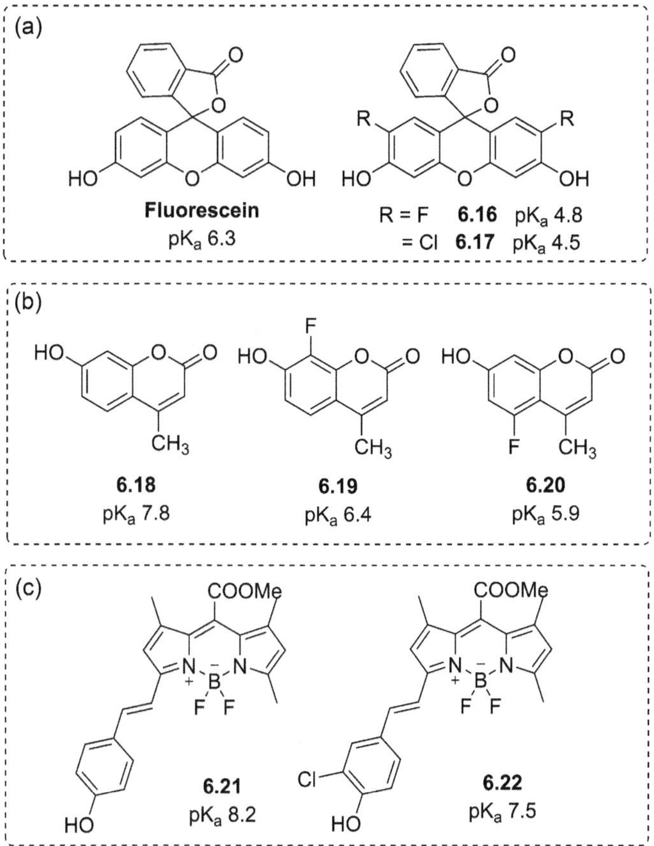

Figure 6.15 Examples of fluorophores **6.16**–**6.22** where the introduction of halogens was used as a method to lower the pK$_a$ (a) fluoresceins [75, 76] (b) coumarins [77] and (c) BODIPYs [78].

Other electron withdrawing groups such as 1-phenyl-piperidine and phenyl piperazine can also decrease the pK$_a$, for example as shown in the 3′-mono-substituted BODIPYs **6.23** and **6.24** (Figure 6.16a) [78]. In addition to the inductive effect, the electron-withdrawing ability of such groups means that electrons can be delocalised over more atoms, therefore stabilising the conjugate base of the acid. This promotes the dissociation of the acid, which is characteristic of a stronger acid. Furthermore, adding electron withdrawing substituents on the *meso* phenyl ring of the BODIPY **6.25–6.28** has also been shown to cause a decrease in pK$_a$, which is an effect modulated *via* PET (Figure 6.16b) [79]. Examples of PET-based fluorescent pH sensors suitable for biological imaging will be given in Section 6.2.3.1. Decreasing the ICT donor ability through substitution at the *N*-atom of xanthene core with methylpiperazine (**6.29**) and benzylpiperazine (**6.30**) are also effective at decreasing the pK$_a$ (Figure 6.16c) [80].

Strategies to raise pK$_a$ can also be useful for optimising pK$_a$ values that are too low. Generally, introducing electron-donating groups to the ICT donor raises the pK$_a$. This can be observed in a xanthene-based scaffold in instances where ethyl groups were substituted to the 2′ and 7′ position of the xanthene core of fluorescein **6.31** that slightly raise the pK$_a$ to 6.4 (Figure 6.17a) [81]. Oxygen to carbon substitution at the 4′ position of the xanthene core also raises the pK$_a$ from 6.4 to 7.4 as it raises the electron-donating ability of the ICT donor (**6.32**, Figure 6.17a) [82]. Replacing the 4′ position with a more electron-donating silicon further

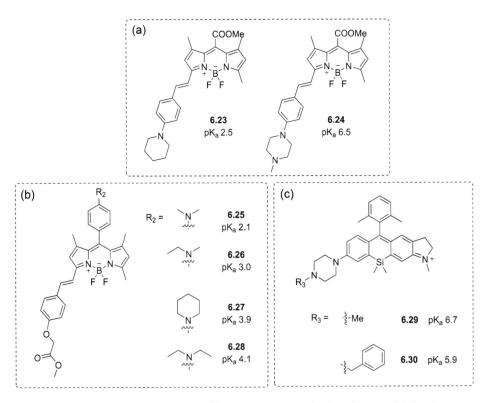

Figure 6.16 Examples of fluorophores **6.23–6.30** where substituting electron-withdrawing groups lowers pK$_a$ of fluorophores in (a) 3′-mono-substituted BODIPYs [78], (b) at the *meso* phenyl ring of BODIPYs [79], and (c) at the *N*-atom of the xanthene core [80].

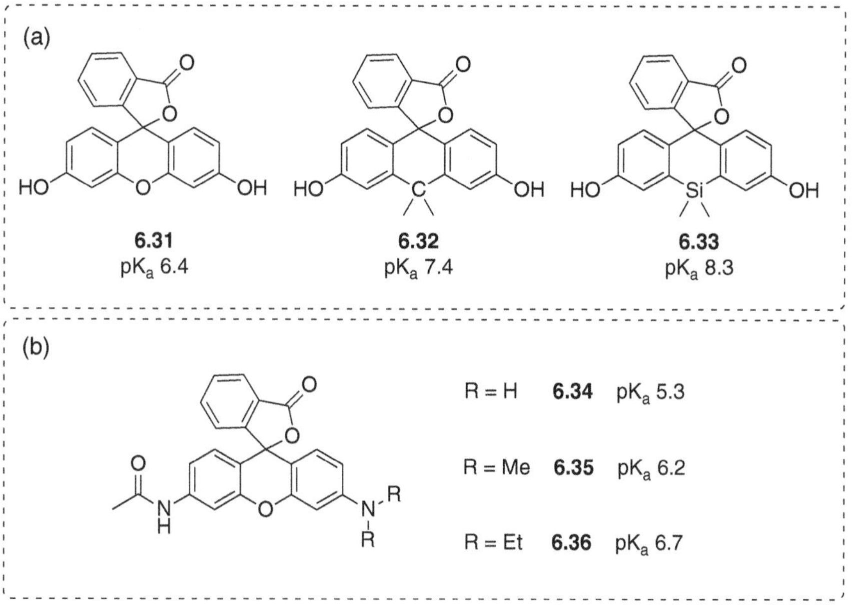

Figure 6.17 Examples of fluorophores **6.31–6.36** where substitution with electron-donating groups raises the pK$_a$ of fluorophores, (a) at the 4′ position of the xanthene core [81, 82] and (b) at the N-atom of the xanthene core with alkyl groups [83].

raises the pK$_a$ higher to 8.3 (**6.33**, Figure 6.17a) [82]. Additionally, mono-acetylated hydroxy-methyl rhodamines subjected to *N*-alkyl substitution with electron-donating groups increases the pK$_a$ from 5.3 to 6.2 (**6.34** *vs.* **6.35**, NH$_2$ *vs.* *N,N*-dimethylamine), and from 5.3 to 6.7 (**6.34** *vs.* **6.35**, NH$_2$ *vs.* *N,N*-diethylamine) (Figure 6.17b) [83].

6.2.3 Examples of pH Sensors

There are many pH-responsive fluorescent sensors that are commercially available or reported in the literature. Here, we review a range of approaches to developing pH sensors, with key examples of sensors used in biological studies.

6.2.3.1 Photoinduced Electron Transfer as a Mechanism for Sensing

Fluorescent pH sensors are often based on reversible protonation events, and these commonly operate *via* a photoinduced electron transfer (PET) mechanism [84]. The amine group is a good PET donor as it has a lone pair of nonbonding electrons that can donate electrons to the fluorophore in its first excited singlet state, effectively preventing the excited electron from undergoing radiative decay back to the ground state and resulting in non-radiative energy loss (Figure 6.18a) [84]. During amine protonation, this lone electron pair accepts a hydrogen ion, and the energy of the lone pair decreases due to a bonding interaction. As a result of this, the energy level of the pH receptor in its protonated state is lower than the HOMO of the fluorophore, hence it can no longer participate in PET and the fluorescence response is restored (Figure 6.18b).

Hall *et al.* reported **6.37** and **6.38**, two PET-based pH sensors containing an *aza*-BOPIDY scaffold (Figure 6.19); **6.37** and **6.38** have a diethylamine and morpholine groups, respectively,

(a) **OFF**

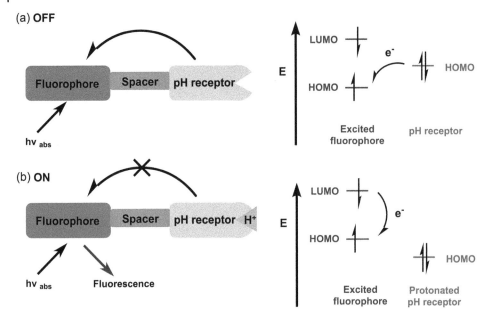

(b) **ON**

Figure 6.18 PET mechanism between a fluorophore and pH receptor in the (a) OFF unprotonated state and (b) ON protonated state.

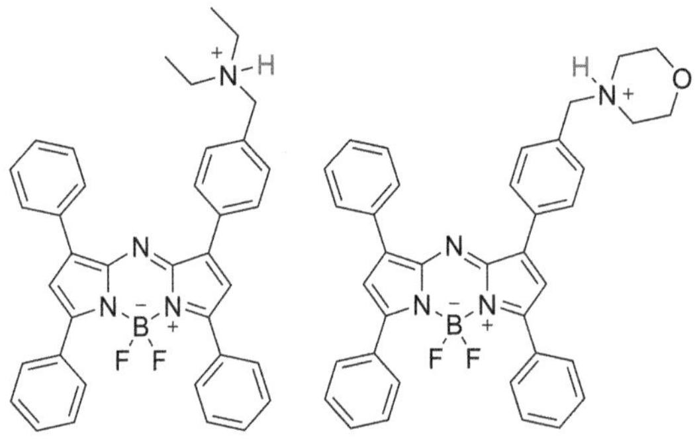

6.37   **6.38**

Figure 6.19 Examples of PET-based pH sensors **6.37** and **6.38** on the *aza*-BODIPY scaffold [85].

that facilitate pH sensing, with pK_a values of 6.9 and 4.8, respectively [85]. The fluorescence process is PET-quenched in alkaline solution, while in acidic solution, PET is inhibited, and the sensors display far-red absorbance and fluorescence. They were used *in vitro* cellular imaging with HeLa cells. The localisation of **6.37** was determined to be in the cytoplasm through Z-stack image analysis.

Tang and co-workers used a *N*-heterocycle called terpyridine as a pH-sensing moiety [86]. Their tricarbocyanine sensor **6.39** has three nitrogen atoms that may participate in PET quenching, where the lone pair electrons of the terpyridine group can quench the fluorescence of the tricarbocyanine (Figure 6.20) [86]. This sensor gives near-infrared fluorescence in acidic solution, where protonation of these nitrogen atoms leads to restoration of the

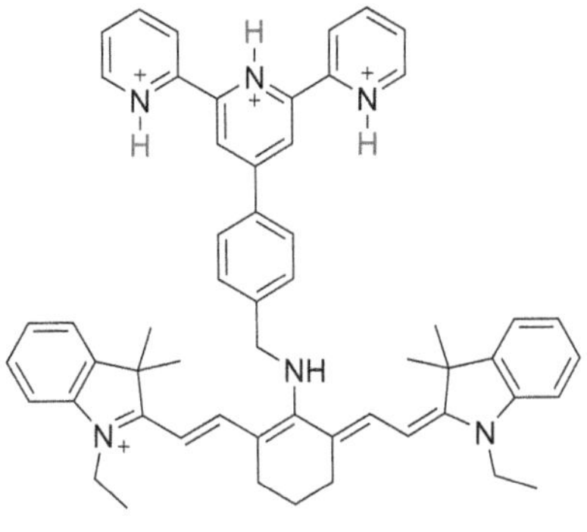

Figure 6.20 PET quenching using terpyridine as pH receptor in **6.39** [86].

6.39

fluorescence; **6.39** has a pK$_a$ of 7.10 and was shown to have a strong linear dependency to pH fluctuations within the range of 6.70–7.90. It was tested in live HepG2 and HL-7702 cells.

6.2.3.2 The Ring Opening of Rhodamines as a Mechanism for pH Sensing

Rhodamine-based sensors are often favoured as fluorophores for their high resistance to photobleaching, high water solubility, and high fluorescence quantum yields (see Chapter 1). In particular, their inherent pH sensitivity makes them suitable for the development of fluorescent pH sensors [84]. In near-neutral aqueous solvent, rhodamines exist in a spirolactam form, which is colourless and non-fluorescent (Figure 6.21a). In acidic aqueous solution, the protonation of rhodamines leads to spirolactam ring opening, resulting in the appearance of a pink colour and a fluorescence turn-on. The pH-dependent ring opening of rhodamines has therefore been exploited over the past decades for the preparation of fluorescent pH sensors for biological imaging.

As an ICT-based fluorophore where the zwitterion is the fluorescent form, the stability of the rhodamine zwitterion determines the pK$_a$. Chi *et al.* has demonstrated that modifying the R$_1$ – R$_4$ substituents can decrease the ICT 'push-pull' effect to favour the ring-closed form of the xanthene, which consequently lowers the rhodamine's pK$_a$ (Figure 6.21b) [87]. R$_1$ and R$_2$ are substituents that serve as electron donors in the push-pull model, while R$_3$ and R$_4$ are the electron acceptors. Substituting R$_1$ and R$_2$ with electron-withdrawing groups will decrease the push-pull effect, lowering the pK$_a$ value. For instance, decreasing the electronegativity of R$_2$ from oxygen to carbon to silicon leads to a decreased pK$_a$ value. Conversely, R$_3$ and R$_4$ positions will need to be substituted with electron-donating group to decrease the push-pull effect and lowering the pK$_a$.

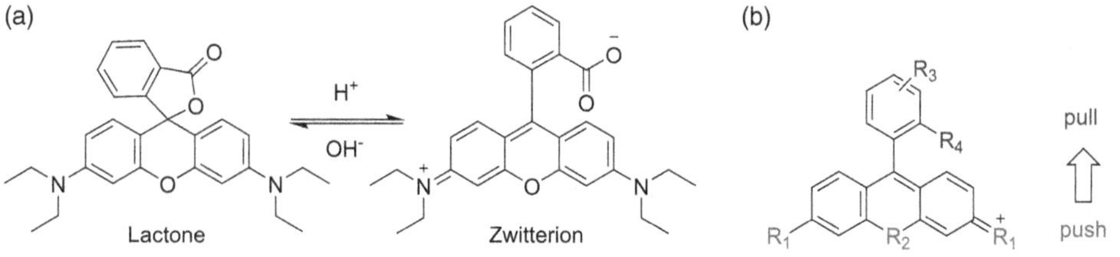

Figure 6.21 (a) Lactone-zwitterion equilibrium of rhodamine, where the zwitterion is the fluorescent form; (b) four representative substitution sites of xanthene derivatives that affect the lactone-zwitterion equilibrium [87].

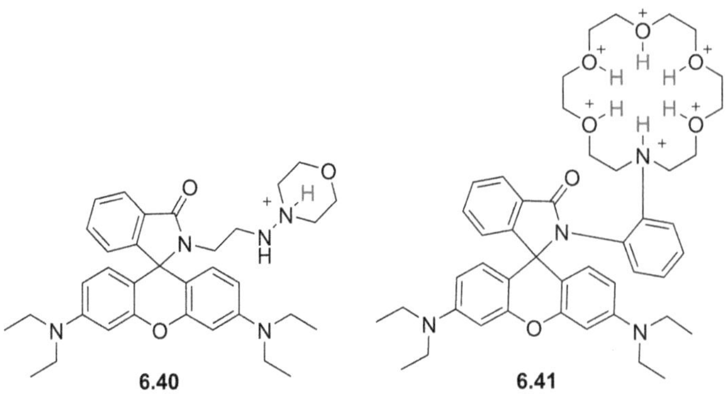

Figure 6.22 Examples of fluorescent pH sensors **6.40** and **6.41** based on the ring-opening of rhodamines [88, 89].

Wang *et al.* reported **6.40**, which employs the rhodamine spirolactam ring-opening process to sense the pH environment (Figure 6.22) [88]. With a pK_a of 5.42, **6.40** demonstrated linear fluorescence response from pH 5.00 to 6.00. The sensor also contains and a lysosome-localising morpholine group, and confocal fluorescence microscopy experiments in HeLa cells showed that lysosomal pH increases during rapamycin-induced mitophagy. Colocalisation experiments with MitoTracker Green and Lysotracker Red DND-99 confirmed that the increase in lysosomal pH occurs after the endocytosis of mitochondria into the lysosome.

Another example of the use of rhodamine ring-opening for pH sensing is **6.41**, which exhibits an extended pH sensing range through the incorporation of *N*-(2-aminophenyl)-*aza*-18-crown-6 group (Figure 6.22) [89]. The multiple proton acceptors in the *aza*-18-crown-6 group are key to the wide sensing range. Although it has a pK_a of 4.10, **6.41** exhibits a linearity of fluorescence response towards pH changes from pH 4.00 to 7.00. Since **6.41** localises in the lysosome, an artesunate-induced decrease in lysosomal pH and dexamethasone-induced increase in lysosomal pH were visualised and quantified *via* confocal imaging.

6.2.3.3 Intramolecular Charge Transfer as a Mechanism for Ratiometric pH Sensing

Fluorescent sensors that operate on a PET-based mechanism and spirolactam ring-opening tend to be turn-on sensors. In contrast, ratiometric sensors offer advantages for cellular imaging applications including concentration-independence of response, as outlined in Chapter 1. Since protonation and/or deprotonation leads to a change in electron density, installation of the pH-sensitive group into the conjugated system of the fluorophore can give rise to a sensor that changes fluorescence emission wavelength when electron density varies with pH.

The protonation of quinoline can also be used to sense pH. Foley and co-workers reported a lysosome-targeted morpholine pH sensor, **6.42**, that displays a ratiometric response to pH changes (Figure 6.23) [90]. Protonation of the quinoline nitrogen of **6.42** creates a quaternary ammonium with a greater electron-withdrawing effect, leading to enhanced ICT and a red-shifted fluorescence. The ratio of fluorescence intensity at 613 and 560 nm (I_{614nm}/I_{560nm}) increases as pH decreases with good linearity in the pH 4.0–6.0 range, making the probe suitable for quantifying lysosomal pH. Because of its low cytotoxicity and good photostability,

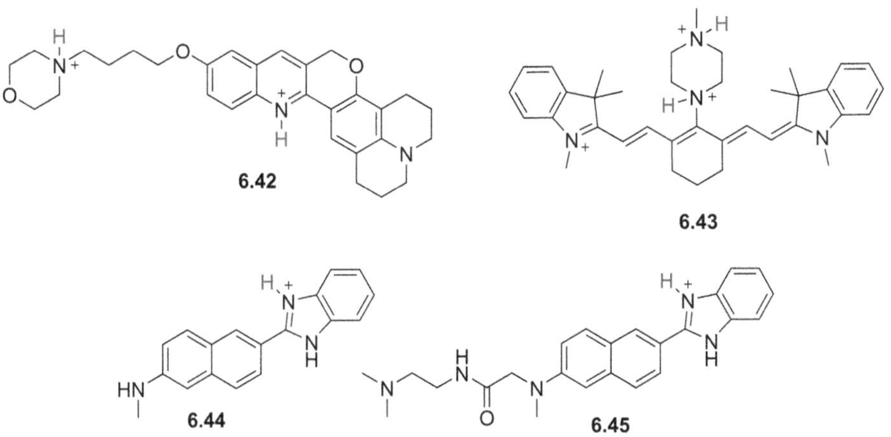

Figure 6.23   Examples of ratiometric ICT-based pH sensors **6.42–6.45** [90–92].

6.42 was applied for lysosomal detection in HeLa cells equilibrated to different pH levels using nigericin. Furthermore, it was able to detect the effect of a malaria drug chloroquine in increasing lysosomal pH [91].

Nagano and co-workers developed a series of excitation-ratiometric near-infrared fluorescent pH sensor based on a tricarbocyanine framework, with the methylpiperazine-substituted **6.43** exhibiting the most desirable reversibility and fluorescence turn-on with a pK_a of 6.8 (Figure 6.23) [93]. Under acidic conditions, protonation occurs at the two nitrogen atoms on the piperazine, which decreases their electron-donating ability causing a red-shift of the absorbance peak. The ratio of emission at 780 nm when excited at 750 and 670 nm exhibits linearity from pH 4.5 to 6.0. The treatment of **6.43**-dosed HeLa cells with basic ammonium chloride revealed that the pH value of lysosome changed from 4.7 to 6.2 in confocal microscopy.

Kim and co-workers reported a benzimidazole-naphthalene-based, two-photon, ratiometric fluorescent sensor for acidic pH in live cells and tissues, **6.44** (Figure 6.23) [92]. Protonation of the nitrogen atom on the benzimidazole leads to red-shifted emission, giving a ratiometric response; **6.44** was found to localise in the cytosol and exhibits a pK_a of 5.34, with potential to measure acidic membrane compartments of the cytosol. A lysosomally targeted analogue, **6.45**, was also prepared with a pK_a of 5.63 [92]. Both **6.44** and **6.45** were used to visualise pH distribution in rat brain tissues.

6.2.3.4   pH Sensors Based on Addition Reactions

While the previous sections have described pH-sensing mechanisms that involve addition or removal of a proton, another strategy for pH sensing is the addition of a hydroxyl group (–OH). Tang and co-workers reported a ratiometric sensor **6.46** based on tetraphenylethene-cyanine, with claims of a broad pH sensing range (Figure 6.24) [94]. The sensor works through exploiting the abundant hydroxide groups in basic environments, which can react with the carbon-nitrogen double bond on the indolinium unit in an addition reaction. This disrupts the hemicyanine conjugation and displays a turn-on of blue fluorescence at 489 nm between pH 10 and 14. From pH 7 to 10, the sensor is non-fluorescent, possibly due to quenching effect by the anionic sulfonate group. Lowering the pH further leads to protonation of

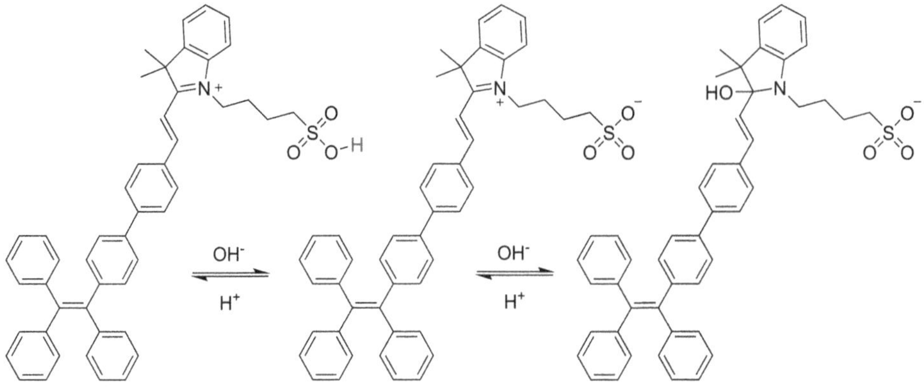

Figure 6.24 Acid–base equilibrium of sensor **6.46**. Under acidic environment, it exhibits red emission, while in basic environment, a blue emission can be observed [94].

the sulfonate group, which alleviates the quenching effect and displays red emission from the highly conjugated molecule. Colocalisation studies of **6.46** with LysoTracker Green confirms its localisation in the lysosome. It was used to visualise the effects of acetic acid and ammonium chloride treatments on lysosomes. While there are multiple protonation sites, only a single pK_a of 6.42 was reported. Despite this, **6.46** was used to measure pH between 4.7 and 8.0 in the lysosomes, in addition to excellent biocompatibility, cell permeability, and photostability. High-throughput analysis using flow cytometry was employed for live cell pH sensing, which further demonstrated the practicality of **6.46** as a lysosomal ratiometric pH sensor.

6.3 Fluorescent Redox Sensors for Biological Studies

6.3.1 The Regulation of Redox State in Health and Disease

The balance of oxidants and antioxidants is essential in regulating cellular signalling processes [93, 95]. Amongst all endogenous sources of reactive oxygen species (ROS), mitochondria are the major sites of ROS production [65]. In mitochondria, the production of ROS begins with the leakage of electrons from the electron transport chain, which can reduce molecular oxygen to produce superoxide as the initial mitochondrial ROS [96, 97]. Superoxide then rapidly undergoes a dismutation reaction by the enzyme superoxide dismutase to form hydrogen peroxide, which inherently is more stable than superoxide and has the ability to cross the cell membrane. In the cytosol, hydrogen peroxide can perform reversible modifications of proteins such as protein phosphatases, protein kinases, and transcription factors.

ROS have been found to regulate important signalling processes, including stem cell differentiation [98], cardiac remodelling [99], maintenance of mitochondrial morphology [100, 101], and apoptosis [102]. While cells have an effective antioxidant system to negate the harmful effects of ROS, excessive ROS can react deleteriously with proteins, membranes, and DNA, leading to impaired function [97, 103]. The dysregulation of redox state is known as oxidative stress and has been shown to be implicated in many diseases including diabetes mellitus [104], cardiovascular disease [104], Alzheimer's disease [105], and certain cancers [106]. Thus, redox regulation and dysregulation have great implications on cellular health. There is an ongoing demand to understand redox processes to study both disease pathology and development of drugs for mitochondrial associated disease.

Irreversible, reaction-based sensors for a specific ROS provides quantitative end-point data as to how much a specific ROS is produced in a biochemical pathway. The design principles and examples of such sensors are discussed in Chapter 5. On the other hand, reversible sensors to probe global redox state can enable study of transient oxidation and reduction events that are typical of physiological processes. Reversible sensors for specific ROS are also desirable to provide information about the individual role of each ROS in biochemical events. A combination of irreversible and reversible sensors can be used in combination to better understand the physiological and pathological roles of cellular redox states. Here, we discuss key design strategies of reversible fluorescent redox sensors and provide key examples of sensors that have been successfully applied to cellular studies.

6.3.2 Design Strategies of Fluorescent Redox Sensors and Key Examples

6.3.2.1 Redox Sensors Based on the Nitroxyl Radical/Hydroxylamine Redox Couple

Nitroxides are stable cyclic radicals and effective antioxidants that can react with ROS [107]. They undergo redox switching between the diamagnetic hydroxylamine and paramagnetic nitroxide-radical states. Fluorescent sensors can be developed by tethering the nitroxide radical to a fluorophore of choice. The nitroxide radical quenches fluorescence, and the reversible reduction to the diamagnetic hydroxylamine will lead to restoration of fluorescence (Figure 6.25). These are often known as profluorescent nitroxide (PFN) sensors, as healthy biological systems are reductive and these sensors would first exist in the fluorescent hydroxylamine form, before reacting with ROS to give a turn-off response.

While many PFN sensors use tetramethyl nitroxide as the redox active moiety, more recently, tetraethyl nitroxides have been shown to have extended lifetimes and lower reduction rates *in vitro* [108] and *in vivo* [109], providing better sensitivity for highly reducing environments such as the mitochondria [110]. A prime example is the rhodamine-based PFN sensor **6.47** (Figure 6.26) [111]. This sensor could report on rotenone- and antimycin-induced ROS production in a dose-dependent matter in hTERT-immortalised cells. The reversible response was confirmed through multiple cycles, using succinate as a respiratory substrate to induce a reducing environment followed by aeration with oxygen gas to oxidise the environment. Furthermore, as the nitroxyl radical and hydroxylamine form display different fluorescence lifetimes, FLIM was proposed as a ratiometric output to monitor

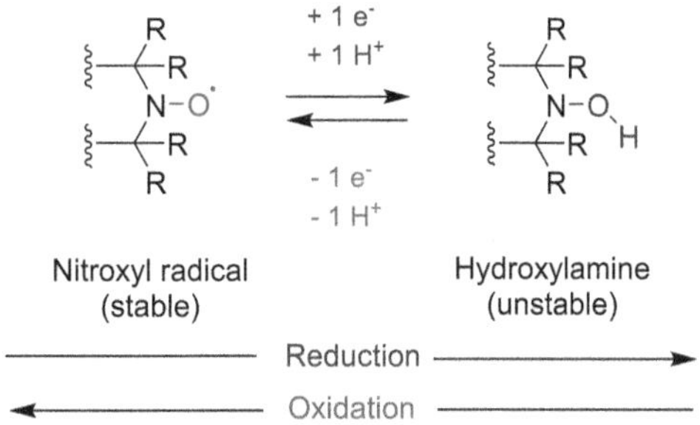

Figure 6.25 The reversible reduction and oxidation of the nitroxyl radical.

Nitroxyl radical (stable)

$+ 1\,e^-$
$+ 1\,H^+$

$- 1\,e^-$
$- 1\,H^+$

Hydroxylamine (unstable)

——————— Reduction ———————→

←——————— Oxidation ———————

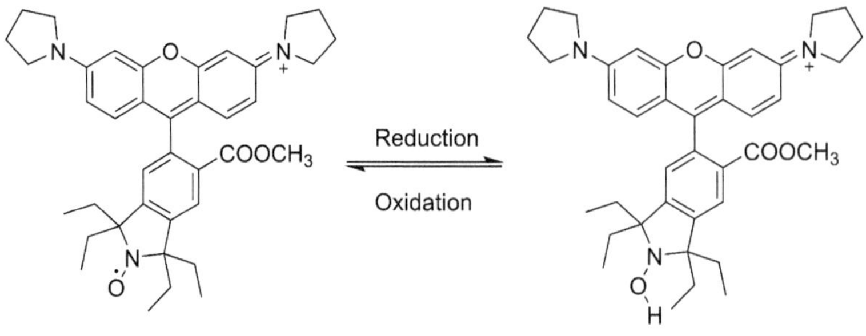

6.47   hydroxylamine form

Figure 6.26 Example of nitroxyl-based sensor in its radical form **6.47** and its fluorescent hydroxylamine form [111].

mitochondrial redox state. The redox response of **6.47** is an example that reflects the overall redox status, rather than being selective for a specific ROS.

6.3.2.2 Redox Sensors Based on the Quinone/Hydroquinone Redox Couple

Quinone/hydroquinone is another reversible redox couple that has been employed as a redox-sensitive group. The fluorescence of a quinone-based redox sensor is modulated through PET quenching [57], with quinone acting as the PET acceptor. The quenching effect is removed upon reduction of the quinone to hydroquinone, which cannot participate in PET quenching.

Tang and co-workers reported a quinone-based fluorescent sensor **6.48** that is selective for superoxide and GSH (Figure 6.27) [112]; **6.48** is oxidised by superoxide to generate the ketone that has 1000-fold greater fluorescence. It could be reduced back to **6.48** using glutathione (GSH) *in vitro*. The reversible sensing of superoxide was demonstrated in HepG2 cells using ROS-inducer phorbol-12-myristate-13-acetate (PMA) and the reductant GSH; **6.48** is also suitable for dynamic tracking of superoxide *in vivo* and was used to visualise changes in superoxide concentration during reperfusion injury in hepatocytes, zebrafish, and mice.

Similarly, Zhang *et al.* reported a red-shifted two-photon fluorescent sensor **6.49** for the reversible imaging of superoxide/glutathione (Figure 6.28) [113]. The sensor utilises a through-bond energy transfer (TBET) mechanism in a naphthalene-BODIPY system to obtain a longer emission wavelength, enabling deeper imaging penetration depths and minimised background fluorescence. There is a 19-fold fluorescence enhancement upon oxidation. Reversible cycling was achieved with PMA and GSH both *in vitro* and *in vivo* with RAW264.7 cells and in rat liver tissue slices.

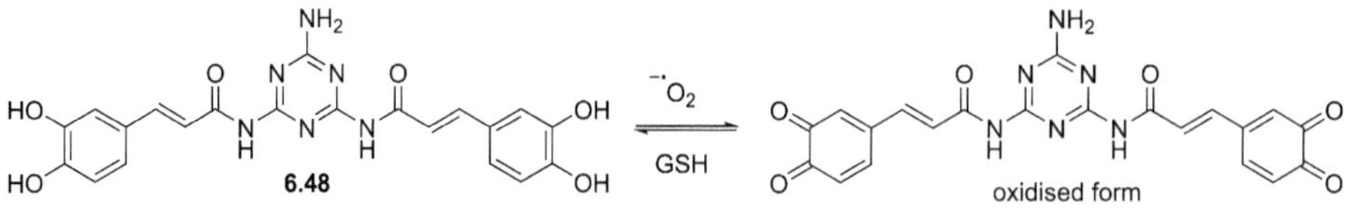

Figure 6.27 Example of quinone/hydroxylamine-based sensor; **6.48** is the reduced non-fluorescent form while the oxidised form is fluorescent [112].

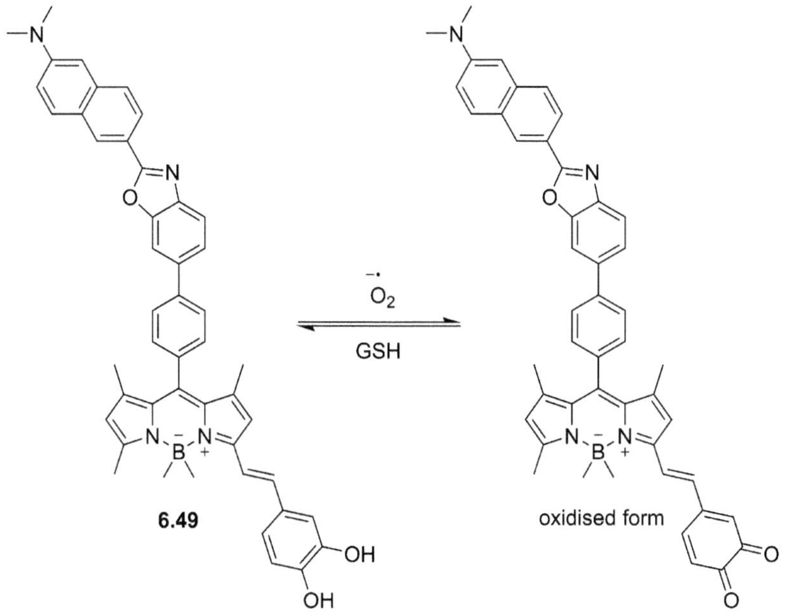

Figure 6.28 An example of two-photon quinone-based sensor **6.49** and its fluorescent oxidised form [113].

6.3.2.3 Redox Sensors Based on Chalcogens

Chalcogens (sulfur, selenium, and tellurium) comprise an important class of redox sensing groups. The ebselen moiety is a glutathione peroxidase mimic that uses glutathione to catalyse the breakdown of hydrogen peroxide [114]. It is a five-membered ring containing a redox active Se-N bond, which is cleaved upon reduction. This property is ideal in fluorescent redox sensors as switching between on and off states often requires the forming and breaking of conjugated systems. For instance, in **6.50**, a cyanine dye is tethered to ebselen, with the cyanine acting as both the fluorescent reporter and mitochondrial-targeting group, and the ebselen as a reversible redox sensor (Figure 6.29) [115]. The reduced sensor is non-fluorescent, with a proposed mechanism to have generated a radical selenium species that can participate in PET quenching of the cyanine. The oxidised sensor prevents this quenching and is fluorescent; **6.50** was able to report on mitochondrial redox state in HepG2

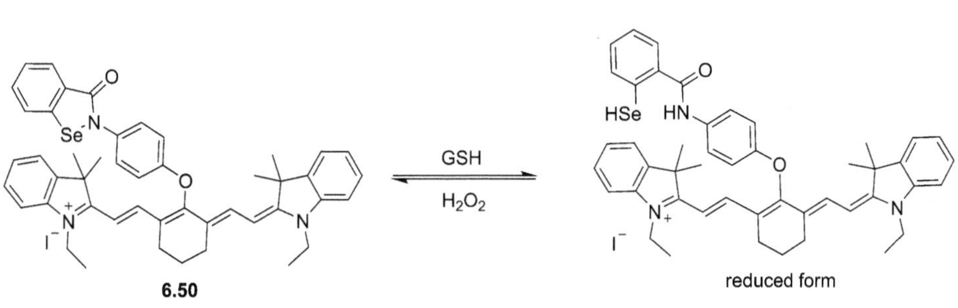

Figure 6.29 An example of chalcogen-based sensor **6.50** and its non-fluorescent reduced form [115].

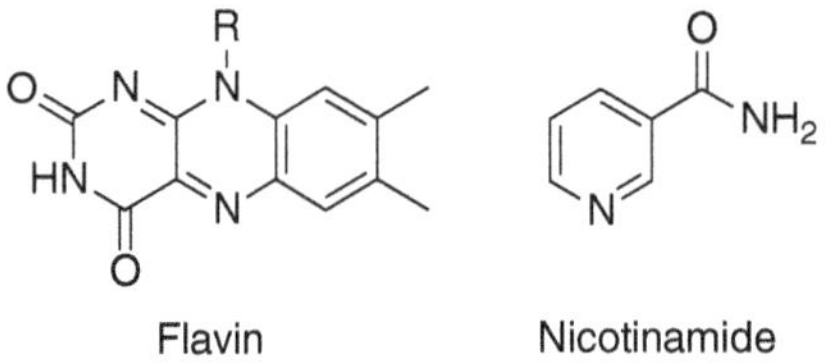

Figure 6.30 Generic structures of flavin and nicotinamide.

Flavin   Nicotinamide

cells under buthionine sulfoximine (BSO)-induced oxidative stress. When BSO was removed, the fluorescence intensity decreased to pre-BSO treatment levels, indicative of the recovery of HepG2 cell from oxidative stress.

6.3.2.4 Redox Sensors Based on Flavins and Nicotinamides

Using redox active molecules that are already employed in biological systems is another method for the design of redox sensors. Flavins and nicotinamides (Figure 6.30) act as redox-active co-factors and co-enzymes in cellular redox reactions and are ubiquitous in biological systems in their nucleotide derivatives form, flavin adenine dinucleotide (FAD) and nicotinamide adenine dinucleotide (NAD), respectively [116]. Both redox active centres can be reversibly oxidised and reduced, but with different reduction potentials (−316 and −219 mV *vs.* SHE, respectively) [117]. Sensors based on flavins and nicotinamides will tend to be oxidised by a broad range of ROS and therefore act as global redox sensors, rather than showing specificity to individual ROS.

New and co-workers have contributed to this field by developing a set of fluorescent sensors based on flavin, utilising the fact that flavins adopt a planar structure in oxidised form, which is fluorescent, while reduction gives rise to a bent, non-fluorescent form. They first reported **6.51**, a naphthalimide-flavin conjugate with a red-shifted fluorescence emission (545 *vs.* 531 nm) compared to native riboflavin (Figure 6.31) [118]; **6.51** exhibited a greater than 100-fold decrease in fluorescence intensity upon reduction along with excellent reversibility. It was used to monitor oxidative stress in 3T3-L1 adipocytes treated with different concentrations of glucose; **6.52** is an analogue that uses a triphenylphosphonium group for mitochondrial targeting [119]. In addition to its utility in confocal imaging, it was able to demonstrate through flow cytometry analysis that different oxidative capacity exists in the bone marrow, thymus, and spleen of mice.

A ratiometric redox sensor **6.53** was also reported (Figure 6.32) [120]. It comprised of a coumarin and flavin fluorophore as the donor–acceptor FRET-pair. In its reduced form **6.53**, the FRET process is inhibited, and blue donor emission is observed; while in its oxidised

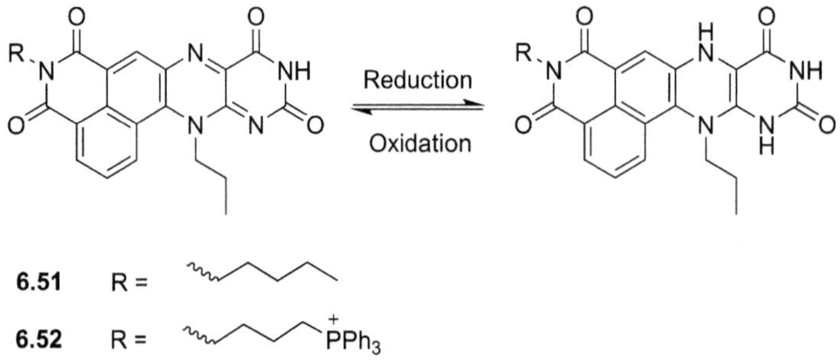

Figure 6.31 Flavin-based redox sensors **6.51** and its mitochondrial analogue **6.52**. The oxidised form is fluorescent [118, 119].

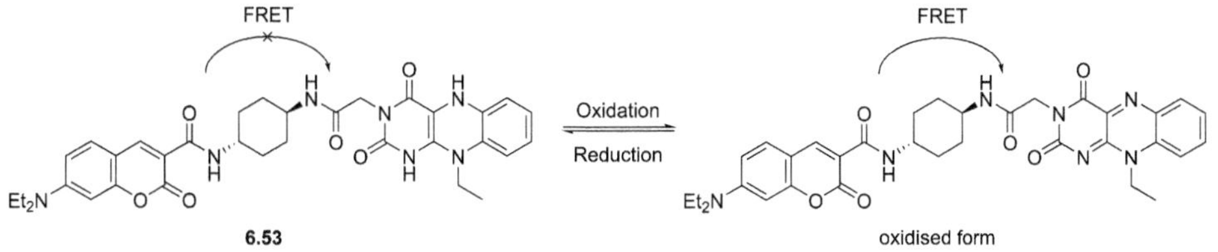

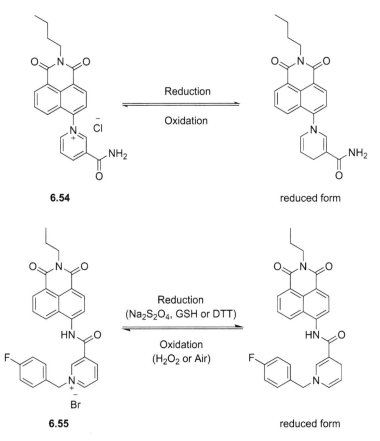

Figure 6.32 Flavin-based redox sensors **6.53** that exhibits ratiometricity. The maximal wavelength of the reduced and oxidised form are 470 and 520 nm, respectively [120].

form, FRET is allowed, and green acceptor emission is observed; **6.53** was able to monitor cellular oxidative capacity in H_2O_2 and *N*-acetyl cysteine treated HeLa cells using confocal microscopy, FLIM, and flow cytometry.

Sensor **6.54** is a naphthalimide with a nicotinamide at the 4-position, displaying a ratiometric response upon chemical and electrochemical reduction to the reduced form (Figure 6.33) [117]. On the reduction of **6.54**, there is no change to the fluorescence intensity at 460 nm and an increase at 550 nm. The fluorescence ratio of **6.54** was measured in A549 cells incubated under hypoxia condition and normoxia conditions, revealing a significantly higher fluorescence ratio in hypoxic cells after a 3-hour incubation. Sensor **6.55**, reported by

Figure 6.33 Nicotinamide redox sensor **6.54** with ratiometric emission at 460 and 550 nm [117]; **6.55** is another nicotinamide-based redox sensor that is fluorescent when oxidised [121].

Pfeffer and co-workers, is an 'on-off' naphthalimide-based NADH mimic that can reversibly detect of intracellular redox states (Figure 6.33) [121]. It demonstrates excellent reversibility between the oxidised form and reduced form. It has been shown to localise in the lipids of A549 cells and responds to redox changes after H_2O_2, sodium dithionite, and *N*-acetyl cysteine treatment. A drawback of **6.54** and **6.55** is that conjugation to the fluorophore gives rise to a marked change in the reduction potential compared to natural nicotinamide derivatives, and so they are unlikely to be sensitive to subtly redox changes within cells.

6.4 Conclusions

Reversible fluorescent sensors have been shown to be effective tools for monitoring fluctuations in cellular environments. Most sensors for cellular environment introduced in this chapter are reversible and able to detect changes within cells. In general, the influences on excited states, especially energy levels and dipole moments of excited states, are the key considerations for the design of sensors for polarity. Fluorophores that undergo ICT or TICT are widely used for constructing sensors for polarity, due to their excited states' sensitivity to polarity. ESIPT molecules that can form hydrogen bonds with surrounding solvents can also be used as sensors for polarity, with PET-based BODIPY derivatives also sensitive to polarity.

Molecular rotors are commonly used as sensors for viscosity, because their fluorescence emission is tightly influenced by the surrounding viscosity. Three types of molecular rotors were discussed in this chapter. Some molecular rotors are also sensitive to surrounding temperature, limiting their usage in detection of viscosity to cellular environments with relatively stable temperature. More efforts are needed in developing selective sensors for viscosity that can be applied in complex cellular environments.

For pH sensors, consideration of the pK_a is crucial. The pK_a of sensors restricts the detection range of which they can effectively be used and hence dictates their suitability for sensing in cells. A pH sensor with a pK_a of 5 would be suitable to monitor lysosomal acidification, while another pH sensor with a pK_a of 7 would deem appropriate for measuring pH fluxes in the near-neutral cytosolic environment. This chapter summarised some of the strategies for sensing pH using PET, rhodamine ring-opening, and ICT mechanisms. Ratiometric fluorescence output is commonly seen in pH sensors that utilise ICT mechanisms, while sensors that operate through PET mechanisms tend to be intensity based. However, a shortcoming of current fluorescent pH probes are its narrow detection range of ±1.0 pH unit. More efforts to expand this sensing range to encompass the full biological relevant pH range (pH 4–8) are needed. This may be achieved by incorporating multiple sensing groups onto one or two fluorophores. Constructing a ratiometric pH sensor may also help to realise this goal.

Biological redox systems are made of many oxidants and reductants. Sensors that have specificity to certain ROS exists, but these tend to be reaction-based and their reversibility require specific reductant such as GSH. Techniques to provide specificity include the nitroxyl radical/hydroxylamine redox couple and the quinone/hydroquinone redox couple. On the other hand, sensors to monitor general redox state has also been reported, using flavins and nicotinamides that are tuned to biological relevant reduction potential. These are reversible and responsive to a wide range of oxidants and reductants. Further research into these reversible, global redox sensors across a broader range of wavelengths are desirable for live cell studies when multiple dyes are involved.

References

1 Song, W., Dong, B., Lu, Y. et al. (2019). *New J. Chem.* 43 (30): 12103–12108.

2 Xiao, H., Li, P., Zhang, W., and Tang, B. (2016). *Chem. Sci.* 7 (2): 1588–1593.

3 Drubin, D.G. and Nelson, W.J. (1996). *Cell* 84 (3): 335–344.

4 Kuimova, M.K., Botchway, S.W., Parker, A.W. et al. (2009). *Nat. Chem.* 1 (1): 69–73.

5 Kuimova, M.K. (2012). *Phys. Chem. Chem. Phys.* 14 (37): 12671–12686.

6 Liu, T., Liu, X., Spring, D.R. et al. (2014). *Sci. Rep.* 4: 5418.

7 Su, D., Teoh, C.L., Wang, L. et al. (2017). *Chem. Soc. Rev.* 46 (16): 4833–4844.

8 Yang, Z., He, Y., Lee, J.H. et al. (2013). *J. Am. Chem. Soc.* 135 (24): 9181–9185.

9 Nadiv, O., Shinitzky, M., Manu, H., and Hecht, D. (1994). *Biochem. J.* 298 (2): 443–450.

10 Deliconstantinos, G., Villiotou, V., and Stavrides, J.C. (1995). *Biochem. Pharmacol.* 49 (11): 1589–1600.

11 Gleason, M.M., Medow, M.S., and Tulenko, T.N. (1991). *Circ. Res.* 69 (1): 216–227.

12 Zubenko, G.S., Kopp, U., Seto, T., and Firestone, L.L. (1999). *Psychopharmacology* 145: 175–180.

13 Grabowski, Z.R. and Rotkiewicz, K. (2003). *Chem. Rev.* 103 (10): 3899–4031.

14 Sasaki, S. and Drummen, G.P.C. (2016). Konishi G-i. *J. Mater. Chem. C* 4 (14): 2731–2743.

15 Grabowski, Z.R. (1992). *Pure Appl. Chem.* 64 (9): 1249–1255.

16 Lu, Z., Liu, N., Lord, S.J. et al. (2009). *Chem. Mater.* 21 (5): 797–810.

17 Mata, G. and Luedtke, N.W. (2013). *Org. Lett.* 15 (10): 2462–2465.

18 Haidekker, M.A., Brady, T.P., Lichlyter, D., and Theodorakis, E.A. (2005). *Bioorg. Chem.* 33 (6): 415–425.

19 Haidekker, M.A. and Theodorakis, E.A. (2010). *J. Biol. Eng.* 4: 11.

20 Haidekker, M.A. and Theodorakis, E.A. (2007). *Org. Biomol. Chem.* 5 (11): 1669–1678.

21 Haidekker, M.A., Ling, T., Anglo, M. et al. (2001). *Chem. Biol.* 8 (2): 123–131.

22 Köhler, G., Wolschann, P., and Rotkiewicz, K. (1992). *Chem. Sci.* 104 (2): 197–207.

23 Haidekker, M.A., L'Heureux, N., and Frangos, J.A. (2000). *Am. J. Physiol. Heart Circ. Physiol.* 278 (4): H1401–H1406.

24 Su, D., Teoh, C.L., Gao, N. et al. (2016). *Sensors* 16 (9): 1397.

25 Kuimova, M.K., Yahioglu, G., Levitt, J.A., and Suhling, K. (2008). *J. Am. Chem. Soc.* 130 (21): 6672–6673.

26 Levitt, J.A., Kuimova, M.K., Yahioglu, G. et al. (2009). *J. Phys. Chem. C* 113 (27): 11634–11642.

27 Yin, X., Li, Y., Zhu, Y. et al. (2010). *Dalton Trans.* 39 (41): 9929–9935.

28 Liu, X., Chi, W., Qiao, Q. et al. (2020). *ACS Sens.* 5 (3): 731–739.

29 Peng, X., Yang, Z., Wang, J. et al. (2011). *J. Am. Chem. Soc.* 133 (17): 6626–6635.

30 Karpenko, I.A., Niko, Y., Yakubovskyi, V.P. et al. (2016). *J. Mater. Chem. C* 4 (14): 3002–3009.

31 Vu, T.T., Méallet-Renault, R., Clavier, G. et al. (2016). *J. Mater. Chem. C* 4 (14): 2828–2833.

32 Howell, S., Dakanali, M., Theodorakis, E.A., and Haidekker, M.A. (2012). *J. Fluoresc.* 22 (1): 457–465.

33 Vysniauskas, A., Qurashi, M., Gallop, N. et al. (2015). *Chem. Sci.* 6 (10): 5773–5778.

34 Sasaki, S., Niko, Y., Klymchenko, A.S., and Konishi, G.-i. (2014). *Tetrahedron* 70 (41): 7551–7559.

35 Marini, A., Muñoz-Losa, A., Biancardi, A., and Mennucci, B. (2010). *J. Phys. Chem. B* 114 (51): 17128–17135.

36 Reichardt, C. (1994). *Chem. Rev.* 94 (8): 2319–2358.

37 Krasnowska, E.K., Gratton, E., and Parasassi, T. (1998). *Biophys. J.* 74 (4): 1984–1993.

38 Parisio, G., Marini, A., Biancardi, A. et al. (2011). *J. Phys. Chem. B* 115 (33): 9980–9989.

39 Kucherak, O.A., Didier, P., Mély, Y., and Klymchenko, A.S. (2010). *J. Phys. Chem. Lett.* 1 (3): 616–620.

40 Ren, B., Gao, F., Tong, Z., and Yan, Y. (1999). *Chem. Phys. Lett.* 307 (1–2): 55–61.

41 Saroja, G., Soujanya, T., Ramachandram, B., and Samanta, A. (1998). *J. Fluoresc.* 8 (4): 405–410.

42 Signore, G., Nifosì, R., Albertazzi, L. et al. (2010). *J. Am. Chem. Soc.* 132 (4): 1276–1288.

43 Norris, S.R., Warner, C.C., Lampkin, B.J. et al. (2019). *Org. Lett.* 21 (10): 3817–3821.

44 Zhuang, Y.D., Chiang, P.Y., Wang, C.W., and Tan, K.T. (2013). *Angew. Chem. Int. Ed.* 52 (31): 8124–8128.

45 Prioli, S., Reinholdt, P., Hornum, M., and Kongsted, J. (2019). *J. Phys. Chem. B* 123 (49): 10424–10432.

46 Demchenko, A.P., Tang, K.C., and Chou, P.T. (2013). *Chem. Soc. Rev.* 42 (3): 1379–1408.

47 Padalkar, V.S. and Seki, S. (2016). *Chem. Soc. Rev.* 45 (1): 169–202.

48 Li, C., Yang, Y., Ma, C., and Liu, Y. (2016). *RSC Adv.* 6 (6): 5134–5140.

49 Schmidtke, S.J., Underwood, D.F., and Blank, D.A. (2004). *J. Am. Chem. Soc.* 126 (28): 8620–8621.

50 Paterson, M.J., Robb, M.A., Blancafort, L., and AD, D.B. (2004). *J. Am. Chem. Soc.* 126 (9): 2912–2922.

51 Klymchenko, A.S., Pivovarenko, V.G., Ozturk, T., and Demchenko, A.P. (2003). *New J. Chem.* 27 (9): 1336–1343.

52 Marciniak, H., Hristova, S., Deneva, V. et al. (2017). *Phys. Chem. Chem. Phys.* 19 (39): 26621–26629.

53 Chou, P.-T., Pu, S.-C., Cheng, Y.-M. et al. (2005). *J. Phys. Chem. A* 109 (17): 3777–3787.

54 Kwon, J.E. and Park, S.Y. (2011). *Adv. Mater.* 23 (32): 3615–3642.

55 Sedgwick, A.C., Wu, L., Han, H.H. et al. (2018). *Chem. Soc. Rev.* 47 (23): 8842–8880.

56 Klymchenko, A.S., Ozturk, T., Pivovarenkob, V.G., and Demchenko, A.P. (2001). *Tetrahedron Lett.* 42 (45): 7967–7970.

57 Nunomura, A., Perry, G., Aliev, G. et al. (2001). *J. Neuropathol. Exp. Neurol.* 60: 759–767.

58 Duan, X., Li, P., Li, P. et al. (2011). *Dyes Pigments* 89 (3): 217–222.

59 Kim, S.Y., Cho, Y.J., Son, H.J. et al. (2018). *J. Phys. Chem. A* 122 (13): 3391–3397.

60 Zhu, H., Fan, J., Mu, H. et al. (2016). *Sci. Rep.* 6: 35627.

61 Sunahara, H., Urano, Y., Kojima, H., and Nagano, T. (2007). *J. Am. Chem. Soc.* 129 (17): 5597–5604.

62 Casey, J.R., Grinstein, S., and Orlowski, J. (2010). *Nat. Rev. Mol. Cell Biol.* 11 (1): 50–61.

63 Zhang, J., Li, J., Chen, B. et al. (2019). *Dyes Pigments* 170: 107620.

64 Cooper, G.M. (2000). *The Cell: A Molecular Approach*, 2nde. Sunderland: Sinauer Associates.

65 Alberts, B., Johnson, A., and Lewis, J. (2002). The mitochondrion. In: *In Molecular Biology of the Cell*, 4the. New York: Garland Science https://www.ncbi.nlm.nih.gov/books/NBK26894/.

66 Han, J. and Burgess, K. (2010). *Chem. Rev.* 110 (5): 2709–2728.

67 Paradise, R.K., Lauffenburger, D.A., and Van Vliet, K.J. (2011). *PLoS One* 6 (1): 15746.

68 Nicolson, G.L. (2014). *Integr. Med. (Encinitas)* 13 (4): 35–43.

69 Rodríguez, S.J. (2002). *J. Am. Soc. Nephrol.* 13 (8): 2160.

70 Corbet, C. and Feron, O. (2017). *Nat. Rev. Cancer* 17 (10): 577–593.

71 Pastorekova, S. and Gillies, R.J. (2019). *Cancer Metastasis Rev.* 38 (1): 65–77.

72 ThermoFisher Scientific (2010). pH Indicators. In: *Molecular Probes Handbook: A Guide to Fluorescent Probes and Labeling Technologies*, 11ee, 885. ThermoFisher Scientific.

73 Collection, H.R. Bordwell pKa Table 2017. https://organicchemistrydata.org/hansreich/resources/pka/ (accessed 27 September 2017)

74 Li, X., Gao, X., Shi, W., and Ma, H. (2014). *Chem. Rev.* 114 (1): 590–659.

75 Mottram, L.F., Boonyarattanakalin, S., Kovel, R.E., and Peterson, B.R. (2006). *Org. Lett.* 8 (4): 581–584.

76 Minta, A., Kao, J.P.Y., and Tsien, R.Y. (1989). *J. Biol. Chem.* 264 (14): 8171–8178.

77 Sun, W.-C., Gee, K.R., and Haugland, R.P. (1998). *Bioorg. Med. Chem. Lett.* 8 (22): 3107–3110.

78 Miao, F., Uchinomiya, S., Ni, Y. et al. (2016). *ChemPlusChem* 81 (11): 1209–1215.

79 Hoogendoorn, S., Blom, A.E.M., Willems, L.I. et al. (2011). *Org. Lett.* 13 (20): 5656–5659.

80 Takahashi, S., Kagami, Y., Hanaoka, K. et al. (2018). *J. Am. Chem. Soc.* 140 (18): 5925–5933.

81 Lavis, L.D., Rutkoski, T.J., and Raines, R.T. (2007). *Anal. Chem.* 79 (17): 6775–6782.

82 Grimm, J.B., English, B.P., Chen, J. et al. (2015). *Nat. Methods* 12 (3): 244–250.

83 Sakabe, M., Asanuma, D., Kamiya, M. et al. (2013). *J. Am. Chem. Soc.* 135 (1): 409–414.

84 Yin, J., Hu, Y., and Yoon, J. (2015). *Chem. Soc. Rev.* 44 (14): 4619–4644.

85 Hall, M.J., Allen, L.T., and O'Shea, D.F. (2006). *Org. Biomol. Chem.* 4 (5): 776–780.

86 Tang, B., Yu, F., Li, P. et al. (2009). *J. Am. Chem. Soc.* 131 (8): 3016–3023.

87 Chi, W., Qi, Q., Lee, R. et al. (2020). *J. Phys. Chem. C* 124 (6): 3793–3801.

88 Wang, X., Fan, L., Zhang, X. et al. (2020). *Analyst* 145 (21): 7018–7024.

89 Lee, D., Swamy, K.M.K., Hong, J. et al. (2018). *Sensors Actuators B Chem.* 266: 416–421.

90 Liu, X., Su, Y., Tian, H. et al. (2017). *Anal. Biochem.* 89 (13): 7038–7045.

91 Chen, P.M., Gombart, Z.J., and Chen, J.W. (2011). *Cell Biosci.* 1 (1): 10.

92 Kim, H.J., Heo, C.H., and Kim, H.M. (2013). *J. Am. Chem. Soc.* 135 (47): 17969–17977.

93 Myochin, T., Kiyose, K., Hanaoka, K. et al. (2011). *J. Am. Chem. Soc.* 133 (10): 3401–3409.

94 Chen, S., Hong, Y., Liu, Y. et al. (2013). *J. Am. Chem. Soc.* 135 (13): 4926–4929.

95 Handy, D.E. and Loscalzo, J. (2012). *Antioxid. Redox Signal.* 16 (11): 1323–1367.

96 Jastroch, M., Divakaruni, A.S., Mookerjee, S. et al. (2010). *Essays Biochem.* 47: 53–67.

97 Murphy, M.P. (2009). *Biochem. J.* 417 (1): 1–13.

98 Bigarella, C.L., Liang, R., and Ghaffari, S. (2014). *Development* 141 (22): 4206–4218.

99 Tsutsui, H., Kinugawa, S., and Matsushima, S. (2009). *Cardiovasc. Res.* 81 (3): 449–456.

100 Yu, T., Robotham, J.L., and Yoon, Y. (2006). *Proc. Natl. Acad. Sci. U. S. A.* 103 (8): 2653.

101 Willems Peter, H.G.M., Rossignol, R., Dieteren Cindy, E.J. et al. (2015). *Cell Metab.* 22 (2): 207–218.

102 Lee, J. and Song, C.-H. (2021). *Antioxidants* 10 (6): 872.

103 Birben, E., Sahiner, U.M., Sackesen, C. et al. (2012). *World Allergy Organ. J.* 5 (1): 9–19.

104 Teresa Vanessa, F., Annamaria, P., Pengou, Z., and Franco, F. (2013). *Curr. Pharm. Des.* 19 (32): 5695–5703.

105 Wang, X., Wang, W., Li, L. et al. (2014). *Biochim. Biophys. Acta* 1842 (8): 1240–1247.

106 Tafani, M., Sansone, L., Limana, F. et al. (2016). *Oxidative Med. Cell. Longev.* 2016: 3907147.

107 Blinco, J.P., Fairfull-Smith, K.E., Morrow, B.J., and Bottle, S.E. (2011). *Aust. J. Chem.* 64 (4): 373–389.

108 Kajer, T.B., Fairfull-Smith, K.E., Yamasaki, T. et al. (2014). *Free Radic. Biol. Med.* 70: 96–105.

109 Kinoshita, Y., Yamada, K.-i., Yamasaki, T. et al. (2010). *Free Radic. Biol. Med.* 49 (11): 1703–1709.

110 Go, Y.-M. and Jones, D.P. (2008). *Biochim. Biophys. Acta* 1780 (11): 1273–1290.

111 Chong, K.L., Chalmers, B.A., Cullen, J.K. et al. (2018). *Free Radic. Biol. Med.* 128: 97–110.

112 Zhang, W., Li, P., Yang, F. et al. (2013). *J. Am. Chem. Soc.* 135 (40): 14956–14959.

113 Liu, H.-W., Zhu, X., Zhang, J. et al. (2016). *Analyst* 141 (20): 5893–5899.

114 Sies, H. (1993). *Free Radic. Biol. Med.* 14 (3): 313–323.

115 Xu, K., Qiang, M., Gao, W. et al. (2013). *Chem. Sci.* 4 (3): 1079–1086.

116 Berg, J.M., Tymoczko, J.L., and Stryer, L. (2002). Nucleotide biosynthesis. In: *Biochemistry*, 5the, 1051–1052. New York: W H Freeman.

117 Leslie, K.G., Kolanowski, J.L., Trinh, N. et al. (2019). *Aust. J. Chem.* 73 (10): 895–902.

118 Yeow, J., Kaur, A., Anscomb, M.D., and New, E.J. (2014). *Chem. Commun.* 50 (60): 8181–8184.

119 Kaur, A., Brigden, K.W.L., Cashman, T.F. et al. (2015). *Org. Biomol. Chem.* 13 (24): 6686–6689.

120 Kaur, A., Haghighatbin, M.A., Hogan, C.F., and New, E.J. (2015). *Chem. Commun.* 51 (52): 10510–10513.

121 Sharma, H., Tan, N.K., Trinh, N. et al. (2020). *Chem. Commun.* 56 (15): 2240–2243.

7

Labelling Proteins and Biomolecules with Small Fluorescent Sensors

Joy Ghrayche[1,2], Marcus E. Graziotto[1,2], Paris I. Jeffcoat[1,3], and Elizabeth J. New[1,2,3]

[1] *School of Chemistry, The University of Sydney, NSW, Australia*
[2] *Australian Research Council Centre of Excellence for Innovations in Peptide and Protein Science, The University of Sydney, NSW, Australia*
[3] *The University of Sydney Nano Institute (Sydney Nano), The University of Sydney, NSW, Australia*

Whilst most fluorescent sensors can be successfully targeted to cellular regions, targeting individual biomolecules such as proteins, lipids, enzymes, and sugars is more challenging. Traditional methods for visualisation involve fluorescently labelled antibodies or fluorescent proteins. Fluorophore-antibody conjugates require permeabilisation of the cell membrane for binding and cannot be used to monitor intracellular dynamic events in living cells [1]. Fluorescent proteins can be fused to a protein of interest with genetic modification and can have emission wavelengths spanning the visible spectrum [2]. Whilst they remain a useful imaging tool, their invariant fluorescence signals are typically less bright and less photostable than organic dyes [3], and it is challenging to modify these proteins for sensing applications [4]. In addition, the large size of fluorescent proteins is known to perturb the native structure and function of the biomolecules of interest [5].

Small-molecule fluorophores overcome these limitations as they can have a fluorescent response specific to the cellular experiment while minimising disruption to the native structure or function of a protein [6]. They can be made ratiometric or fluorogenic and can respond to a variety of different analytes or biomolecules.

This chapter gives an overview of the different techniques that can be used to image biomolecules in living cells. The chapter begins by outlining the main classes of bioorthogonal reactions that can be used to label biomolecules generally. The second part of the chapter focuses on developments in techniques that involve larger modifications and can be used for the imaging of proteins and biomolecules.

7.1 Labelling Biomolecules in Cells with Fluorescent Sensors

The general strategy for chemoselective labelling of proteins and biomolecules with small-molecule fluorophores involves two components (Figure 7.1).

Molecular Fluorescent Sensors for Cellular Studies, First Edition. Edited by Elizabeth J. New.
© 2022 John Wiley & Sons Ltd. Published 2022 by John Wiley & Sons Ltd.

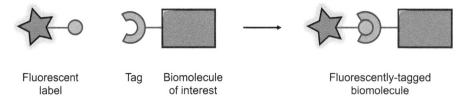

Fluorescent   Tag   Biomolecule   Fluorescently-tagged
label   of interest   biomolecule

Figure 7.1 Schematic of a biomolecule labelling strategy. The fluorophore and tagged biomolecule each contain a group that selectively reacts with the other to form a stable linkage between fluorophore (orange star) and biomolecule (purple rectangle).

1) **Tagged biomolecule of interest:** The tag is a non-natural group covalently attached to the biomolecule under investigation that is designed to selectively react with a group installed on an exogenous label. The tag can be a single amino acid modification, or a larger enzyme or protein as described below. If the molecule under investigation is a protein, then it is termed a protein of interest (POI).

2) **Label:** The small-molecule fluorophore, known as the fluorescent label, is an exogenous molecule synthesised to have the desired fluorescence and sensing properties. A group that selectively reacts with the tag is appended to the small molecule to allow for conjugation.

In the figures for this chapter, a purple rectangle represents the biomolecule of interest and an orange star represents the appended label, as shown in Figure 7.1.

In order to label a biomolecule in a cellular sample, the tag is introduced first, either by exogenous addition or genetic modification [7]. For biomolecules such as sugars, lipids, and nucleic acids, modified versions of these compounds can be synthesised externally in the chemical laboratory and then introduced to cells by incubation. The cells' own enzymatic machinery recognises the natural part of the modified biomolecule and incorporates the modification within the cell [8].

There is growing demand for robust methods to introduce fluorescent sensors into cellular environments, in particular, where specific proteins need to be visualised. A number of methods have been utilised to meet this need:

- *Orthogonal tRNA-aaRS pairs* – This allows for the incorporation of a non-natural amino acid with a modified side chain into a protein of interest. In this technique, a random or rare codon (such as the amber-stop codon) is inserted into the gene of interest to express the modification. An unnatural amino acid with a group on the side chain is added and recognised by an aminoacyl-tRNA-synthetase enzyme (aaRS) that aminoacylates orthogonal transfer RNA (tRNA) with the modification [9, 10]. The modified amino acid is then incorporated into the gene of interest at the ribosome where the codon is recognised [11]. The typical reactions used with these modifications are discussed in Section 7.2.

- *Short peptide recognition sequences* – These involve adding a six amino acid recognition sequence to the protein with genetic encoding. This strategy is used in the protein tagging systems discussed in Section 7.3.

- *Fusion proteins* – Provided that a suitable plasmid can be obtained, a cell can be transfected with DNA that instructs the cell to produce the protein of interest fused to a modified enzyme that possesses an unnatural reactive site that can selectively react with a fluorescent label [9]. This strategy is used in the protein tagging systems discussed in Section 7.4.

- *Engineered enzymes* – This method utilises modified ligase enzymes to append small-molecule functionalities to acceptor peptides and proteins [10]. This strategy is used in the tagging systems discussed in Section 7.5.

After the tag has been introduced, the fluorescent label is then incubated with the cell sample, and this results in a covalent bond between fluorophore and biomolecule. Most fluorescent labels are 'always-on', meaning that the fluorescence does not change after being introduced into the cellular system (Figure 7.2). A major disadvantage of 'always-on' fluorophores is the need for extensive washing steps to remove any unbound dye. Inefficient sample washing can lead to lower signal-to-noise ratios and thus greater background fluorescence (Figure 7.2). Fluorogenic probes, where the fluorescence changes upon binding inside cellular systems, are often preferred as they minimise background signal from off-target labelling and reduce the need for washing [11].

This chapter provides an overview of the main strategies to label proteins and biomolecules in cells for fluorescent imaging. It begins by outlining the general chemical reactions that can be used to label any modified biomolecule. The latter parts of the chapter focus on the range of protein-specific labelling technologies that have been developed using short peptide tags, fusion proteins, and enzymatic modifications. While the majority of systems reported to date use non-responsive fluorophores to label biomolecules, the strategies can be applied to the incorporation of fluorescent sensors to report on specific sub-cellular micro-environments. Key examples of biomolecule labelling with fluorescent sensors are highlighted throughout.

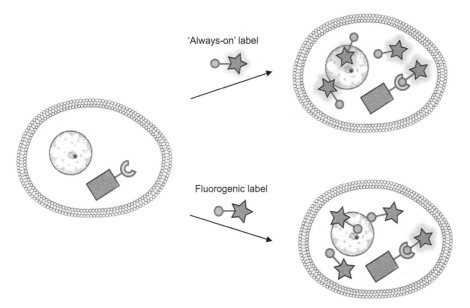

Figure 7.2 Comparison of 'always-on' and fluorogenic labels for tagging biomolecules. For 'always-on' labelling, fluorescent signal can be observed if the label localises in off-target locations, leading to false-positive readouts. If the label is fluorogenic, the fluorescent response is only observed when the label binds to the desired tagged biomolecule (purple rectangle).

7.2 Small-molecule Modifications and Bioorthogonal Reactions

The most useful chemical reactions for attaching small fluorophore labels to tagged biomolecules in biological systems are bioorthogonal reactions, which are defined as chemical reactions that can proceed in biological environments without perturbing the native state and function of the system. These types of reactions are also commonly called 'click' reactions, where clicking refers to the formation of bonds between the two reagents involved. In this approach, one reactant is appended to the label and one reactant is used as a tag on a biomolecule, resulting in selective labelling. These types of reactions require the tag to be added on a single amino acid in a protein or on an exogenous copy of the biomolecule itself. With many bioorthogonal reactions developed in the last 20 years, chemists and biologists are spoilt for choice in fluorescently labelling biomolecules. This section outlines the main bioorthogonal reactions that have been used to tag and label biomolecules, highlighting the advantages and disadvantages of each. Selected examples where the reaction has been used to generate a fluorogenic label have been included. The figures in this section present the typical way in which the reaction is used, but they can be installed on either the fluorescent reporter or biomolecule.

7.2.1 Polar Ketone and Aldehyde Condensations

The first known bioorthogonal reactions were polar reactions, namely ketone and aldehyde condensations (Figure 7.3a). With the initial report that ketones could be incorporated into proteins with unnatural alanine and phenylalanine residues [12], as well as later reports on the efficient incorporation of aldehydes into proteins [13], the field flourished [14]. In this reaction, a hydrazine- or aminooxy-containing nucleophile reacts with an aldehyde or ketone in acidic conditions forming an imine. The most appealing features of this reaction

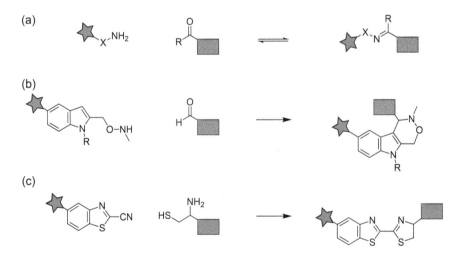

Figure 7.3 Examples of bioorthogonal polar addition reactions that are used to label biomolecules. (a) Aldehyde/ketone condensations, (b) Pictet-Spengler ligation, and (c) cyanobenzothiazole condensation.

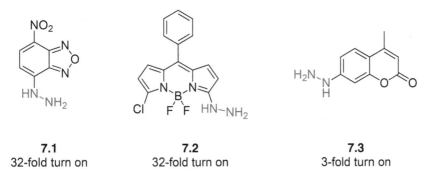

7.1	**7.2**	**7.3**
32-fold turn on	32-fold turn on	3-fold turn on

Figure 7.4 Selected examples of fluorescent labels containing hydrazines. The reactive moiety is shown in red.

are the small size of aldehydes and ketones and their chemical selectivity for hydrazines and aminooxy derivatives amongst the amines and thiols that are abundant in cells. Aldehydes and ketones are rare in proteins but can be found in mammalian cells: for example, aldehydes are present in open chain sugars; and ketones are found in hormones [15]. However, these molecules are not expressed on the cell surface [16], so aldehydes and ketones are most suitable for labelling cell surface proteins and biomolecules [17]. Additionally, in the buffered environment of the cell at pH 7.4, the equilibrium reaction disfavours the formation of the imine, and hydrazone products can be susceptible to hydrolysis. Using anilines as catalysts can overcome the requirement of an acidic pH and increase the rate of reaction at pH 7 [18].

Fluorogenic labels can be developed using hydrazines, since they contain a lone pair that is believed to quench intramolecular charge transfer (ICT) fluorescence [16]. As shown in Figure 7.4, hydrazines were installed on an oxadiazole derivative (**7.1**) in order to develop an *in vitro* fluorogenic assay for lipases [19]. Hydrazines were also appended to BODIPY (**7.2**) and coumarin (**7.3**) dyes, resulting in up to a 32-fold increase in fluorescence upon reaction with aldehydes [20]; **7.3** was used for live cell imaging of α-tubulin structures [21]. Since these developments, fluorescent labels bearing hydrazine and aminooxy moieties have been made commercially available.

More recent developments in this field have taken advantage of a Pictet-Spengler-type reaction involving *N*-aminooxytryptamine (Figure 7.3b) [22]. This reaction product was resistant to hydrolysis and formed faster than the above condensation products but has not been extensively applied to image biomolecules. Other polar condensations, such as the 1,2-aminothiol-cyanobenzothiazole ligation (Figure 7.3c), have been tried and tested *in vitro* [23] and in cells [24]. The reaction proceeds faster ($9\,M^{-1}\,s^{-1}$) than other polar reactions and neither of the bioorthogonal partners are found in cells [25]. This ligation has been used to label thiols on proteins, cell membranes, and brain tissue [26] and for the fluorescent imaging of glycoside hydrolase and protease activity in live cells [27].

7.2.2 Azide Bioorthogonal Chemistry

Azides are one of the most useful groups in bioorthogonal chemistry. Their biocompatibility, absence from eurkaryotic cells, ease of synthesis, and tolerance of chemical reaction conditions have elevated them to being one of the most diversely used bioorthogonal

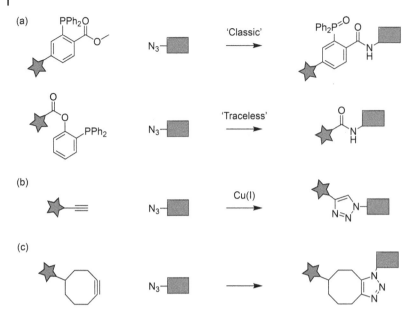

Figure 7.5 Common bioorthogonal azide additions used for fluorescently labelling tagged biomolecules in cells. (a) Staudinger ligations, (b) copper azide-alkyne click (CuAAC), and (c) strain-promoted azide-alkyne click (SPAAC).

groups [28]. Chemically, azides are resistant to oxidation, stable in aqueous solution, and not highly polar, meaning that hydrogen bonding does not affect these groups *in cellulo* [29]. Azide modifications do not typically disrupt the structure or function of the biomolecule due to their small size, meaning that azides are used for many different reactions as shown in Figure 7.5.

7.2.2.1 Staudinger Ligations

The Bertozzi group first proposed the use of bioorthogonal azides in the Staudinger ligation (Figure 7.5a) [30]. This reaction occurs between an azide and a triarylphosphine containing an acyl-electrophilic trap on one of the aryl rings, with both groups are absent from biological systems. In this reaction, the phosphine reacts with the azide, and the acyl-electrophilic trap stabilises the intermediate allowing the formation of a biologically stable amide linkage and a phosphine oxide [31]. The drawbacks of the Staudinger ligation are the slow rate of reaction ($\sim 10^{-3}$ M^{-1} s^{-1}), the susceptibility of some phosphines to oxidation by air and enzymes within the cell, and the fact that these labels are not yet commercially available [7]. However, in terms of selectivity, toxicity, and bioorthogonality considerations, the Staudinger ligation ranks among the best of all bioorthogonal reactions developed to date [15].

Various fluorescent probes bearing aryphosphines have been developed for labelling azides in cells (Figure 7.6a). The stability and selectivity of this ligation permits fluorescent labelling of *in vitro* proteins, for example, using the coumarin derivative **7.4**, which has a diphenylphosphine at its 3-position [32, 33]. Suitable phosphines were appended to several common dyes such as cyanine (**7.5**), fluorescein (**7.6**), and rhodamine (**7.7**) for live cell fluorescence imaging, with **7.5** determined to be the best candidate for detecting glycans in living

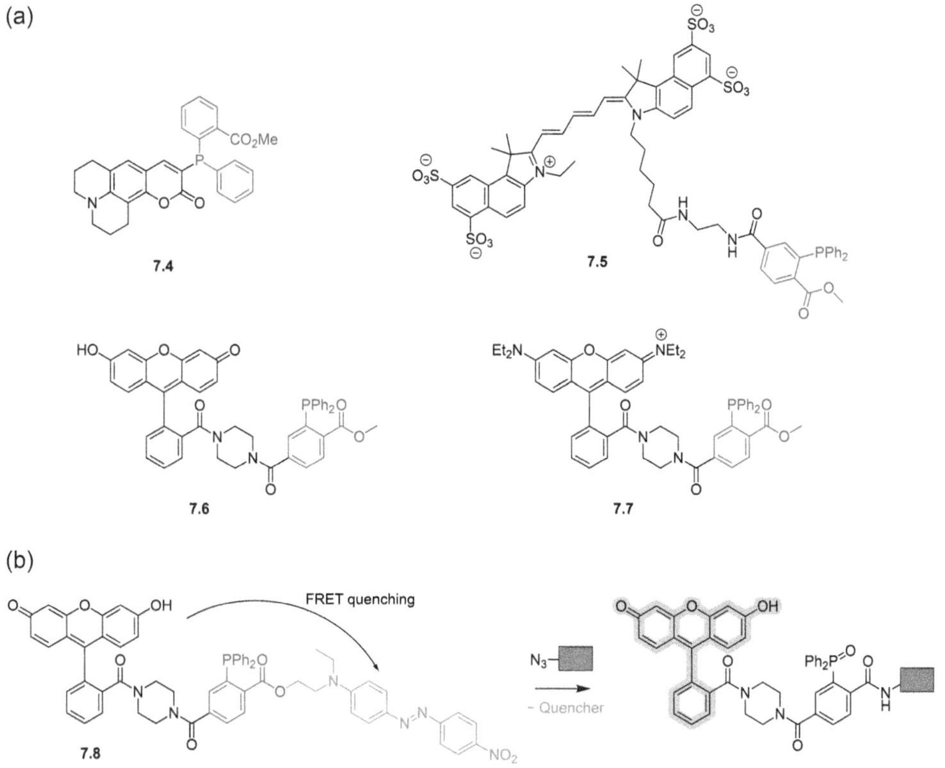

(a)

(b)

Figure 7.6 (a) Examples of phosphine-conjugated fluorophores used for Staudinger ligation reactions to label modified biomolecules in cells. (b) Example of a fluorogenic probe for the Staudinger ligation where the fluorescein fluorophore is quenched by an attached FRET quencher (green). Upon click reaction with the azide-tagged biomolecule, the quencher is cleaved and fluorescence restored to the fluorescein. The reactive moiety is shown in red.

cells, providing the least background fluorescence of the series [34]. A fluorogenic probe (**7.8**) was developed by exploiting Förster resonance energy transfer (FRET) between a fluorescein and an azobenzene quencher (Figure 7.6b). The bioorthogonal reaction of the arylphosphine with exogenous azide-tagged sugars inside living and fixed cells displaced the quencher, providing fluorescent signals that indicated cell-surface and Golgi apparatus-localisation of the tagged sugars [35].

In the typical ligation, the phosphine oxide remains attached between the label and biomolecule, but it can also be removed during the bioorthogonal reaction to minimise interference with the biomolecule of interest. This is achieved by installing the reporter on the carbonyl and upon rearrangement only an amide is left behind in the 'traceless' Staudinger ligation (Figure 7.5a) [36]. This variant of the ligation has great potential for the fluorescent labelling of biomolecules in cells.

7.2.2.2 Copper Azide–Alkyne Cycloadditions

The azide functionality also undergoes copper-click reactions, that is, the reaction of azides and alkynes to form triazoles, catalysed by copper(I), as shown in Figure 7.5b. First proposed by Sharpless, the copper azide-alkyne cycloaddition (CuAAC) reaction proceeds in mild, aqueous conditions [37] and was considered to be compatible with biological systems as the triazole product is stable in water [38]. Indeed, the term 'click reaction' has become synonymous with this class of cycloaddition. One advantage of this reaction is the relatively fast rate of reaction $(10^1$–$10^2 \, M^{-1} \, s^{-1})$ compared to the polar additions mentioned above. However, the rate of labelling is dependent on the concentration of the catalyst, which is hard to control in cells and animals [15]. Thus, while CuAAC can be used as a bioorthogonal chemical reaction, it is typically used *in vitro*. This is due to the requirement of a Cu(I) catalyst, which is toxic to cells as excess Cu(I) can increase the concentration of reactive oxygen species, inducing oxidative stress [39]. There are examples of the CuAAC reaction being used to label cell surface glycans [40], but these have been superseded by newer strain promoted additions.

7.2.2.3 Strain-promoted Cycloadditions

To make such azide additions more suitable for use in cells, Bertozzi and colleagues proposed that incorporating ring-strain to the alkyne partner would lower the activation energy of the cycloaddition so that no catalyst was required. The resultant strain-promoted [3 + 2] azide-alkyne cycloaddition (SPAAC) was first used to fluorescently label proteins in cells for flow cytometry [41]. Chemically, the reaction proceeds similarly to CuAAC, but the ring strain permits the formation of the triazole to occur catalyst-free (Figure 7.5c). The rate of reaction $(10^{-3}$–$10^{-1} \, M^{-1} \, s^{-1})$ is typically slower than for CuAAC, but the removal of a potentially toxic catalyst is a considerable advantage, allowing for the widespread use of this bioorthogonal reaction for the labelling of membrane glycans in both live cells [42] and zebrafish [43]. It remains one of the most widely used bioorthogonal reactions as these compounds are commercially available with reactive groups for facile conjugation to biomolecules. One limitation of this reaction is that the triple bond on the cyclooctyne is susceptible to attack from reduced cysteines such as glutathione [44] as well as sulfenic acids [45], which may degrade the reagent before labelling. Strategies to avoid this include the pre-alkylation of cysteines before labelling [46] or the addition of low concentrations of β-mercaptoethanol, which is proposed to scavenge sulfanyl radicals and/or minimise post-translational modifications of cysteines that can react with cyclooctynes [47].

Iterations to improve the selectivity and reactivity of the cycloalkynes used for SPAAC have been prominent, and several notable examples are presented in Figure 7.7. Adding ring strain to the first reported bioorthogonal cyclooctyne (OCT) increases reaction rate, which has led to many reported derivatives with improved reaction kinetics [48]. Modifications that increase the rate of reaction with azides have resulted in widely used derivatives such as the difluorine-substituted DIFO and the cyclopropane-substituted analogue BCN. Later-generation variants such as DIBO, BARAC, and DBCO (also sold as ADIBO and DIBAC) have additional phenyl rings to invoke further ring strain and increase reactivity. One issue with the development of new cyclooctynes is that the addition of groups to induce additional ring strain typically increases the hydrophobicity of the molecule. This may result in non-specific membrane interactions that produce false-positive results in biological experiments [7]. In addition, as the ring strain is increased, the stability of a cyclooctyne tag typically decreases in the biological system [49].

7.2.2.4 Fluorogenic Dyes for Azide–Alkyne Labelling

With many examples of azide-modified biomolecules present in the literature, researchers have developed fluorogenic probes that contain alkynes (Figure 7.8). The advantage of this

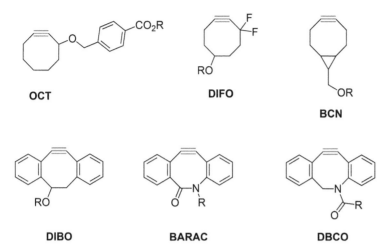

OCT   **DIFO**   **BCN**

DIBO   **BARAC**   **DBCO**

Figure 7.7 Examples of commonly used strained cycloalkynes for labelling of modified biomolecules in cells. R groups represent attachment points for labels or tags. The reactive moiety is shown in red.

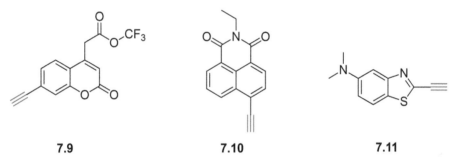

7.9   **7.10**   **7.11**

Figure 7.8 Examples of alkyne-conjugated fluorophores for fluorogenic labelling of modified biomolecules in cells. The reactive moiety is shown in red.

approach is that the azide is the partner installed on the biomolecule of interest, minimising perturbation of the native structure and function of the system being investigated. Installing the alkyne at a position where electron donation is required for fluorescence is a general strategy that allows for the development of small-molecule fluorogenic labels. When the alkyne is installed at the 7-position of a coumarin, the coumarin is not fluorescent, and the click reaction restores fluorescence to the molecule. Using this strategy, the fluorogenic response of coumarin **7.9** was used to detect azide-labelled glycoproteins in live cells [50]. Similarly, installation of the alkyne at the 4-position of a naphthalimide dye (**7.10**) resulted in a probe that could also be utilised for imaging intracellular azide-tagged glycoproteins [51]. This process was also applied to benzothiazoles such as **7.11** as a fluorogenic probe for imaging azide-modified nucleosides in cells [52]. This strategy has been extended and applied to a range of fluorescent scaffolds, including fluoresceins, rhodamines, and all provided a fluorogenic response *in vitro*, with great potential for imaging in mammalian systems [53]. The main issue with these alkyne dyes is that they require the addition of a copper salt to instigate the CuAAC click reaction, meaning that the native biological levels of copper are perturbed during imaging. Fluorogenic strained alkyne systems have also been developed by fusing coumarin with a BARAC derivative and fusing cyclopropenes and DIBO derivatives [54, 55]. However, these alkyne systems are yet to be utilised for cellular imaging.

In the examples of azide-alkyne reactions in this section, the azide was installed on the biomolecule of interest as it is the smaller bioorthogonal partner, but alkynes and cycloalkynes can be appended to proteins and biomolecules. As shown in Figure 7.9a, fluorogenic labels can be made when azides are attached to fluorophores, as they block ICT-based fluorescence mechanisms. The lone pair on the azide quenches the ICT excited state, and subsequent click reaction allows for the donation of electron density from the triazole, restoring ICT fluorescence. Installing an azide at the 3-position of the ICT fluorophore 7-hydroxycoumarin gave a fluorogenic probe amenable to cellular imaging (**7.12**) [56]. Probe **7.12** was used to label and image methionine-tagged proteins in a range of cell lines [57]. A similar fluorogenic effect can be achieved with naphthalimides when the electron donor at the 4-position is replaced by an azide. Upon reaction, the fluorescence is restored and so **7.13** was used for imaging glycoproteins in cells [51]. ICT-based quenching is also believed to be responsible for the fluorogenic nature of the BODIPY-azide **7.14**, which was used to label folate receptors in live cells [58]. One limitation of these fluorogenic dyes is that aryl azides can be reduced by intracellular hydrogen sulfide, meaning that some interferences could be observed in quantitative experiments [59]. However, alkyl azides, such as those on the side chains of modified amino acids, are not typically reduced *in cellulo*.

Azides can also quench xanthene dyes through photoinduced electron transfer (PET) processes (Figure 7.9b). The triazole product does not cause PET quenching, meaning that a fluorogenic response is observed after click reaction of the azide. The first reports of fluorogenic xanthene dyes were fluorescein derivatives, with compound **7.15** used to image glycoproteins on the surface of living cells [60]. Interestingly, the authors also confirmed that the reduction of the azide group on fluorescein derivatives by hydrogen sulfide does not produce a fluorescent product that interferes with imaging. The silicon rhodamine derivative **7.16** was also used for the fluorogenic imaging of peptidoglycans in bacterial cells [61]. This strategy was also utilised to develop four azide-xanthene dyes, known as the **CalFluor** series, that include fluorescein, rhodamine, and silicon rhodamine derivatives. Upon click reaction with alkynes, these dyes exhibited 24–173-fold turn-ons in fluorescence *in vitro*. All four dyes were utilised to image modified cell surface glycoproteins and labelled DNA in live and fixed cell samples, as well as for tracking tagged sialic acids in zebrafish [62]. **CalFluor 488, 555, 580,** and **647** are now commercially available.

7.2.3 Tetrazine Ligation

One of the most significant developments in the field of bioorthogonal reactions was the introduction of the tetrazine ligation. In this reaction, a tetrazine undergoes an inverse electron demand Diels-Alder (IEDDA) cycloaddition reaction with a strained *trans*-cyclooctene (TCO) to form a stable product with the elimination of nitrogen as a by-product (Figure 7.10a) [63]. This reaction proceeds rapidly in aqueous cellular media and is one of the fastest bioorthogonal ligations known to date, with reaction rates up to $10^6 \, M^{-1} \, s^{-1}$, depending on the reaction partner. This permits the collection of images with good signal-to-noise ratios so that minimal concentrations of reagent are required for an experiment. In

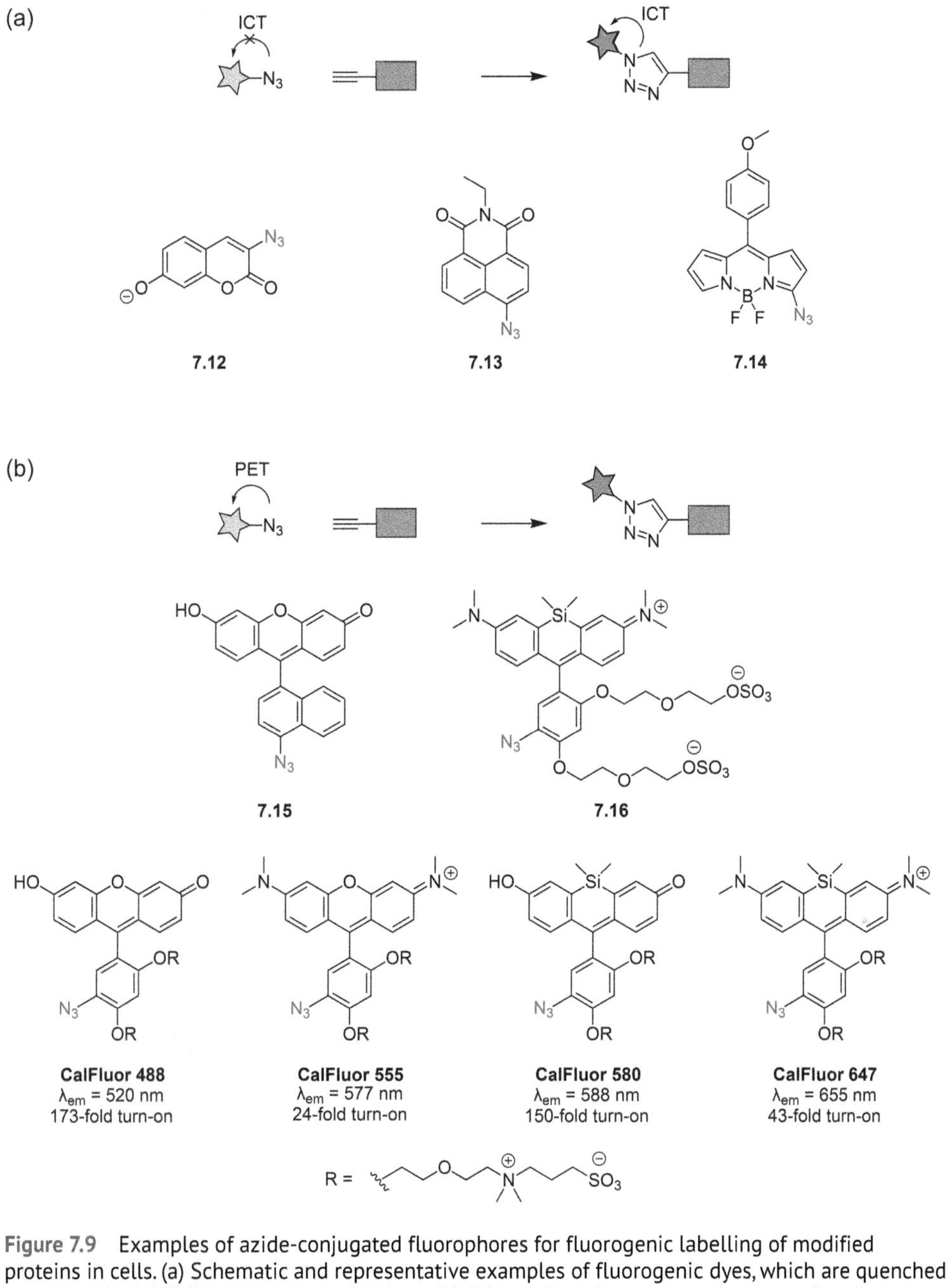

Figure 7.9 Examples of azide-conjugated fluorophores for fluorogenic labelling of modified proteins in cells. (a) Schematic and representative examples of fluorogenic dyes, which are quenched as azides block ICT fluorescence mechanisms. (b) Schematic and representative examples of fluorogenic xanthene-type dyes, which exploit PET quenching from azides. Fold turn-on refers to the fluorescence enhancement after copper-catalysed click reactions with alkynes *in vitro*. The reactive moiety is shown in red.

addition, the tetrazine moiety can quench fluorescence by FRET and through-bond energy transfer (TBET) processes. Upon reaction with a suitable partner, the quenching processes are halted, restoring fluorescence to the fluorophore. Thus, tetrazines present an excellent strategy for the development of fluorogenic labels.

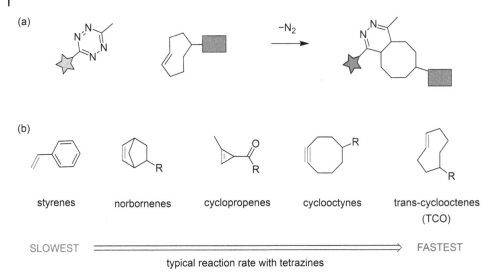

Figure 7.10 (a) Representative tetrazine and a *trans*-cyclooctene in an inverse electron demand Diels Alder cycloaddition reaction. (b) Examples of partners that can be used for the tetrazine ligation, arranged by approximate rates of reaction with tetrazines.

Tetrazines can also undergo the IEDDA reaction with cyclooctynes (see Section 7.2.2.3), cyclopropenes, norbornadienes, and even terminal alkenes. The choice of reaction partner for a tetrazine ligation will depend on the experiment (Figure 7.10b) [64]. This makes it simpler to find a compatible partner than for many of the other bioorthogonal reactions but comes at the cost of potential side reactivity when using more than one bioorthogonal reaction in a sample. One drawback of the ligation is that neither reagent is as small as an azide, and strained cycloalkenes have similar hydrophobicity to the cyclooctynes used in SPAAC. In addition, the fastest reacting *trans*-cyclooctene moieties have been known to hydrolyse and react with thiols in cells [7, 15].

The tetrazine ligation is compatible with fixed and live cell fluorescence imaging [65–67], can fluorescently label organelles for super-resolution microscopy [68], and has been used to radiolabel tumours for multimodal fluorescence and nuclear imaging [69, 70]. Modifications to both the tetrazine and dienophile partner allow for the elimination of groups during the ligation, which has great potential for the development of targeted fluorescent sensors. The diversity of applications using the tetrazine-ligation demonstrated in recent years demonstrates the utility of this technique for the development of fluorogenic labels in cellular imaging.

There are extensive reports of tetrazines being used as fluorescent labels for biomolecules, and Figure 7.11a presents representative examples where tetrazines have been conjugated to the most common dyes and used for imaging. To date, coumarin-tetrazines conjugates have demonstrated the highest fluorescence fold turn-on after reaction with a bioorthogonal tag. The coumarin derivative **7.17** exhibited an 11 000-fold turn-on upon reaction with TCO and was used to label TCO-tagged mitochondrial proteins and actin [71]. However, recent studies of coumarin derivatives have indicated that the fold turn-on values determined in cuvette studies do not always correlate to the fluorescence enhancement observed in cells [72]. In addition, autofluorescence was observed in these imaging conditions, necessitating the

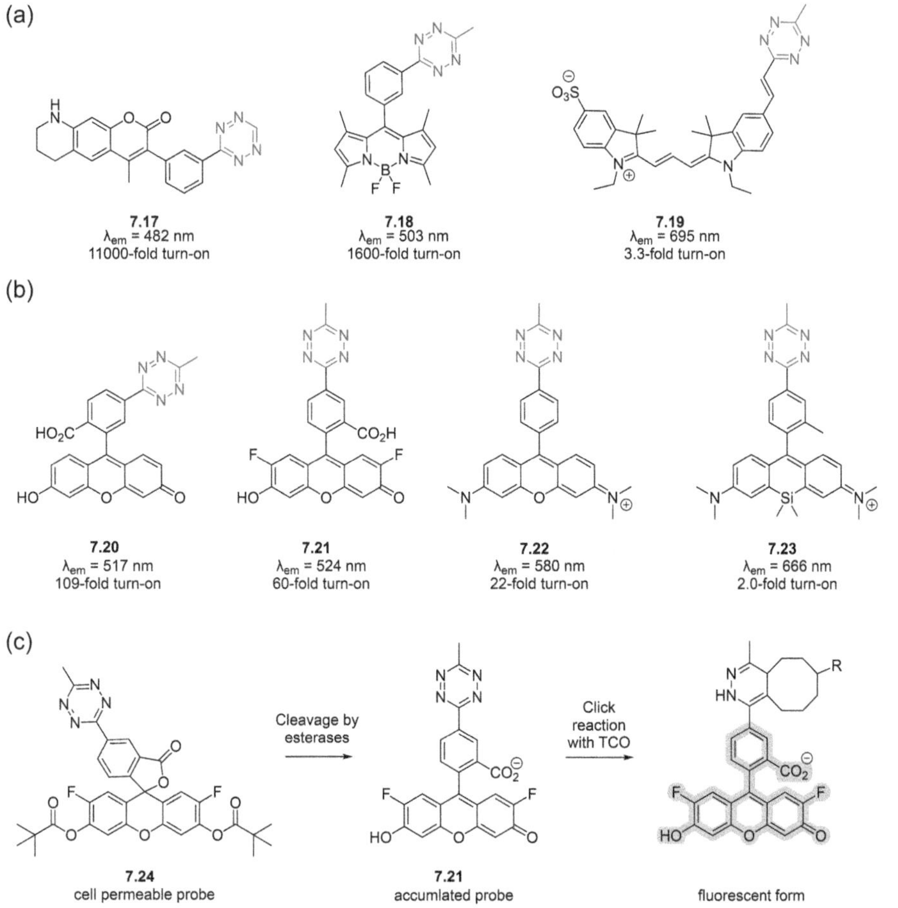

(a)

7.17
λ_{em} = 482 nm
11000-fold turn-on

7.18
λ_{em} = 503 nm
1600-fold turn-on

7.19
λ_{em} = 695 nm
3.3-fold turn-on

(b)

7.20
λ_{em} = 517 nm
109-fold turn-on

7.21
λ_{em} = 524 nm
60-fold turn-on

7.22
λ_{em} = 580 nm
22-fold turn-on

7.23
λ_{em} = 666 nm
2.0-fold turn-on

(c)

Cleavage by esterases

Click reaction with TCO

7.24
cell permeable probe

7.21
accumlated probe

fluorescent form

Figure 7.11 (a) Examples of fluorophore-tetrazine conjugates used for fluorogenic labelling of tagged biomolecules in cells. Fold turn-on values refer to the fluorescence enhancement after reaction with TCO *in vitro*. The reactive moiety is shown in red. (b) Scheme demonstrating an ester protection strategy to increase cell retention of tetrazines in cells.

search for green- and red-emitting fluorogenic tetrazine derivatives. As such, BODIPY derivatives have also been used for bioorthogonal labelling. The fluorescence of BODIPY **7.18** increased 1600-fold upon reaction with TCO and was used for the imaging of epidermal growth factor receptors (EGFR) in fixed cells [73]. These experiments required the introduction of a TCO-EGFR-antibody, meaning that these receptors could not be imaged dynamically in living cells. Cyanines were also conjugated to tetrazines, resulting in fluorogenic labels with red emission near 700 nm. Cyanine **7.19** was used for the no-wash, super-resolution imaging of TCO-tagged actin in fixed cells [74]. Despite this cyanine derivative only possessing a 3.3-fold turn-on in fluorescence after bioorthogonal reaction, the compatibility with super-resolution microscopy and red-shifted emission means that it would be the most suitable for *in cellulo* applications.

Numerous xanthene derivatives have been conjugated to tetrazines, including fluorescein (**7.20**), oregon green (**7.21**), rhodamine (**7.22**), and silicon rhodamine (**7.23**). The fluorescence emission of these dyes spanned 517–666 nm, with decreasing fluorogenic responses

(a)

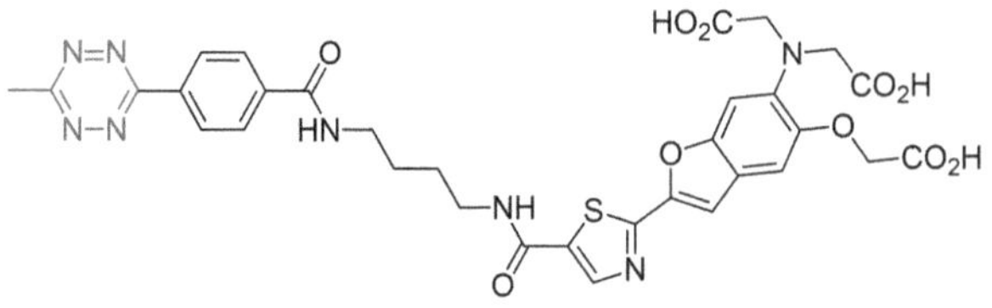

7.25

(b)

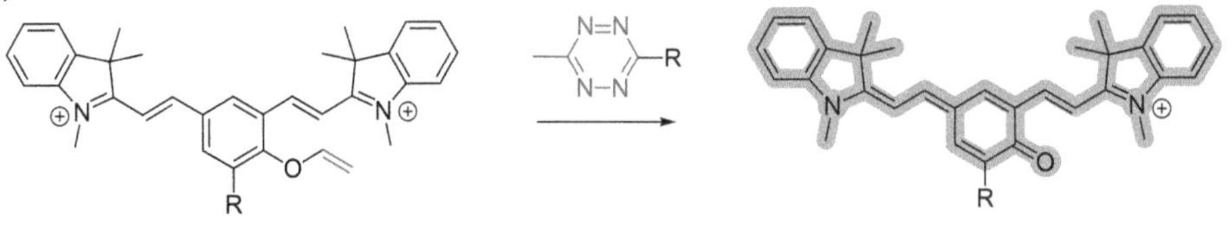

7.26

Figure 7.12 Examples of tetrazine-based fluorescent labels that have been used for sensing in cells. (a) A Mg^{2+} fluorescent sensor that is localised to the nucleus and Golgi using the tetrazine ligation. (b) Detection of mRNA using the tetrazine ligation between a fluorogenic vinyl ether and tetrazine. The mRNA (R groups) localises in only in cells transfected with the complementary mRNA sequence.

observed as the emission wavelength of the dye increased. All of these dyes were biocompatible and were used to image TCO-tagged actin in living cells. To increase cell permeability and retention, derivative **7.22** was also protected with pivalate esters that can be cleaved by esterases within the cell to give derivative **7.24** (Figure 7.11b). The cleavage of these esters resulted in increased cell retention and thus **7.24** was used to image TCO-tagged nuclear and mitochondrial proteins [75]. One challenge that this section highlights is that the ability of the tetrazine to quench is wavelength dependent, and probes that emit in the 400–500 nm range are better quenched than the red-shifted probes [76].

There are limited examples of fluorescent sensors that have been localised to specific sites within cells using the tetrazine ligation. The Buccella group ambitiously combined their group's existing small-molecule Mg(II) sensor with a tetrazine (**7.25**) in order to achieve in cell localised metal sensing [77] (Figure 7.12). The tetrazine moiety quenched the fluorescent output of their sensor, meaning that no fluorescence output was observed unless the probe had clicked at the targeted location. The BCN group targeted to nuclear and Golgi-localised proteins using a HaloTag (Section 7.4), permitting localised Mg^{2+} sensing. To date, this remains one of the only examples where tetrazines have been used to activate a small-molecule fluorescent sensor, and there is great potential for localised analyte sensing using this strategy. The only other report of fluorescent sensing of biomolecules involved the detection of mRNA and quantification by fluorescence. Devaraj and colleagues demonstrated that a cyanine dye could be formed *in situ* after reaction of a vinyl ether masking group and a tetrazine moiety. Sensor **7.26** was subsequently used to detect and quantify a specific sequence of mRNA that had been transfected into live CHO cells by conjugating the tetrazine group and the probe to complementary mRNA sequences [78].

Table 7.1 Cross compatibilities of the bioorthogonal reagents discussed in this chapter.

	Aldehyde/Ketone	Arylphosphine	Alkyne	Strained cyclooctyne	DIBO	DBCO	TCO	BCN	Cyclopropene	Norbornene	Styrene
Hydrazine/aminooxy	✓										
Azide		✓	✓	✓	✓	✓		✓			
Tetrazine				✓	✓	✓	✓	✓	✓	✓	✓

7.2.4 Commercial Fluorescent Labels

With many different reactions explored in the literature, the choice of the right bioorthogonal reactants depends various factors including sample type, size of modification, stability of reactants, rate of reaction, concentrations used, and the commercial availability of the reactants [15]. In response to this need, many companies have commercialised the core bioorthogonal groups. Typically, these groups bear pendant activated esters or amines that can be readily attached to biomolecules. Table 7.1 has been included in order to summarise the cross compatibilities of the bioorthogonal groups presented in this section.

In the above discussion, the best dyes that have been reported for fluorescent labelling would require considerable synthetic expertise for preparation. In response to this need, several companies have developed suites of 'clickable' fluorophores, containing hydrazide, azide, alkyne, and tetrazine moieties for conjugation to suitable bioorthogonal partners. Table 7.2 below summarises the availability of common commercially available dyes where bioorthogonal groups have been appended. Many of the companies also sell other less well-known dyes, but this table focuses on the dyes that are sold by multiple companies. Recent studies have confirmed that many of these dyes that bear tetrazines are suitable for super-resolution microscopy [79].

7.3 Short peptide Recognition Sequences

While the previously discussed section details methods of labelling any biomolecule with fluorophores, the following sections focus on the various techniques available for labelling proteins. One of the earliest developments in protein labelling is known as FlAsH, which requires a short peptide recognition sequence that can be readily appended to both *in vitro* and *in cellulo* proteins [80]. The designed peptide consists of a tetracysteine motif in the sequence Cys-Cys-X-X-Cys-Cys (where X is any amino acid except cysteine) that is genetically incorporated into the protein of interest. The short peptide motif adopts an alpha helical structure in which the four thiol groups are arranged on one side [81].

The FlAsH ligand, FlAsH-EDT$_2$, is a fluorescein-based, membrane-permeable small molecule containing two arsenic atoms complexed by 1,2-ethanedithiol (EDT) (Figure 7.13).

Table 7.2 Selected examples of commercially available fluorophores available for fluorescent bioorthogonal labelling.

Dye	λ_{em} (nm)	Hydrazine/aminoxy	Azide	Alkyne	DIBO	DBCO	Tetrazine
7-Hydroxycoumarin	407		✓[a,b]				
ATTO425	484					✓[b]	
BODIPY FL	510	✓[c]	✓[b,d]	✓[b,d]		✓[b]	✓[b,d]
5-Fluorescein (5-FAM)	512		✓[b,d]	✓[b,d]			✓[b,d]
6-Fluorescein (6-FAM)	512		✓[a,b,d]	✓[b,d]			✓[d]
ATTO488	520					✓[b]	✓[b]
Alexa Fluor 488	525	✓[c]	✓[b,c]	✓[b,c]	✓[c]	✓[b]	✓[b]
Oregon Green	526		✓[c]	✓[c]			
5/6-Carboxyrhodamine 110	527		✓[a,b,d]	✓[a,b]		✓[b]	
ATTO532	552						✓[b]
Alexa Fluor 546	573	✓[c]	✓[b]	✓[b]		✓[b]	
TAMRA	578		✓[a,b,c]	✓[a,b,c,d]	✓[c]		✓[b]
Alexa Fluor 555	580	✓[c]	✓[b,c]	✓[b,c]	✓[c]	✓[b]	✓[b]
Cy3	591		✓[a,b,d]	✓[a,b,d]		✓[a,b]	✓[b,d]
Alexa Fluor 568	603	✓[c]					
Texas Red	603	✓[c]	✓[b]	✓[b]			
Alexa Fluor 594	617	✓[c]	✓[b,c]	✓[b,c]	✓[c]	✓[b]	
Alexa Fluor 633	639	✓[c]					
ATTO 643	665						✓[b]
Alexa Fluor 647	665	✓[c]	✓[b,c]	✓[b,c]	✓[c]	✓[b]	
ATTO 647N	669						✓[b]
Cy5	670		✓[a,b,d]	✓[a,b,d]		✓[a,b]	✓[b,d]
Cy7	779		✓[b,d]	✓[b,d]			✓[d]

Footnotes indicate companies, which listed product in March 2021.
[a] Sigma Aldrich/Merck.
[b] Jena Biosciences.
[c] Thermo Fisher Scientific.
[d] Lumiprobe

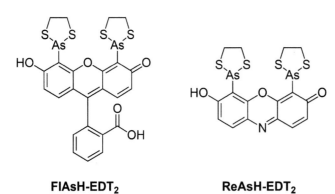

FlAsH-EDT$_2$ **ReAsH-EDT$_2$**

Figure 7.13 Chemical structures of the biarsenic ligands FlAsH-EDT$_2$ and ReAsH-EDT$_2$.

Protein-specific labelling by this system is based on the reversible, high-affinity binding of arsenic to thiol pairs: indeed, the toxicity of arsenic has been attributed to its ability to bind tightly to proteins that contain closely spaced thiol pairs [82]. This binding can be displaced by small vicinal dithiols such as EDT as they have a higher binding affinity for arsenic than the cellular dithiols [83]. Tsien *et al.* envisioned designing a short dithiol-containing peptide domain with a higher affinity for the FlAsH ligand than EDT ligands. The EDT ligands remain complexed to any unreacted probe, ensuring negligible arsenic toxicity.

The appropriately spaced arsenic centres of FlAsH-EDT$_2$ bind to the tetracysteine motif in an entropically favourable manner, as shown in Figure 7.14. The advantage of using the FlAsH system is the small size (<1 kDa) of the protein modification that minimises potential for interference with the function of the protein of interest when compared to fluorescent proteins (~30 kDa) [84].

FlAsH-EDT$_2$ has been found to react with off-target endogenous cysteine-rich proteins, resulting in substantial background fluorescence [85]. However, the tetracysteine motif has since been modified in attempt to maximise its affinity for FlAsH-EDT$_2$, thus decreasing the likelihood of non-specific background staining [86].

Flash-EDT$_2$ is non-fluorescent until bound to the tetracysteine motif, after which fluorescence emission can be observed at 528 nm. It is proposed that the EDT ligands allow for the free rotation of the aryl-arsenic bonds, resulting in quenching of the excited state through PET. Once complexed to the tetracysteine, rotation is restricted, which halts the quenching process and fluorescence is restored [87]. This fluorogenicity is advantageous in that it allows the protein of interest to be imaged without the need for extensive washing to remove unreacted dye before imaging.

The success of the FlAsH protein labelling system has led to the development of other FlAsH analogues such as resorufin-based, ReAsH, with a red fluorescence emission [88]. FlAsH and ReAsH are both commercially available, but can also be readily synthesised in two steps from inexpensive starting materials [89].

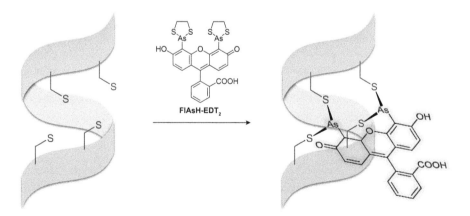

Figure 7.14 Four cysteines arranged on one side of the alpha helix such that the thiol pairs are appropriately spaced for favourable binding to the biarsenic FlAsH-EDT$_2$ ligand. One must also consider the isometric configuration in which the arsenic binds in a manner perpendicular to the alpha helix.

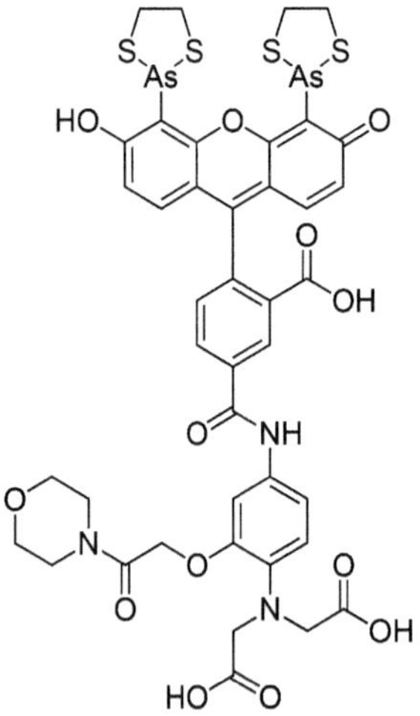

7.27

Figure 7.15 Ca^{2+}-specific sensing label based on the biarsenic FlAsH ligand.

FlAsH and ReAsH can be utilised as FRET acceptors and have been combined with traditional fluorescent protein FRET donors such as CFP and YFP, allowing for ratiometric and quantitative imaging of the interaction between proteins within the cell. For example, the CFP/FlAsH FRET pair was used to study the activation of G-coupled protein receptors and was successfully used to visualise the activation of α_{2A}-adrenegic receptors in real-time [90]. Similarly, ReAsH used as a FRET acceptor to GFP was used for live-cell studies of protein folding and stability [91].

The intermolecular FRET response between FlAsH and ReAsH has also been used in some protein studies. Jovin *et al.* used this FlAsH/ReAsH FRET pair to study the self-association of α-synuclein-C4, highlighting the potential for utilising this system to study the aggregation of amyloid proteins [92]. While the spectral overlap between FlAsH and ReAsH is not significant, increasing the amount of acceptor fluorophore relative to the donor can be used to achieve more favourable FRET conditions.

Other biarsenic compounds have been developed for use with the tetracysteine motif. A modular targeting approach, SplAsH, was introduced as a non-fluorescent targeting moiety, which would allow a variety of fluorophores, such as Alexa Fluor 594, to be used. SplAsH-Alexa594 was shown to have high specificity and greater sensitivity for the tetracysteine motif than ReAsH [93]. To the best of our knowledge, the utility of SplAsH-based labels in live-cell imaging has not been extensively explored. However, there is great potential for its application in *in vivo* and *in cellulo* studies. The main disadvantage of using SplAsH-based labels is that they are not inherently fluorogenic.

While the biarsenic compounds described above act as protein labels, an interesting set of compounds have been prepared for use in fluorescent sensing. Tsien *et al.* developed a Ca(II) sensor, **7.27**, based on the FlAsH-EDT$_2$ core tethered to a BAPTA-like chelator (Figure 7.15); **7.27** displayed a 10-fold fluorescence increase when bound to Ca(II) and was successfully used to image Ca(II) dynamics at the connexin 43 protein in live HeLa cells [94]. Given the abundance of fluorescein-based fluorescent sensors as described throughout this book, it is possible to imagine that many other responsive FlAsH agents could be prepared.

7.4 Fusion Protein Tagging Systems

Fluorescent sensors can be introduced into cellular environments by way of fusion proteins in a variety of methods that leverage the non-covalent interactions between protein-receptor-binding domains and their specific substrates. In these applications, fusion proteins consisting of the POI fused to a proteinogenic tag are expressed in cells. A small-molecule substrate with an appended fluorescent label is then introduced to the system, where it selectively binds to the receptor-binding domain of the tag *in cellulo* and labels the POI (Figure 7.16). The tag, substrate, and label can be selected based on the imaging needs

Figure 7.16 General strategy for site-specific labelling of proteins *in cellulo*. Step 1: Cells are engineered to express a fusion protein, which consists of the POI fused to a proteinogenic tag. Step 2: A small-molecule substrate with appended fluorescent label is introduced. Step 3: The substrate selectively binds to the receptor-binding domain of the tag *in cellulo*, fluorescently labelling the protein of interest.

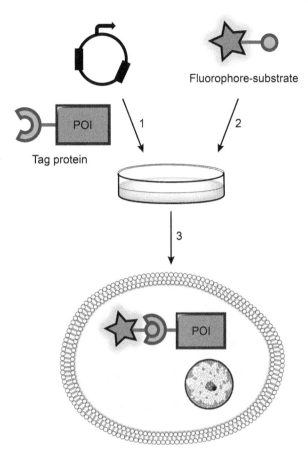

and the specific protein of interest, and can be engineered to optimise certain features such as fluorogenicity or substrate binding [95]. In this section, six prominent fusion protein tagging systems and their applications *in cellulo* are described. These include the FKBP tag, the TMP-eDHFR pair, and the PYP-tag systems, and then the SNAP-tag, CLIP-tag, and HaloTag systems.

7.4.1 FKBP Tag

FK506 binding proteins (FKBPs) are a family of immunophilin proteins that have high binding affinity to immunosuppressive drugs. There are at least 15 immunophilin proteins within the FKBP family, all of which are present in most eukaryotic tissues. Different types of FKBPs can also be found in various subcellular regions and organelles, including the cytoplasm, endoplasmic reticulum, and the nucleus. Achieving selectivity between FKBPs is challenging but can be attained by exploiting or further engineering the variance in binding pocket environments for substrates. The design of new synthetic substrates with selectivity between FKBPs, or selectivity of mutant FKBPs over wild-type FKBPs, led to the first examples of fusion protein tagging systems.

(a)

SLF'-substrate   FKBP12

(b)

7.28

Figure 7.17   (a) A protein of interest tagged with mutant FKBP12 can selectively bind fluorescently labelled SLF' substrates to enable *in cellulo* visualisation. SLF' substrate made up of SLF' itself (represented by the blue circle) and a fluorophore (represented by orange star). (b) Substrate **7.28** structure: SLF' (in red) conjugated to 5-carboxyfluorescein (in black).

Containing the minimum sequence for immunophilin activity, FKBP12 is a 12 kDa cytosolic protein within the FKBP family. Owing to its small size relative to FPs, FKBP12 is an ideal tag. FKBP12 has a high affinity for its natural substrates, the immunosuppressant drugs rapamycin and the family's namesake FK506, but it is the synthetic substrate, SLF, and its more sterically hindered derivative, SLF', that have found utility in fusion protein tagging systems. The SLF substrate is selective towards FKBP12, binds with high affinity, and lacks the immunosuppressive effects of the natural substrate, FK506. SLF is non-toxic, cell permeable, and can readily be conjugated to a fluorophore label.

The first example of the FKBP tag system for protein labelling *in cellulo* utilised a mutant FKBP12 variant, FKBP12(F36V), as the tag and a fluorescein conjugated SLF' (**7.28**) as the substrate (Figure 7.17) [96]. This system successfully labelled a range of intracellular proteins in a number of cell lines with little background staining. Later, the system was further optimised with a number of novel SLF'-dye conjugates that possessed improved fluorescence properties [97]. Within this optimisation process, synthetic polymers (pluronics) were co-administered to improve solubility and cellular uptake of SLF'-conjugates, and suppression dyes were used to improve the signal-to-noise ratio without the need for washing steps. The system was also validated on over 12 different target POIs fused to FKBP12 (F36V).

7.4.2 eDHFR Tag

Dihydrofolate reductase is an enzyme found in all organisms that plays important roles in cell proliferation and growth. It is an appealing candidate for the role of the tag in fusion protein tagging systems due to its small size relative to FPs (18 kDa) and monomeric nature.

Bacterially produced *Escherichia coli* dihydrofolate reductase (eDHFR), in particular, is suitable for *in cellulo* labelling since it is orthogonal to mammalian systems.

Initial reports of this system utilised fluorescently labelled O^6-alkylguanine-DNA derivatives [98] (see Section 7.4.4) and fluorescently labelled methotrexate (Mtx) as substrates [99]. Such examples successfully demonstrated selective labelling in mammalian cells. However, it is the use of DHFR with 2,4-diamino-5-(3,4,5-trimethoxybenzyl)pyrimidine (TMP) as the substrate that has seen the furthest development in imaging applications. In such applications, fusion proteins of the POI and eDHFR are expressed. The TMP substrate with an appended fluorescent label is then introduced to the system, where it can selectively bind to the receptor-binding domain of eDHFR, labelling the POI (Figure 7.18a).

The Cornish group were amongst the first to report applications of the TMP-eDHFR system, demonstrating that selective labelling of plasma membrane, nucleic, and cytosolic proteins could be achieved with low background fluorescence and relatively fast reaction

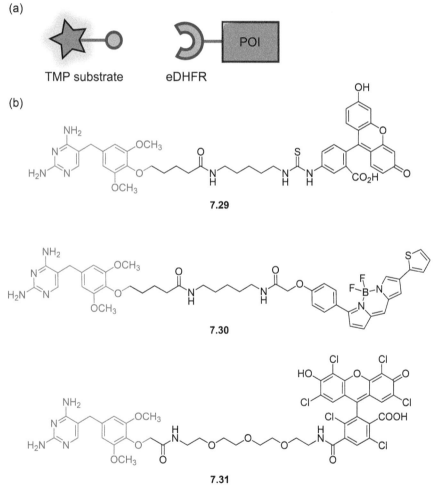

Figure 7.18 (a) A protein of interest tagged with eDHFR can selectively bind fluorescently labelled TMP substrates to enable *in cellulo* visualisation, (b) TMP-fluorescein derivative **7.29**, TMP-BODIPY Texas Red derivative, **7.30**, TMP-Hexachlorofluorescein derivative, **7.31**, with TEG linker for decreased lipophilicity. TMP shown in red.

kinetics [100]. Here, different fluorophores were appended to the TMP substrate, including fluorescein (**7.29**) and BODIPY Texas Red (**7.30**) derivatives (Figure 7.18b). It was demonstrated that changing the fluorophore did not disrupt the binding of the substrate to eDHFR. To further optimise the TMP-eDHFR system, a library of TMP substrate analogues were developed to further increase labelling kinetics, improve the specificity of the substrate binding, reduce lipid partitioning, and red-shift the fluorophores [101]. In addition, a TEG linker was added to one of the substrates (**7.31**) in order to decrease lipophilicity (Figure 7.18b).

The TMP-eDHFR system has been used in super-resolution imaging by utilising the commercially available fluorophore ATTO655 conjugated to the TMP substrate [102]. Using this technique, the dynamics of human histone h2b protein at ~20 nm resolution was visualised. In addition, the TMP-eDHFR system has proved to be compatible with two-photon imaging of live cells when suitable fluorophores were conjugated to the substrate [103].

Fluorogenic labels for the TMP-eDHFR system have been explored to improve signal-to-noise ratio. One such system was shown to rapidly label intracellular nuclear and cytoplasmic proteins with minimal background fluorescence (Figure 7.19) [104]. Here, the TMP substrate was modified to include not only an ATTO 520 fluorophore but also a cleavable quenching moiety (**7.32**). When unbound, the label's fluorescence is quenched, but upon binding, the quencher is cleaved from the substrate scaffold by a proximal cysteine residue, and fluorescence is restored.

Further developments of the TMP-eDHFR labelling system have seen the introduction of a 'tagging-then-labelling' approach in which SPAAC (see Section 7.2.2.3) is utilised to label the fusion-POI. This approach has been used to selectively label nuclear proteins *in cellulo* [105]. In this approach, the TMP substrate is modified with an azide, which acts as a bioorthogonal handle for SPAAC (Figure 7.20). The binding of the substrate to the fusion POI is termed the 'tagging' step. Next, a fluorophore with a free alkyne, which acts as the complimentary bioorthogonal handle for SPAAC, is introduced to the system. Once the two bioorthogonal handles react *via* SPAAC *in cellulo*, the 'labelling' of the POI is complete.

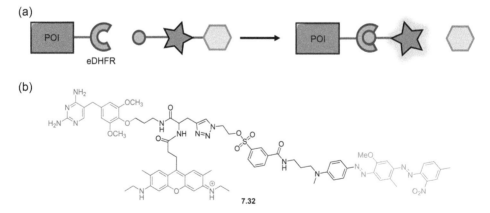

Figure 7.19 (a) Schematic of the fluorogenic TMP substrate binding to an eDHFR fusion protein. Once bound, a 20-fold increase in fluorescence occurs with the displacement of the quenching moiety. Fluorogenic substrate contains TMP (blue circle), quenched fluorophore (grey star), and quencher (yellow hexagon). (b) Actual structure of substrate **7.32**, TMP (shown in red) with fluorophore (commercially available ATTO520, shown in blue), cleavable group (sulfonate), and quencher (commercially available BHQ1, shown in green).

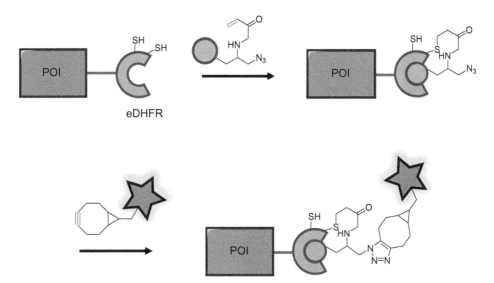

Figure 7.20 Schematic of the 'tagging-then-labelling' approach, which utilises a bioorthogonal strain-promoted azide-alkyne cycloaddition (SPAAC).

7.4.3 PYP Tag

Photoactive yellow protein (PYP) is a bacterial blue-light photosensor protein, which plays a role in the proteinogenic signalling pathways of bacteria. As for eDHFR, PYP is an ideal tag for use in fusion protein tagging systems due to its small size (14 kDa) relative to FPs, orthogonality to mammalian systems, and high substrate affinity and reaction kinetics. PYP binds to a natural, coumarin-based substrate, the Coenzyme A thioester of 4-hydroxycinnamic acid, and also binds to the thioester derivative of 7-hydroxycoumarin-3-carboxylic acid (**7.33**) (Figure 7.21).

The Kikuchi group introduced fusion protein tagging systems involving the PYP-tag, with initial reports demonstrating that **7.33** could selectively label cytosolic fusion-POIs

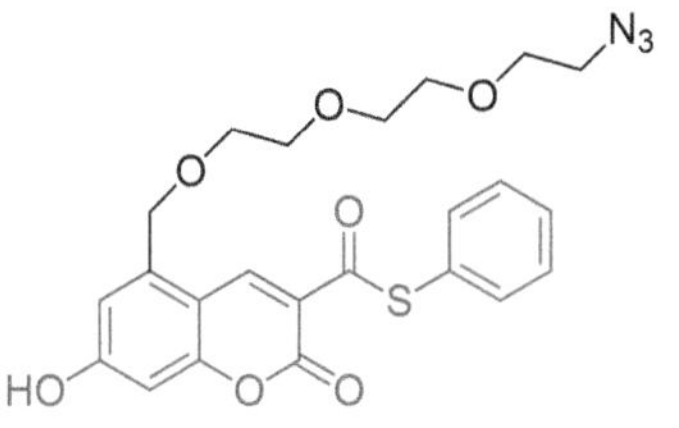

7.33 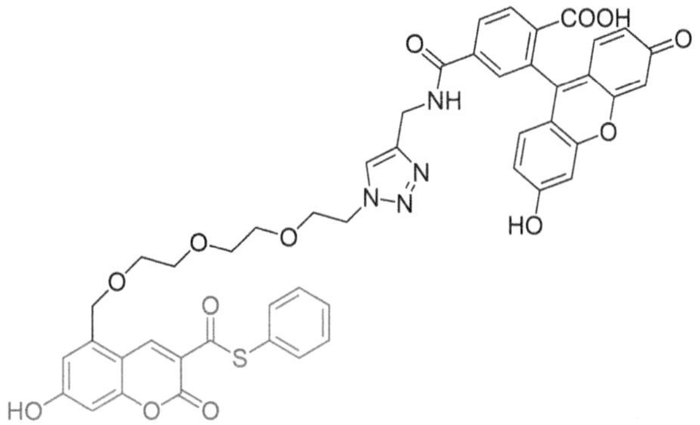   **7.34**

Figure 7.21 Structures of PYP-tag substrates: 7-hydroxycoumarin-3-carboxylic acid, **7.33**, and the fluorescein derivative, **7.34**.

using the PYP-tag *in cellulo* [106]. In comparison to other fusion protein tagging systems where the substrate must be appended to a fluorophore label, here the substrate is the label. A ratiometric version of the system was also described (**7.34**), with a FRET quenching fluorescein moiety added to the substrate (Figure 7.21). While this system permitted the labelling of cell membrane proteins, labelling was only successfully achieved after a 24-hour incubation period, which presents a serious limitation in the further applicability of the system.

Fluorogenic labels for the PYP tagging system have also been explored [107]. The fluorogenic scaffold **7.35** utilises a cinnamic acid thioester as the substrate. As this substrate is not inherently fluorescent, it is necessary to append a fluorophore label to it. The **7.35** substrate also includes a cleavable nitrobenzene quenching moiety such that once bound to the fusion protein, the quenching moiety can be displaced and fluorescence reinstated (Figure 7.22a). While this system was demonstrated to selectively label cell membrane proteins, the labelling of intracellular proteins was not possible since the charged **7.35** substrate is not cell permeable. To overcome this limitation and allow intracellular labelling, an acetylated derivative of the **7.35** substrate, **7.36**, was developed (Figure 7.22b) [106]. Here, the neutrally charged substrate is able to cross the cell membrane and enter the cytosolic space where it is subsequently cleaved by cellular esterases and returned to the anionic **7.35** form.

Coumarin-based substrates for PYP-tag systems have been developed for rapid intracellular cytosolic and nuclear labelling [108]. The fluorogenic and solvatochromatic substrates, **7.37** and **7.38**, exhibit sensitivity to the polarity of the solvent, so turn on when they enter the low polarity environment of the PYP-tag receptor-binding domain (Figure 7.23).

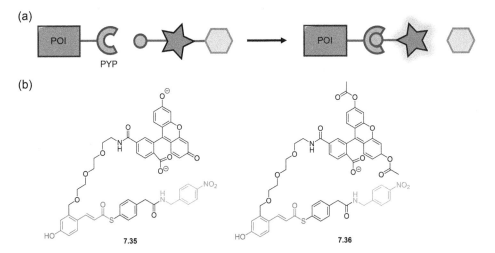

Figure 7.22 (a) Schematic of the fluorogenic substrate-receptor pair labelling of a PYP-tagged fusion-POI. Fluorogenic substrate contains PYP substrate (blue circle), quenched fluorophore (grey star), and quencher (yellow hexagon). (b) Actual structure of the fluorogenic PYP-tag substrate, **7.35**, which includes a cinnamic acid thioester as the PYP-substrate (shown in red), a quenched fluorescein moiety, and a cleavable nitrobenzene quenching moiety (shown in green). Acetylated derivative of the **7.35** substrate, **7.36**, was developed so as to allow intracellular labelling.

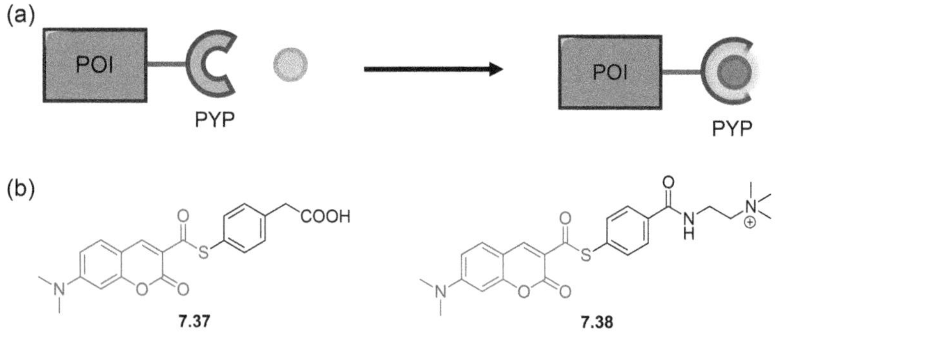

(a)

POI — PYP ○ → POI — PYP

(b)

7.37   7.38

Figure 7.23 (a) The fluorescence of the substrate is 'turned on' within the low polarity environment of the PYP-tag receptor-binding domain. (b) Structures of fluorogenic and solvatochromatic PYP substrates utilised in PYP-tag system for intracellular protein labelling, **7.37** and **7.38**.

7.4.4 SNAP-Tag and CLIP-Tag

The SNAP-tag system is a self-labelling system based on a modified version of the human DNA repair protein, O^6-alkylguanine-DNA alkyltransferase (hAGT), which reacts covalently with O^6-benzylguanine (BG) derivatives [98]. The modified hAGT works by dealkylating the BG derivative through one of its cysteine residues to form a stable thioether bond between the tag and label, releasing a guanine in the process (Figure 7.24a). The SNAP-tag enzyme was modified from native hAGT using a directed evolution approach [109–111]. Compared to wild-type hAGT, SNAP-tag possesses a 52-fold higher activity towards BG derivatives, lower susceptibility to proteolytic degradation, greater stability post-labelling, and does not bind to DNA [112].

A close relative of SNAP-tag, CLIP-tag utilises a different variant of hAGT, which reacts irreversibly to O^2-benzylcytosine (BC) derivatives. The mechanism occurs in a similar fashion to SNAP-tag, where a cysteine residue dealkylates the BC to form a thioether bond, releasing a cytosine (Figure 7.24b) [113]. CLIP-tag was designed to be orthogonal to SNAP-tag, allowing them both to be used for simultaneous labelling of two different proteins in the

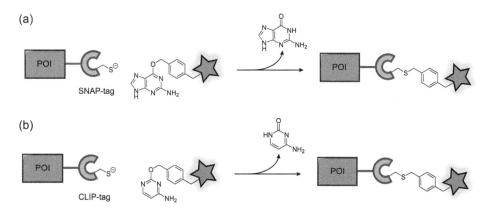

Figure 7.24 Schematic representation of labelling with the (a) SNAP-tag and (b) CLIP-tag systems.

same cell without cross-reacting. An array of fluorescent SNAP-tag and CLIP-tag ligands have been made commercially available. These ligands are based on a variety of core 'always-on' fluorophores, which cover a range of the blue to near infrared spectrum.

Fluorogenic labels for SNAP-tag and CLIP-tag are not yet commercially available. Fortunately, synthesis of fluorescent SNAP-tag and CLIP-tag ligand derivatives is relatively straight forward. Early development of fluorogenic labels involved the use of a fluorescence quenching moiety, which is cleaved upon binding to the tag, leading to a significant increase in the fluorescence output. In the case of SNAP-tag, BG is a known PET donor, which can act as a quencher to PET responsive fluorophores [114]. Herten *et al.* exploited this phenomenon to create a ATTO700-derived SNAP-tag label, **7.39** which exhibits a 30-fold fluorescence enhancement upon binding to the SNAP-tag (Figure 7.25) [115]. FRET-based fluorogenic probes have also been reported for SNAP-tag. Urano *et al.* designed a fluorescein-based BG derivative functionalised with a FRET accepting moiety (**7.40**) (Figure 7.25). Once bound to the SNAP-tag enzyme, the quencher is released along with the guanine, resulting in a 300-fold turn on. Sensor **7.40** was used to study the dynamics of the EGF protein receptor during cell migration [116]. The disadvantage of these fluorogenic labels is that they can be difficult to synthesise. However, it is possible to exploit commercially available quencher/fluorophore FRET pairs to generate fluorogenic SNAP-tag labels of varying emission wavelengths. An example of such a label is **7.41**, which was synthesised using the commercially available QSY 7 and CBG-NH$_2$ (Figure 7.25) [117].

Other fluorogenic enzyme-tag labels have been designed based on silicon-rhodamine dyes (Si-R). Si-Rs have a higher propensity to exist in the non-fluorescent ring-closed form in higher polarity solvents such as water. As such, the fluorogenic mechanism is thought to arise from the formation of the ring-opened dye in the hydrophobic binding pockets of the tag, while unreacted label remains in the ring-closed state in the aqueous cellular environment (Figure 7.26) [118]. This Si-R-tag was adapted to SNAP and CLIP utilising their respective linkers.

Recently, environmentally sensitive fluorophores have been used to create fluorogenic labels. Solvatochromic dyes are a well-known class of fluorophores, which vary in fluorescence intensity based on the polarity or viscosity of their surroundings [119]. Johnsson and co-workers were first to exploit environmental sensitivity for the development of a fluorogenic Nile Red-based SNAP-tag label. Nile Red is known for its strong fluorescence in less polar environments and spontaneous insertion into the cell membrane due to its hydrophobic nature [120]. The Nile Red SNAP-tag label, **7.42**, was designed such that membrane insertion was only possible once bound to the SNAP-tag enzyme (Figure 7.27). The use of this probe is limited to imaging membrane proteins [121].

Crystallographic data of the SNAP-tag enzyme has revealed that the binding pocket is narrow and hydrophobic in nature [122]. As such, a small label that responds to the environmental polarity may be utilised as a probe that changes in fluorescence once bound deep into the enzyme-binding pocket. In order to achieve this, the dye must be small enough to fit inside the narrow binding pocket. Tan *et al.* developed a probe based on this principle by utilising an environmentally sensitive small-molecule sensor, 4-sulfamonyl-7-aminobenzoxadiazole (SBD). The BG-bound fluorophore, **7.43**, had a 280-fold turn-on upon binding to the SNAP-tag enzyme (Figure 7.27). The probe was used to successfully image both intracellular and cell-surface proteins in HeLa cells using a no-wash method [123].

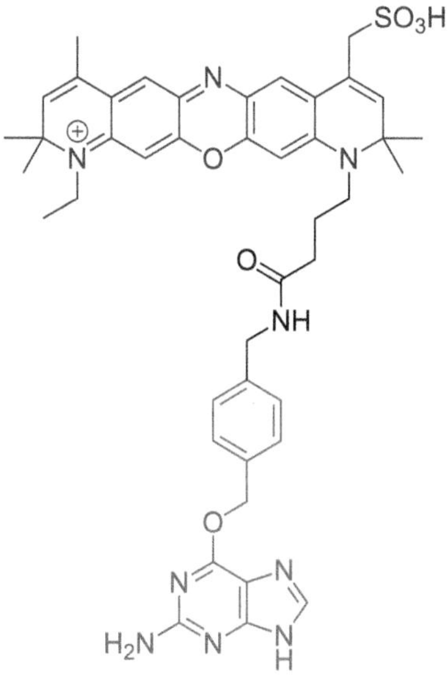

7.39

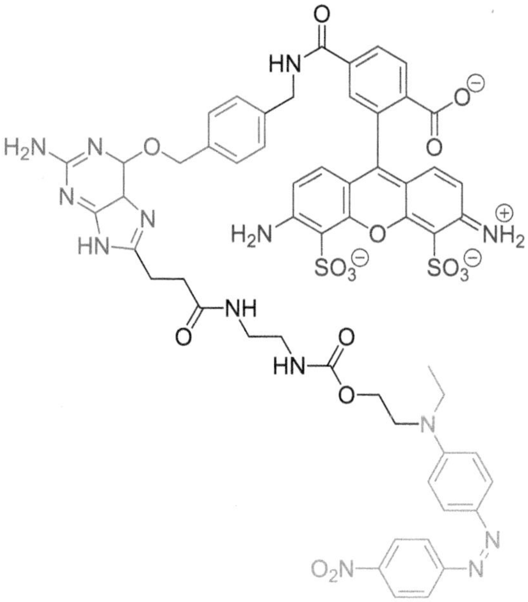

7.40

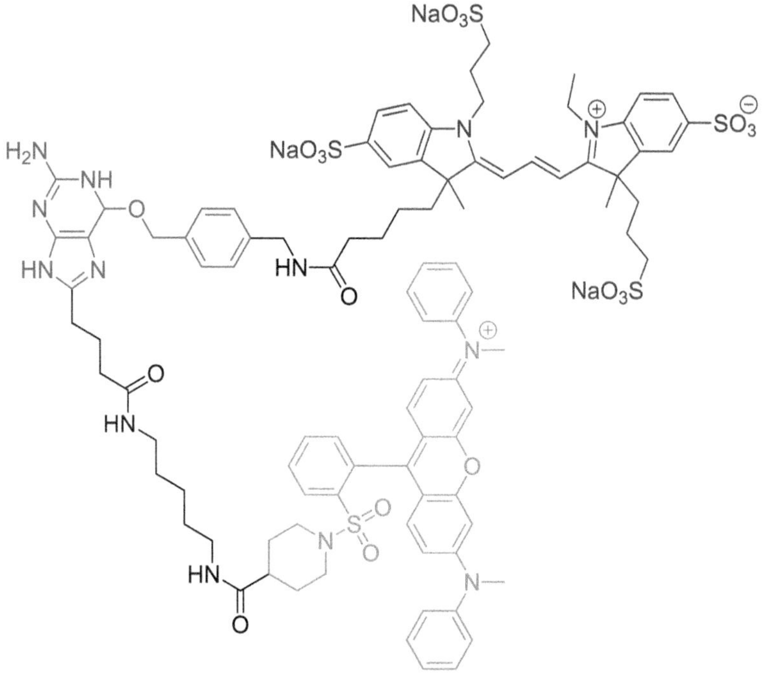

7.41

Figure 7.25 Quencher-release-based fluorogenic SNAP-tag probes highlighting the quenching moiety (green) and the fluorophore (blue). The SNAP-tag substrate is shown in red.

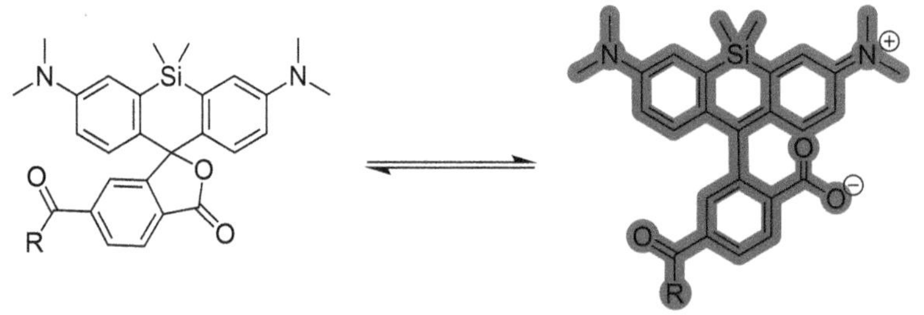

Figure 7.26 Si-R based enzyme-tag substrate in the non-fluorescent ring-closed state and the ring-opened fluorescent form. R is the core substrate for either SNAP-tag, CLIP-tag, or HaloTag.

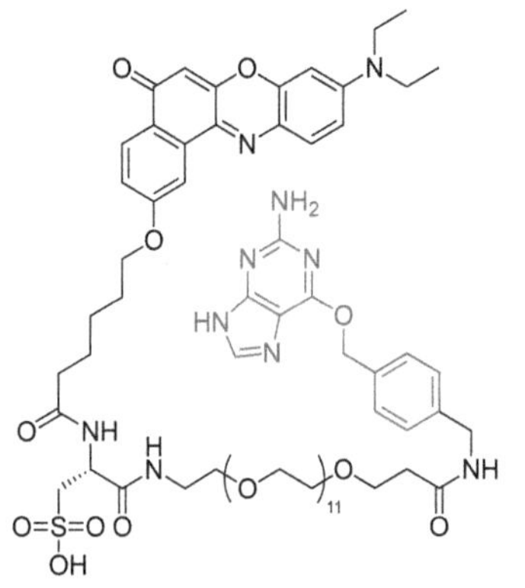

7.42 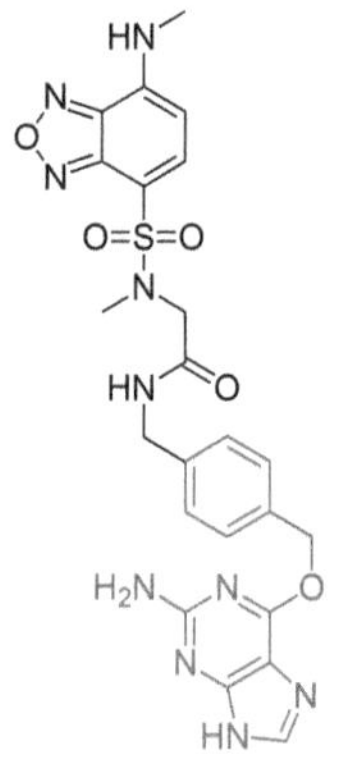　　　　　　　7.43 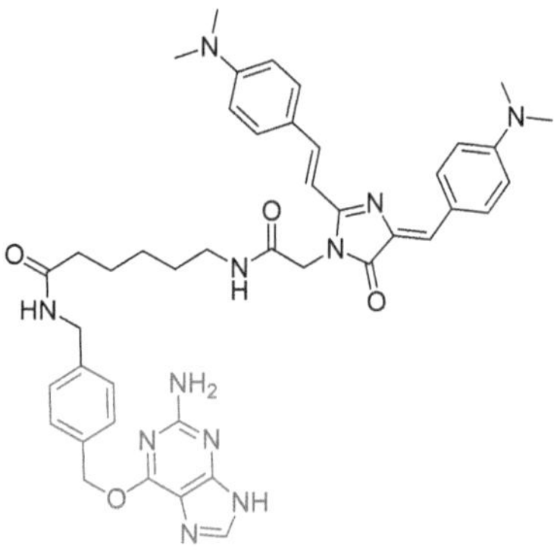　　　　　　　7.44

Figure 7.27 Environmentally sensitive fluorogenic SNAP-tag labels. The SNAP-tag substrate is shown in red.

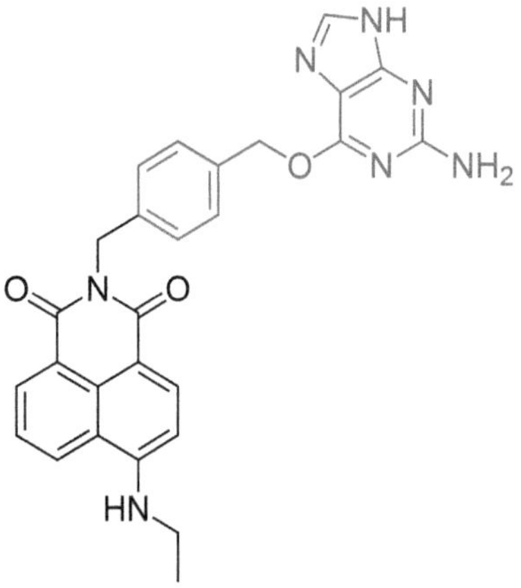

7.45

Figure 7.28 Environmentally sensitive and PET quencher release based fluorogenic SNAP-tag substrate. The SNAP-tag substrate is shown in red.

Other fluorescence quenching mechanisms such as twisted internal charge transfer (TICT) have been exploited to create fluorogenic SNAP-tag labels. For example, Zhang *et al.* designed a fluorogenic label based on the chromophore of the red fluorescent protein, Kaede. This fluorophore is quenched via a TICT mechanism. Once bound to the SNAP-tag enzyme, rotation becomes restricted, which, in turn, restores its fluorescence. Label **7.44** showed a 90-fold increase in fluorescence intensity upon binding to the SNAP-tag enzyme and was used to study Tar DNA-binding protein 43 in a wash-free imaging setup (Figure 7.27) [124].

Xu *et al.* utilised a combination of PET quenching by BG and the environmental sensitivity of naphthalimides to generate a fluorogenic SNAP-tag label (**7.45**, Figure 7.28). Naphthalimides have higher fluorescence intensities in hydrophobic environments, are small enough to fit into the narrow SNAP-tag binding pocket, and are quenched by PET mechanisms. Probe **7.45** was found to have a 36-fold increase in fluorescence when bound to SNAP-tag and was used for wash-free imaging of cytosolic, mitochondrial, and nuclear proteins [125].

7.4.5 HaloTag

The HaloTag system is based on a modified bacterial haloalkane dehalogenase enzyme, designed to link covalently to synthetic haloalkane-bearing ligands. The enzyme cleaves the terminal halogen *via* an S_N2 reaction, forming an ester bond between the alkane, and an aspartate residue located deep within the enzyme-binding pocket (Figure 7.29). In the wild-type enzyme, this ester intermediate is readily cleaved through base-mediated hydrolysis, catalysed by a nearby histidine residue. The modified HaloTag enzyme is designed such that this histidine is replaced with a phenylalanine residue, which is ineffective as a base. This inhibits the ester hydrolysis, thereby trapping the intermediate as a stable covalent adduct [126]. Several fluorescent HaloTag substrates ranging from the blue to near-infrared spectrum, including some fluorogenic substrates, have been made commercially available.

Fluorogenic HaloTag labels have been designed based on environmentally sensitive fluorophores. This was a difficult task to achieve due to the narrow, rigid, and hydrophobic binding pocket of the HaloTag. Kool *et al.* were the first to report an environmentally sensitive fluorogenic dye for the HaloTag system (**7.46**). This was achieved by using a truncated version of the original HaloTag substrate, tethered to a linear fluorophore. The fluorophore is quenched by TICT, which is relieved once bound within the binding pocket as bond rotation is restricted. The substrate exhibited a 27-fold increase in fluorescence output and was successfully utilised in bacterial imaging, with the potential for use in mammalian cells (Figure 7.30) [127].

A novel class of fluorogenic probes developed by Zhang *et al.* operate by cation–π interaction between the electron-donating group of the dye and the tryptophan residue found in the HaloTag enzyme binding pocket. It is believed that this interaction promotes excited-state separation and suppresses TICT, both of which enhance fluorescence. The label **7.47**

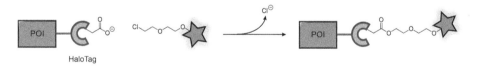

Figure 7.29 Schematic representation of labelling with the HaloTag system.

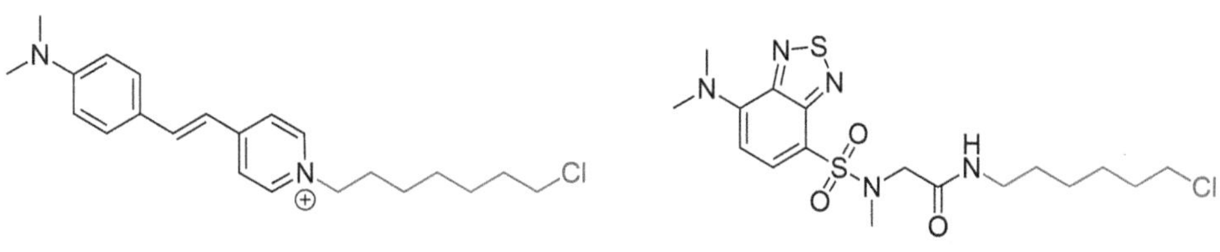

7.46   **7.47**

Figure 7.30 Environmentally sensitive fluorogenic HaloTag substrate based on TICT mechanism.

(Figure 7.30) exhibited a 1000-fold increase in fluorescence intensity when bound to the HaloTag enzyme and was utilised successfully in no-wash imaging of superoxide dismutase proteins in live cells [128].

As shown above. Si-R based substrates were shown to be effective as a fluorogenic probe for the HaloTag system by utilising the HaloTag linker (Figure 7.27). All three Si-R-tags were shown to be effective at tagging actin proteins in live cells [118].

Fluorogenic probes have yet to be designed based on quencher-release-based mechanism for the HaloTag system. This can be explained by the difficulty of introducing a cleavable quenching moiety to the linear substrate, and the very narrow nature of the HaloTag enzyme-binding pocket. This remains a niche to be filled.

7.5 Enzymatic Modifications for Labelling Proteins

Across all approaches to protein labelling, there is a necessary trade-off between labelling specificity and tag size. Enzymatic modifications aim to provide the best of both worlds: the specificity of protein machinery and the modular, bioorthogonal nature of small-molecule labels. For the purpose of protein labelling, enzymatic machinery can be exploited to attach small-molecule fluorophores onto a POI. A range of such systems exist, including approaches which exploit Coenzyme A derivatives [129], biotin ligase [130], and Sortase A [131]. Here, two enzymatic modifications for protein labelling are explored in detail: the LAP-tag system and protein *trans*-splicing. Both systems have been demonstrated to label membrane and intracellular proteins.

7.5.1 The LAP-tag System

As with all enzyme modifications used for the purpose of fluorescent protein labelling, the LAP-tag system utilises enzymatic machinery to attach small-molecule fluorophores onto a POI. In the LAP-tag system, the enzyme exploited for probe conjugation is lipoic acid ligase (LplA), which in natural systems catalyses the ATP-dependent ligation of lipoic acid onto various proteins involved in oxidative metabolism (Figure 7.31a). In the LAP-tag system, target POIs are modified to express short peptide tags, which act as specific substrates for LplA. In contrast to fluorescent proteins, these LplA accepter peptide (LAP) tags introduce minimal perturbation to the native system being studied. The LAP-tags act as highly specific recognition points for LplA, which can direct the ligation of small-molecule handles for fluorescent labelling [132] or even small-molecule fluorophores [133] onto the mutant POI.

Initial reports of the LAP-tag labelling system utilised wild-type LplA and demonstrated labelling of cell surface proteins only [134]. The system involved POIs transfected to express a 22-amino acid recognition sequence (LAP1), which directed the LplA-assisted ligation of an alkyl azide handle for SPAAC reactions onto the POI (Figure 7.32b). When reacted in cells with a cyclooctyne-appended fluorophore, the LAP-tagged POI becomes fluorescently labelled. An optimised LAP-tag has been developed for this system using yeast surface display, which demonstrated superior reaction kinetics for the labelling of cell surface proteins [135]. The optimised LAP-tag, LAP2, consisted of a 13-amino acid recognition sequence that had increased binding kinetics of over 70-fold compared to the first-generation LAP1

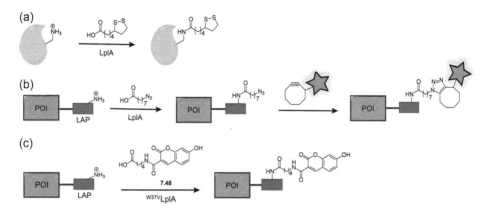

Figure 7.31 (a) In natural systems lipoic acid ligase (LplA) catalyses the ATP-dependent ligation of lipoic acid onto various proteins involved in oxidative metabolism, including the E2p domain of *E. coli* pyruvate dehydrogenase. (b) General scheme for LAP-tag system: The POI is engineered to express an LplA accepter peptide (LAP) tag. Upon recognition of this tag, wild-type LplA ligates an alkyl azide to the POI, which can be further derivatised by a fluorophore to complete the tagging mechanism. (c) The POI is engineered to express an LplA accepter peptide (LAP) tag. Upon recognition of this tag, mutant LplA ligates the fluorophore substrate **7.48** directly onto the POI completing the tagging mechanism.

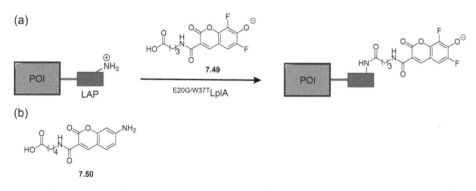

Figure 7.32 (a) A double mutant LplA can direct the ligation of the fluorescent substrate **7.48** directly onto a LAP-tagged POI using an extension of the PRIME method of fluorescent protein labelling. (b) Structure of **7.49**. ^{W37V}LplA can direct the ligation of the **7.49** fluorescent substrate directly onto a LAP-tagged POI.

recognition sequence. While labelling in mammalian cells was achieved with these systems using a number of fluorophore scaffolds with no background labelling of endogenous proteins, the relatively slow reaction kinetics of SPAAC alongside competing biological reactions in a cellular environment limits the systems applicability to cell surface proteins.

Efforts to overcome this limitation resulted in the development of the novel PRobe Incorporation Mediated by Enzymes (PRIME) fluorescent labelling strategy [132]. In PRIME, a mutant LplA (^{W37V}LplA) is expressed, which selectively recognises and directs the ligation of a 7-hydroxycoumarin (**7.48**) substrate onto a LAP-tagged POI (Figure 7.31c). The LAP-tag used is LAP2 [135], a 13-amino acid recognition sequence. With PRIME, fluorophore ligation is rapid, occurring within 10 minutes, and is highly selective. The

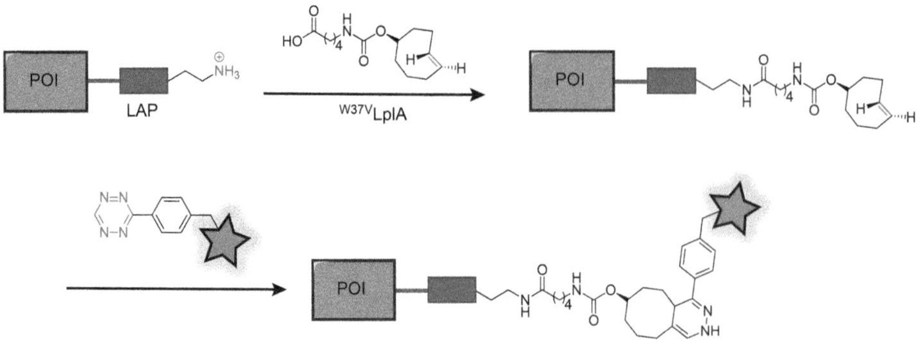

Figure 7.33 General scheme for fluorescent labelling using Diels-Alder cycloaddition. The POI is engineered to express a LAP-tag. Mutant LplA directs the ligation of a trans-cyclooctene handle onto the LAP-tagged POI. This handle can subsequently react in a bioorthogonal manner with a fluorophore-conjugated tetrazine moiety to fluorescently label the POI.

system was demonstrated to label both cell surface and intracellular nuclear proteins. A one-step ligation also reduces the number of excess wash steps required for the other two-step ligation systems.

While the PRIME method is useful in that it allows for intracellular proteins to be labelled with minimal washing steps and background fluorescence, substrate **7.48** has pH-dependent fluorescence and is only fluorescent in its anionic form. This factor limits cellular applications, as acidic cellular environments or organelles, such as endosomes, cannot be imaged as **7.48** would be non-emissive in these environments. To overcome this limitation, efforts to expand the recognition capability of PRIME have been made, including a system that recognises the pH-independent substrate Pacific Blue (**7.49**) [133]. While wild-type LplA and ^{W37V}LplA do not recognise substrate **7.49**, a double-mutant LplA ($^{E20G/W37T}$LplA) was developed, which could direct the ligation of **7.49** onto a LAP-tagged POI (Figure 7.32a). The **7.49** system was demonstrated to selectively label proteins on the cell surface, within intracellular environments, and inside acidic endosomes. Further, a 7-aminocoumarin (**7.50**) derivative of the PRIME fluorescent substrate was developed, which was fluorescent over a wide range of pH values and demonstrated selective labelling for both cell surface and intracellular proteins (Figure 7.32b) [136]. While **7.49** and **7.50** are not limited by any pH dependence, **7.48** remains the brightest fluorophore substrate for application with the PRIME system.

Novel two-step LAP-tag systems were later introduced, including one in 2011 [67], which utilised the IEEDA cycloaddition between TCO and tetrazine to label both cell surface and intracellular nuclear proteins (Figure 7.33). In this system, the POI is engineered to express a LAP-tag. Mutant LplA directs the ligation of a TCO moiety onto the LAP-tagged POI. TCO subsequently reacts in a bioorthogonal manner with a tetrazine-conjugated fluorophore to fluorescently label the POI. A number of fluorophores, including tetramethylrhodamine and Alexa Fluor 647, and a number of IEEDA cycloaddition partners were compatible with the system. The labelling kinetics for this system were demonstrated to be exceptionally fast.

7.5.2 Protein *Trans*-splicing

Discovered in 1990, protein splicing is a multi-step, autocatalytic, post-translational process thought to have been involved in enzyme evolution [137]. In general, a precursor protein

Figure 7.34 General schematic of protein splicing: a precursor protein is converted into its mature, functional form by self-removal of an internal polypeptide sequence, termed the intein.

is converted into its mature, functional form by self-removal of an internal polypeptide sequence (Figure 7.34).

While a number of protein splicing mechanisms exist [138], the mechanism most commonly used as a biotechnological platform for applications such as protein synthesis and protein labelling is protein *trans*-splicing. Protein *trans*-splicing is catalysed by *trans*-splicing (or split) inteins, one of the three structural types of inteins. *Trans*-splicing inteins do not have a peptide bond connecting the N- and C-termini of the intein splicing domain, but instead the domain is split in two, with the N-terminal sequence dubbed the N-intein (IntN) and the C-terminus known as the C-intein (IntC). IntN and IntC are encoded by separate genes and must associate and re-fold for the intein to be catalytically active and trigger the *trans*-splicing process (Figure 7.35).

Trans-splicing inteins can not only catalyse the ligation of endogenous proteins; they have also been demonstrated to carry out splicing in exogenous, engineered settings. This presents an opportunity for biotechnology and protein engineering applications, including protein labelling, since targeting moieties and small-molecule fluorophores can be

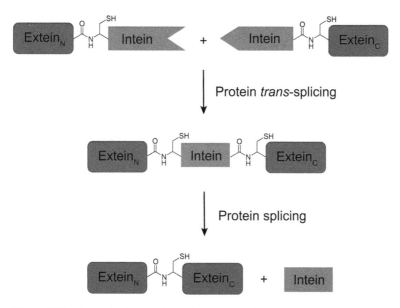

Figure 7.35 General scheme of protein *trans*-splicing, which occurs over three key steps. Step 1: Formation of the intein splicing domain – IntN and IntC associate and fold into the active conformation. Step 2: Self-removal of the internal polypeptide sequence: an intervening polypeptide sequence, the intein, is excised from the precursor protein. The mature protein is formed: the flanking polypeptide sequences, termed the exteins, autocatalytically ligate forming a new peptide bond.

incorporated into extein fragments and ligated onto a POI by way of *trans*-splicing inteins and the *trans*-splicing mechanism. *Trans*-splicing inteins can mediate the site-specific fluorescent labelling of a protein of interest *in cellulo* where fluorophores are conjugated onto one of the flanking extein sequences. The natural biorthogonality of the *trans*-splicing mechanism and the relatively high affinity and selectivity that the intein fragments have for each other makes the use of *trans*-splicing inteins for protein labelling advantageous relative to other protein labelling options. However, some key challenges with the approach remain, namely some issues around the solubility of the semi-synthetic protein fragments [139] and around the temperature dependence of some naturally occurring intein's catalytic activity [140].

Nevertheless, *trans*-splicing inteins have found great utility as protein labelling tools since their discovery 30 years ago. One of the first examples of the use of protein *trans*-splicing for the labelling of proteins *in cellulo* involved the naturally occurring Ssp DnaE intein, which was utilised to ligate a short peptide FLAG label onto the target protein, GFP [141]. In this strategy, the IntN sequence was endogenously expressed in the target POI, while the IntC sequence was linked to both the FLAG label and a cell-penetrating peptide (CPP) sequence to facilitate crossing of the extracellular membrane (Figure 7.36). Once inside the cell, the intein fragments associate and are excised, leaving the labelled POI. Similar approaches, where intein fragments directly ligate a fluorophore onto a POI *in cellulo*, have been demonstrated for the labelling of thioredoxin and b-lactamase proteins [142], hexahistidine [143], maltose binding protein, and human growth hormone (hGH) [144]. Dual-fluorescent labelling for FRET studies has also been achieved using protein *trans*-splicing in cases where two fluorophores were ligated onto the target protein by *trans*-splicing inteins [145].

Additionally, cell surface proteins have been labelled using a range of *trans*-splicing approaches [146]. In one approach the transmembrane POI, monomeric red fluorescent protein, was engineered to express the IntC segment of the Ssp GyrB intein [147]. In addition, the IntC segment was conjugated to the bacterial eDHFR protein (Figure 7.37). eDHFR's natural substrate, trimethoprim (TMP), was conjugated to the IntN segment, along with a biotin label. Directed by the high-affinity binding between DHFR and TMP, the intein segments meet *in cellulo* and excise themselves from the transmembrane POI, leaving the proteins labelled.

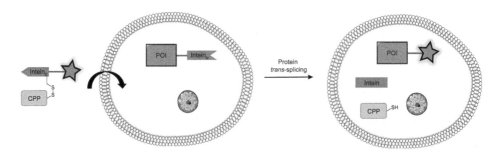

Figure 7.36 General scheme for labelling of POI *in cellulo* via protein *trans*-splicing method. The IntN sequence is endogenously expressed in the target POI, GFP, while the exogenous IntC sequence is linked to both a label and a cell-penetrating peptide (CPP) sequence. Once inside the cell, the intein fragments associate and are excised, leaving the POI labelled.

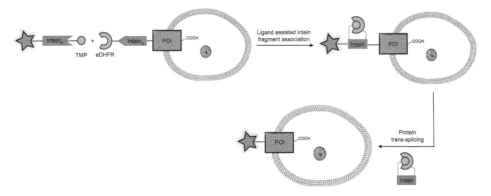

Figure 7.37 Example of the labelling of cell surface proteins using protein *trans*-splicing where the high-affinity binding between bacterial protein DHFR and its substrate, TMP, directs the conjugation between IntN and IntC.

7.6 Future Developments

A diverse group of fluorescent labels have been used to image the localisation and activity of proteins and biomolecules in cells. Through introducing tags to proteins and biomolecules, the label can be localised to the molecule of interest. Despite great progress in this field, there are many examples mentioned throughout this chapter that have only been applied *in vitro* or to bacterial systems and still have great potential for application in mammalian systems. There is also a large scope for the incorporation of new fluorescent sensors that bind specifically to the biomolecule of interest and provide a fluorescence output that correlates to the levels of an analyte in that particular location.

This chapter presents a wide range of labelling strategies. In particular, bioorthogonal reactions have emerged as a method that introduces the smallest modification to the biomolecule of interest. These bioorthogonal reactions can be used in complement with many of the protein labelling systems discussed in the later parts of the chapter. This will permit more ambitious imaging experiments to be performed, with multiple fluorescent signals independently reporting on the location, activity, or chemical environment surrounding the biomolecule of interest. These experiments will unravel the vast unknowns of how proteins and biomolecules change and interact during health and disease.

References

1 Suzuki, T., Matsuzaki, T., Hagiwara, H. et al. (2007). *Acta Histochem. Cytochem.* 40 (5): 131–137.

2 Wiedenmann, J., Oswald, F., and Nienhaus, G.U. (2009). *IUBMB Life.* 61 (11): 1029–1042.

3 Crivat, G. and Taraska, J.W. (2012). *Trends Biotechnol.* 30 (1): 8–16.

4 Frommer, W.B., Davidson, M.W., and Campbell, R.E. (2009). *Chem. Soc. Rev.* 38 (10): 2833–2841.

5 Brock, R., Hamelers, I.H.L., and Jovin, T.M. (1999). *Cytometry.* 35 (4): 353–362.

6 Wysocki, L.M. and Lavis, L.D. (2011). *Curr. Opin. Chem. Biol.* 15 (6): 752–759.

7 McKay, C.S. and Finn, M.G. (2014). *Chem. Biol.* 21 (9): 1075–1101.

8 Prescher, J.A. and Bertozzi, C.R. (2005). *Nat. Chem. Biol.* 1 (1): 13–21.

9 Liss, V., Barlag, B., Nietschke, M., and Hensel, M. (2015). *Sci. Rep.* 5: 1–13.

10 Rashidian, M., Dozier, J.K., and Distefano, M.D. (2013). *Bioconjug. Chem.* 24 (8): 1277–1294.

11 Nadler, A. and Schultz, C. (2013). *Angew. Chem. Int. Ed.* 52 (9): 2408–2410.

12 Cornish, V.W., Hahn, K.M., and Schultz, P.G. (1996). *J. Am. Chem. Soc.* 118 (34): 8150–8151.

13 Carrico, I.S., Carlson, B.L., and Bertozzi, C.R. (2007). *Nat. Chem. Biol.* 3 (6): 321–322.

14 Sletten, E.M. and Bertozzi, C.R. (2009). *Angew. Chem. Int. Ed.* 48 (38): 6974–6998.

15 Patterson, D.M., Nazarova, L.A., and Prescher, J.A. (2014). *ACS Chem. Biol.* 9 (3): 592–605.

16 Shieh, P. and Bertozzi, C.R. (2014). *Org. Biomol. Chem.* 12 (46): 9307–9320.

17 Tuley, A., Lee, Y.J., Wu, B. et al. (2014). *Chem. Commun.* 50 (56): 7424–7426.

18 Dirksen, A., Hackeng, T.M., and Dawson, P.E. (2006). *Angew. Chem. Int. Ed.* 45 (45): 7581–7584.

19 Konarzycka-Bessler, M. and Bornscheuer, U.T. (2003). *Angew. Chem. Int. Ed.* 42 (12): 1418–1420.

20 Dilek, O. and Bane, S.L. (2011). *J. Fluoresc.* 21 (1): 347–354.

21 Banerjee, A., Panosian, T.D., Mukherjee, K. et al. (2010). *ACS Chem. Biol.* 5 (8): 777–785.

22 Agarwal, P., Van Der Weijden, J., Sletten, E.M. et al. (2013). *Proc. Natl. Acad. Sci. U.S.A.* 110 (1): 46–51.

23 Ren, H., Xiao, F., Zhan, K. et al. (2009). *Angew. Chem. Int. Ed.* 48 (51): 9658–9662.

24 Liang, G., Ren, H., and Rao, J. (2010). *Nat. Chem.* 2 (1): 54–60.

25 Nguyen, D.P., Elliott, T., Holt, M. et al. (2011). *J. Am. Chem. Soc.* 133 (30): 11418–11421.

26 Yuan, Y., Wang, X., Mei, B. et al. (2013). *Sci. Rep.* 3: 1–7.

27 Chen, Z., Chen, M., Cheng, Y. et al. (2020). *Angew. Chem. Int. Ed.* 59 (8): 3272–3279.

28 Debets, M.F., Van Der Doelen, C.W.J., Rutjes, F.P.J.T., and Van Delft, F.L. (2010). *ChemBioChem.* 11 (9): 1168–1184.

29 Griffin, R.J. (1994). *Prog. Med. Chem.* 31 (C): 121–232.

30 Saxon, E., Armstrong, J.I., and Bertozzi, C.R. (2000). *Org. Lett.* 2 (14): 2141–2143.

31 Saxon, E. and Bertozzi, C.R. (2000). *Science* 287 (5460): 2007–2010.

32 Agard, N.J., Baskin, J.M., Prescher, J.A. et al. (2006). *ACS Chem. Biol.* 1 (10): 644–648.

33 Lemieux, G.A., De Graffenried, C.L., and Bertozzi, C.R. (2003). *J. Am. Chem. Soc.* 125 (16): 4708–4709.

34 Chang, P.V., Prescher, J.A., Hangauer, M.J., and Bertozzi, C.R. (2007). *J. Am. Chem. Soc.* 129 (27): 8400–8401.

35 Hangauer, M.J. and Bertozzi, C.R. (2008). *Angew. Chem. Int. Ed.* 47 (13): 2394–2397.

36 Nilsson, B.L., Kiessling, L.L., and Raines, R.T. (2000). *Org. Lett.* 2 (13): 1939–1941.

37 Rostovtsev, V.V., Green, L.G., Fokin, V.V., and Sharpless, K.B. (2002). *Angew. Chem. Int. Ed.* 41 (14): 2596–2599.

38 Tornøe, C.W., Christensen, C., and Meldal, M. (2002). *J. Org. Chem.* 67 (9): 3057–3064.

39 Hong, V., Steinmetz, N.F., Manchester, M., and Finn, M.G. (2010). *Bioconjug. Chem.* 21 (10): 1912–1916.

40 Breidenbach, M.A., Gallagher, J.E.G., King, D.S. et al. (2010). *Proc. Natl. Acad. Sci. U.S.A.* 107 (9): 3988–3993.

41 Agard, N.J., Prescher, J.A., and Bertozzi, C.R. (2004). *J. Am. Chem. Soc.* 126 (46): 15046–15047.

42 Baskin, J.M., Prescher, J.A., Laughlin, S.T. et al. (2007). *Proc. Natl. Acad. Sci. U.S.A.* 104 (43): 16793–16797.

43 Laughlin, S.T., Baskin, J.M., Amacher, S.L., and Bertozzi, C.R. (2008). *Science.* 320 (5876): 664–667.

44 Beatty, K.E., Fisk, J.D., Smart, B.P. et al. (2010). *ChemBioChem.* 11 (15): 2092–2095.

45 Poole, T.H., Reisz, J.A., Zhao, W. et al. (2014). *J. Am. Chem. Soc.* 136 (17): 6167–6170.

46 Van Geel, R., Pruijn, G.J.M., Van Delft, F.L., and Boelens, W.C. (2012). *Bioconjug. Chem.* 23 (3): 392–398.

47 Tian, H., Sakmar, T.P., and Huber, T. (2016). *Chem. Commun.* 52 (31): 5451–5454.

48 Jewett, J.C., Sletten, E.M., and Bertozzi, C.R. (2010). *J. Am. Chem. Soc.* 132 (11): 3688–3690.

49 Dommerholt, J., Rutjes, F.P.J.T., and van Delft, F.L. (2016). *Top. Curr. Chem.* 374 (2): 1–20.

50 Rong, L., Liu, L.H., Chen, S. et al. (2014). *Chem. Commun.* 50 (6): 667–669.

51 Sawa, M., Hsu, T.L., Itoh, T. et al. (2006). *Proc. Natl. Acad. Sci. U.S.A.* 103 (33): 12371–12376.

52 Qi, J., Han, M.S., Chang, Y.C., and Tung, C.H. (2011). *Bioconjug. Chem.* 22 (9): 1758–1762.

53 Li, J., Hu, M., and Yao, S.Q. (2009). *Org. Lett.* 11 (14): 3008–3011.

54 Jewett, J.C. and Bertozzi, C.R. (2011). *Org. Lett.* 13 (22): 5937–5939.

55 Friscourt, F., Fahrni, C.J., and Boons, G.J. (2015). *Chem. A Eur. J.* 21 (40): 13996–14001.

56 Sivakumar, K., Xie, F., Cash, B.M. et al. (2004). *Org. Lett.* 6 (24): 4603–4606.

57 Beatty, K.E., Liu, J.C., Xie, F. et al. (2006). *Angew. Chem. Int. Ed.* 45 (44): 7364–7367.

58 Wang, C., Xie, F., Suthiwangcharoen, N. et al. (2012). *Sci. China Chem.* 55 (1): 125–130.

59 Lippert, A.R., New, E.J., and Chang, C.J. (2011). *J. Am. Chem. Soc.* 133 (26): 10078–10080.

60 Shieh, P., Hangauer, M.J., and Bertozzi, C.R. (2012). *J. Am. Chem. Soc.* 134 (42): 17428–17431.

61 Shieh, P., Siegrist, M.S., Cullen, A.J., and Bertozzi, C.R. (2014). *Proc. Natl. Acad. Sci. U.S.A.* 111 (15): 5456–5461.

62 Shieh, P., Dien, V.T., Beahm, B.J. et al. (2015). *J. Am. Chem. Soc.* 137 (22): 7145–7151.

63 Blackman, M.L., Royzen, M., and Fox, J.M. (2008). *J. Am. Chem. Soc.* 130 (41): 13518–13519.

64 Oliveira, B.L., Guo, Z., and Bernardes, G.J.L. (2017). *Chem. Soc. Rev.* 46: 4985–4950.

65 Devaraj, N.K., Weissleder, R., and Hilderbrand, S.A. (2008). *Bioconjug. Chem.* 19 (12): 2297–2299.

66 Šečkute, J. and Devaraj, N.K. (2013). *Curr. Opin. Chem. Biol.* 17 (5): 761–767.

67 Liu, D.S., Tangpeerachaikul, A., Selvaraj, R. et al. (2012). *J. Am. Chem. Soc.* 134 (2): 792–795.

68 Erdmann, R.S., Takakura, H., Thompson, A.D. et al. (2014). *Angew. Chem. Int. Ed.* 53 (38): 10242–10246.

69 Rossin, R., Verkerk, P.R., Van Den Bosch, S.M. et al. (2010). *Angew. Chem. Int. Ed.* 49 (19): 3375–3378.

70 Rossin, R., Van Den Bosch, S.M., Ten Hoeve, W. et al. (2013). *Bioconjug. Chem.* 24 (7): 1210–1217.

71 Meimetis, L.G., Carlson, J.C.T., Giedt, R.J. et al. (2014). *Angew. Chem. Int. Ed.* 53 (29): 7531–7534.

72 Galeta, J., Dzijak, R., Oboříl, J. et al. (2020). *Chem. A Eur. J.* 26: 9945–9953.

73 Carlson, J.C.T., Meimetis, L.G., Hilderbrand, S.A., and Weissleder, R. (2013). *Angew. Chem. Int. Ed.* 52 (27): 6917–6920.

74 Knorr, G., Kozma, E., Schaart, J.M. et al. (2018). *Bioconjug. Chem.* 29 (4): 1312–1318.

75 Wieczorek, A., Werther, P., Euchner, J., and Wombacher, R. (2017). *Chem. Sci.* 8: 1506–1510.

76 Chen, L., Li, F., Nandi, M. et al. (2020). *Dye. Pigment.* 177: 108313.

77 Gruskos, J.J., Zhang, G., and Buccella, D. (2016). *J. Am. Chem. Soc.* 138 (44): 14639–14649.

78 Zhou, C.Y., Alexander, S.C., and Devaraj, N. (2017). *Chem. Sci.* 8: 7169–7173.

79 Beliu, G., Kurz, A.J., Kuhlemann, A.C. et al. (2019). *Commun. Biol.* 2: 261.

80 Miller, L.W. and Cornish, V.W. (2005). *Curr. Opin. Chem. Biol.* 9 (1): 56–61.

81 Griffin, B.A., Adams, S.R., and Tsien, R.Y. (1998). *Science.* 281 (5374): 269–272.

82 Webb, J.L. (2011). *Enzyme and Metabolic Inhibitors*, 595–819p. New York: Academic Press.

83 Whittaker, V.P. (1947). *Biochem. J.* 41 (1): 56–62.

84 Andresen, M., Schmitz-Salue, R., and Jakobs, S. (2004). *Mol. Biol. Cell.* 15 (12): 5616–5622.

85 Stroffekova, K., Proenza, C., and Beam, K.G. (2001). *Eur. J. Physiol.* 442 (6): 859–866.

86 Martin, B.R., Giepmans, B.N.G., Adams, S.R., and Tsien, R.Y. (2005). *Nat. Biotechnol.* 23 (10): 1308–1314.

87 Scheck, R.A. and Schepartz, A. (2011). *Acc. Chem. Res.* 44 (9): 654–665.

88 Adams, S.R., Campbell, R.E., Gross, L.A. et al. (2002). *J. Am. Chem. Soc.* 124 (21): 6063–6076.

89 Adams, S.R. and Tsien, R.Y. (2008). *Nat. Protoc.* 3 (9): 1527–1534.

90 Hoffmann, C., Gaietta, G., Bünemann, M. et al. (2005). *Nat. Methods.* 2 (3): 171–176.

91 Gelman, H., Wirth, A.J., and Gruebele, M. (2016). *Biochemistry.* 55 (13): 1968–1976.

92 Roberti, M.J., Bertoncini, C.W., Klement, R. et al. (2007). *Nat. Methods.* 4 (4): 345–351.

93 Bhunia, A.K. and Miller, S.C. (2007). *ChemBioChem.* 8 (14): 1642–1645.

94 Tour, O., Adams, S.R., Kerr, R.A. et al. (2007). *Nat. Chem. Biol.* 3 (7): 423–431.

95 Gao, J., Hori, Y., Nishiura, M. et al. (2020). *Chem. Lett.* 49 (3): 232–235.

96 Marks, K.M., Braun, P.D., and Nolan, G.P. (2004). *Proc. Natl. Acad. Sci. U.S.A.* 101 (27): 9982–9987.

97 Robers, M., Pinson, P., Leong, L. et al. (2009). *Cytom. Part A.* 75 (3): 207–224.

98 Keppler, A., Gendreizig, S., Gronemeyer, T. et al. (2003). *Nat. Biotechnol.* 21 (1): 86–89.

99 Miller, L.W., Sable, J., Goelet, P. et al. (2004). *Angew. Chem. Int. Ed.* 43 (13): 1672–1675.

100 Miller, L.W., Cai, Y., Sheetz, M.P., and Cornish, V.W. (2005). *Nat. Methods.* 2 (4): 255–257.

101 Calloway, N.T., Choob, M., Sanz, A. et al. (2007). *ChemBioChem.* 8 (7): 767–774.

102 Wombacher, R., Heidbreder, M., Van De Linde, S. et al. (2010). *Nat. Methods.* 7 (9): 717–719.

103 Gallaghers, S.S., Jing, C., Peterka, D.S. et al. (2010). *ChemBioChem.* 11 (6): 782–784.

104 Jing, C. and Cornish, V.W. (2013). *ACS Chem. Biol.* 8 (8): 1704–1712.

105 Chen, X., Li, F., and Wu, Y.W. (2015). *Chem. Commun.* 51 (92): 16537–16540.

106 Kamikawa, Y., Hori, Y., Yamashita, K. et al. (2016). *Chem. Sci.* 7 (1): 308–314.

107 Hori, Y., Nakaki, K., Sato, M. et al. (2012). *Angew. Chem. Int. Ed.* 51 (23): 5611–5614.

108 Hori, Y., Norinobu, T., Sato, M. et al. (2013). *J. Am. Chem. Soc.* 135 (33): 12360–12365.

109 Juillerat, A., Gronemeyer, T., Keppler, A. et al. (2003). *Chem. Biol.* 10 (4): 313–317.

110 Juillerat, A., Heinis, C., Sielaff, I. et al. (2005). *ChemBioChem.* 6 (7): 1263–1269.

111 Gronemeyer, T., Chidley, C., Juillerat, A. et al. (2006). *Protein Eng. Des. Sel.* 19 (7): 309–316.

112 Mollwitz, B., Brunk, E., Schmitt, S. et al. (2012). *Biochemistry.* 51 (5): 986–994.

113 Gautier, A., Juillerat, A., Heinis, C. et al. (2008). *Chem. Biol.* 15 (2): 128–136.

114 Seidel, C.A.M., Schulz, A., and Sauer, M.H.M. (1996). *J. Phys. Chem.* 100 (13): 5541–5553.

115 Stöhr, K., Siegberg, D., Ehrhard, T. et al. (2010). *Anal. Chem.* 82 (19): 8186–8193.

116 Komatsu, T., Johnsson, K., Okuno, H. et al. (2011). *J. Am. Chem. Soc.* 133 (17): 6745–6751.

117 Sun, X., Zhang, A., Baker, B. et al. (2011). *ChemBioChem.* 12 (14): 2217–2226.

118 Lukinavičius, G., Umezawa, K., Olivier, N. et al. (2013). *Nat. Chem.* 5 (2): 132–139.

119 Reichardt, C. (1994). *Chem. Rev.* 94 (8): 2319–2358.

120 Greenspan, P. and Fowler, S.D. (1985). *J. Lipid Res.* 26 (7): 781–789.

121 Prifti, E., Reymond, L., Umebayashi, M. et al. (2014). *ACS Chem. Biol.* 9 (3): 606–612.

122 Wibley, J.E.A., Pegg, A.E., and Moody, P.C.E. (2000). *Nucleic Acids Res.* 28 (2): 393–401.

123 Liu, T.K., Hsieh, P.Y., De, Z.Y. et al. (2014). *ACS Chem. Biol.* 9 (10): 2359–2365.

124 Jung, K.H., Fares, M., Grainger, L.S. et al. (2019). *Org. Biomol. Chem. Biomol. Chem.* 17 (7): 1906–1915.

125 Leng, S., Qiao, Q., Miao, L. et al. (2017). *Chem. Commun.* 53 (48): 6448–6451.

126 Los, G.V., Encell, L.P., McDougall, M.G. et al. (2008). *ACS Chem. Biol.* 3 (6): 373–382.

127 Clark, S.A., Singh, V., Vega Mendoza, D. et al. (2016). *Bioconjug. Chem.* 27 (12): 2839–2843.

128 Liu, Y., Miao, K., Dunham, N.P. et al. (2017). *Biochemistry.* 56 (11): 1585–1595.

129 George, N., Pick, H., Vogel, H. et al. (2004). *J. Am. Chem. Soc.* 126 (29): 8896–8897.

130 Chen, I., Howarth, M., Lin, W., and Ting, A.Y. (2005). *Nat. Methods.* 2 (2): 99–104.

131 Popp, M.W., Antos, J.M., Grotenbreg, G.M. et al. (2007). *Nat. Chem. Biol.* 3 (11): 707–708.

132 Uttamapinant, C., White, K.A., Baruah, H. et al. (2010). *Proc. Natl. Acad. Sci. U.S.A.* 107 (24): 10914–10919.

133 Cohen, J.D., Thompson, S., and Ting, A.Y. (2011). *Biochemistry.* 50 (38): 8221–8225.

134 Fernández-Suárez, M., Baruah, H., Martínez-Hernández, L. et al. (2007). *Nat. Biotechnol.* 25 (12): 1483–1487.

135 Puthenveetil, S., Liu, D.S., White, K.A. et al. (2009). *J. Am. Chem. Soc.* 131 (45): 16430–16438.

136 Jin, X., Uttamapinant, C., and Ting, A.Y. (2011). *ChemBioChem.* 12 (1): 65–70.

137 Perler, F.B. (1998). *Cell.* 92 (1): 1–4.

138 Romero-Casañas, A., Gordo, V., Castro, J., and Ribó, M. (2020). *Methods in Molecular Biology*, 15–29. New York: Springer US (*Methods in Molecular Biology*; vol. 2133).

139 Mootz, H.D. (2009). *ChemBioChem.* 10 (16): 2579–2589.

140 Paulus, H. (2000). *Annu. Rev. Biochem.* 69 (1): 447–496.

141 Giriat, I. and Muir, T.W. (2003). *J. Am. Chem. Soc.* 125 (24): 7180–7181.

142 Ludwig, C., Schwarzer, D., and Mootz, H.D. (2008). *J. Biol. Chem.* 283 (37): 25264–25272.

143 Ludwig, C., Pfeiff, M., Linne, U., and Mootz, H.D. (2006). *Angew. Chem. Int. Ed.* 45 (31): 5218–5221.

144 Kurpiers, T. and Mootz, H.D. (2008). *ChemBioChem.* 9 (14): 2317–2325.

145 Yang, J.Y. and Yang, W.Y. (2009). *J. Am. Chem. Soc.* 131 (33): 11644–11645.

146 Schütz, V. and Mootz, H.D. (2014). *Angew. Chem. Int. Ed.* 53 (16): 4113–4117.

147 Ando, T., Tsukiji, S., Tanaka, T., and Nagamune, T. (2007). *Chem. Commun.* 47: 4995–4997.

8

Future Directions of Fluorescence Sensors for Cellular Studies

Jiarun Lin[1,2], Natalie Trinh[1,3], and Elizabeth New[1,2,3]

[1] School of Chemistry, The University of Sydney, NSW, Australia
[2] The University of Sydney Nano Institute (Sydney Nano), The University of Sydney, NSW, Australia
[3] Australian Research Council Centre of Excellence for Innovations in Peptide and Protein Science, The University of Sydney, NSW, Australia

The past few decades have seen the rapid development of fluorescent sensors and their many applications in biological studies. Earlier chapters in this book have discussed common applications of fluorescent sensing, such as metal ion sensing or monitoring of the cellular environment. However, fluorescent sensors can be used in many other applications and applied in several different directions. This chapter will discuss some recent advances in the field of biological fluorescence sensing, highlighting areas for future research. These include lifetime imaging, near-infrared imaging, dual-analyte sensing, super-resolution microscopy, and multimodal imaging.

8.1 Fluorescence Lifetime Imaging Microscopy

8.1.1 Introduction

Fluorescent sensors have greatly advanced our knowledge of health and disease and of many biological processes. However, the visualisation of these processes and conditions have largely relied on the use of fluorescence intensity-based sensors. This presents a problem as there are many factors that can affect the fluorescence intensity of a sensor. For example, the fluorescence intensity is greatly dependent on the concentration of sensor and heterogeneous mediums such as cells and tissues can absorb and scatter light, therefore reducing the fluorescence intensity of a sensor [1]. Emission ratiometric sensors have the added advantage that they are concentration independent and therefore improves the accuracy of quantification [2]; however, they are also susceptible to factors such as light scattering and autofluorescence.

The fluorescence lifetime (τ) of a fluorophore is another measurable property that can be used to visualise sensors within cells. The development of imaging systems that utilise the fluorescence lifetime has allowed for another method for cellular imaging, which overcomes

Molecular Fluorescent Sensors for Cellular Studies, First Edition. Edited by Elizabeth J. New.
© 2022 John Wiley & Sons Ltd. Published 2022 by John Wiley & Sons Ltd.

drawbacks of fluorescence intensity-based imaging. The fluorescence lifetime is the average time the molecule spends in the excited state after excitation before decaying to the ground state [3] and is defined as the inverse of the sum of the rates for each type of decay pathway:

$$\tau = \frac{1}{k_r + k_{nr}}$$

where k_r is the rate of radiative decay and k_{nr} is the rate of non-radiative decay.

The fluorescence lifetimes of small organic fluorophores can range from picoseconds (*e.g.* for cyanine dyes) to nanosecond (*e.g.* for pyrene dyes) [4]. It has been shown in some classes of fluorophores that dyes with higher fluorescence quantum yields generally exhibit longer fluorescence lifetimes [5]. Structurally, more rigid dyes possess longer lifetimes compared to dyes with flexible bonds, as these also have higher quantum yields due to the absence of rotatable bonds, which can act as a pathway for non-radiative decay [4]. As the quantum yield is sensitive to environmental factors such as viscosity, temperature, pH, and solvent, fluorescent lifetimes can also be used to sense the cellular environment.

8.1.2   Advantages of Fluorescence Lifetime Imaging Microscopy

The common fluorescence lifetime imaging microscopy (FLIM) set-up is explained in detail in Chapter 2. FLIM offers many advantages over standard intensity-based fluorescence imaging. Firstly, the fluorescence lifetime of a fluorophore is independent of fluorophore concentration [6]. Therefore, FLIM is able to determine whether a change in fluorescence intensity is due to fluorescence quenching, changes in the fluorophore concentration, or a combination of both [7]. Furthermore, FLIM is less susceptible to inner filter effects, where highly concentrated solutions can result in spectral distortion or loss of signal [8].

Another advantage that FLIM has over intensity-based fluorescence microscopy is that it can discriminate between spectrally overlapping fluorophores. This is particularly advantageous for separating the emission of the fluorescent sensor of interest from that of other naturally occurring fluorescent molecules within cells.

Furthermore, the fluorescence lifetime of a fluorophore is an absolute measurement. This means that regardless of the method of measurement and the instrument configurations (*e.g.* instrument sensitivity or path length), the measurement will be reproducible across different instruments once the instrument response function (IRF) has been taken into account.

8.1.3   Examples of Sensors for FLIM

8.1.3.1   Endogenous Sensors

Many biomolecules within cells are intrinsically fluorescent, including reduced nicotinamide adenine dinucleotide (NADH) and flavin adenine dinucleotide (FAD), and proteins containing aromatic amino acids. These species can be used as endogenous fluorescent sensors to study metabolic interactions. The use of endogenous sensors has the advantage that an external label is not needed, which avoids any problems that arise from the introduction of external sensors to a biological system, such as toxicity and non-specific binding [8].

One of the earliest examples of FLIM using endogenous sensors was reported by Lakowicz and co-workers [9]. The authors used FLIM to successfully discriminate between free and protein-bound NADH. These two species are spectrally identical and are difficult to distinguish using intensity-based fluorescence microscopy. By measuring changes in fluorescence lifetimes of NADH in solution ($\tau = 0.37\,\text{ns}$) and when bound to malate dehydrogenase ($\tau = 0.94\,\text{ns}$), the authors were able to create 2D lifetime images of free and bound NADH. Similarly, fluorescence lifetimes have been used to distinguish between other related metabolic species, such as NADH and NADPH [10], and NADH and FAD [11].

Since these cofactors and coenzymes play a significant role in cellular metabolism and are involved in the healthy functioning of cells [12], there have been numerous studies using the fluorescence lifetime imaging of these species for clinical applications. For example, Skala and co-workers observed that precancerous epithelial cells exhibited greater variability in the redox ratio and the fluorescence lifetime of NADH compared to normal cells [11]. Furthermore, Sun and co-workers demonstrated the potential use of FLIM for image-guided surgery [13], showing that glioblastoma multiforme brain tumours have a non-uniform distribution of fluorescence lifetimes of NAD/NADH compared to normal cortical tissue, and this can be utilised to precisely distinguish between normal cells and tumour cells.

8.1.3.2 Exogenous Sensors

As fluorescent lifetimes are sensitive to the environment, exogenous sensors have been mainly implemented in the sensing of cellular pH, viscosity and temperature. Selected examples of FLIM-based sensors for these physicochemical properties will be discussed below.

8.1.3.2.1 pH Sensing As discussed in Section 6.2, many intensity-based pH sensors utilise a nitrogen or oxygen as the sensing moiety for pH, whereby protonation or deprotonation of the heteroatom results in a change in fluorescence intensity or a change in wavelength. This method can also be applied for fluorescence lifetime imaging, as the different protonation states of the molecule should have different fluorescence lifetimes. Hille *et al.* used this strategy and investigated the suitability of several commercially-available pH sensors for fluorescence lifetime imaging [14]. They found 2′,7′-bis(2-carboxyethyl)-5-(and -6) carboxyfluorescein (BCECF) to exhibit the greatest variation in fluorescence decay times, with a $\tau = 3.0\,\text{ns}$ at pH 6.4 and $\tau = 3.9\,\text{ns}$ at pH 8.4. The authors used **8.1** to successfully image the pH of cockroach salivary duct cells using both one-photon and two-photon FLIM (Figure 8.1).

Another method for pH sensing is to design a reaction-based sensor where acid can cleave a group to unmask fluorescence. Almutairi *et al.* reported **8.1**, a dendritic nanosensor for pH sensing in cells (Figure 8.1) [15]; **8.1** is a turn-on sensor which shows a fluorescence turn-on in acidic environments; **8.1** features a cypate dye as the fluorescent reporter, attached to an aliphatic PEGylated polymeric dendrimer through acid-labile bonds. In neutral and basic conditions, the cypate dyes can aggregate and stack, which causes fluorescence quenching and a decrease in fluorescence lifetimes. In acidic conditions, the cypate dye is cleaved from the polymeric backbone, resulting in an increase in both fluorescence intensity and fluorescence lifetime. Although no cell studies were performed with this sensor, one can imagine the potential that this sensor has for biological imaging and the other types of pH sensors that can be derived using this system.

8.1.3.2.2 Viscosity As discussed in Section 6.1, viscosity within cells is typically measured using molecular rotors, where a fluorophore contains a moiety that is able to freely rotate and the viscosity of the medium affects this ability [16]. Molecular motors are able to form twisted intramolecular charge transfer (TICT) states when excited, and relaxation occurs non-radiatively without any photon emission [17]. When the free rotation of the molecule is

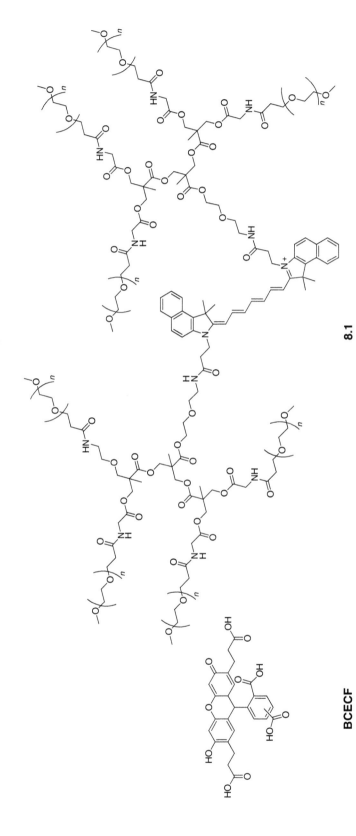

BCECF

8.1

Figure 8.1 Structures of sensors for lifetime imaging of intracellular pH [14, 15].

inhibited, the excited molecule is more likely to return to the ground state through fluorescence emission rather than non-radiative decay, and therefore the fluorescence intensity will increase. Concomitantly, the fluorescence lifetimes will also increase with restricted free rotation.

Kuimova *et al.* reported one of the first fluorescent sensors for the detection of viscosity using FLIM [18]. Their sensor **8.2** is based on a BODIPY fluorophore with a meso-substituted alkoxy phenyl group that is able to freely rotate. The fluorescence intensity of **8.2** increased with viscosity. This was due to the restriction in free rotation of the phenyl group, preventing the non-radiative decay and population of the dark state. The sensor was also shown to have an increase in fluorescence lifetime from 0.7 to 3.8 ns as the viscosity of the medium

Figure 8.2 Structure of a sensor for lifetime imaging of cellular viscosity [18].

increased from 28 to 950 cP. The authors demonstrated that their sensor was able to report on the viscosity within SK-OV-3 cells using FLIM, with the viscosity values confirmed using time-resolved fluorescence anisotropy (Figure 8.2).

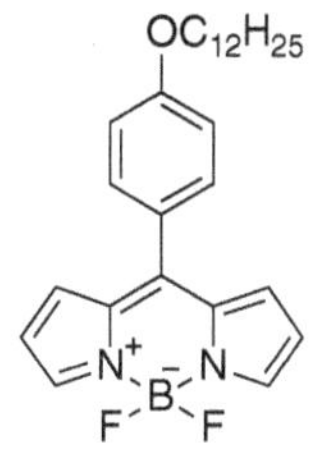

8.1.3.2.3 Temperature

All biochemical reactions involved with proper cell function produce energy, and therefore, temperature is a measurable output of cellular function. Changes in temperature, whether at the cellular or organelle level, have been linked to pathological diseases [19]. Similar to viscosity sensing, many fluorescent sensors exploit conformational changes arising from changes in temperature as a method for temperature sensing. As temperature and viscosity are related parameters, many temperature sensors also rely on the free rotation around a single bond. Through small structural changes, such as the introduction of sterically hindering groups, one can selectively sense temperature over viscosity [19].

Dmitriev and co-workers reported a temperature sensor, **8.3**, based on a sulforhodamine [20]. The authors chose a rhodamine fluorophore due to its well-known temperature-sensing ability, with the rotation of the amine groups disrupting the fluorescence and the non-radiative decay [21]. Sensor **8.3** also features a lipophilic tail for ease of incorporation into a polymeric nanoparticle to increase the cell-penetrating ability and also stabilise the microenvironment of the sensor, shielding it from other factors such as viscosity, which can affect the readout of the sensor. Compound **8.3** was applied to HTC116 cells, and using FLIM, the sensor was shown to have the same fluorescence lifetime across the whole cell. The cells were also subjected to different temperatures and **8.3** was shown to have a change in fluorescence lifetime from 2.8 to 2.2 ns as the temperature increased from 25 to 40 °C. The authors also used **8.3** to study temperature distribution across 3D spheroids using FLIM, determining that the spheroid core had a higher temperature compared to the periphery (Figure 8.3).

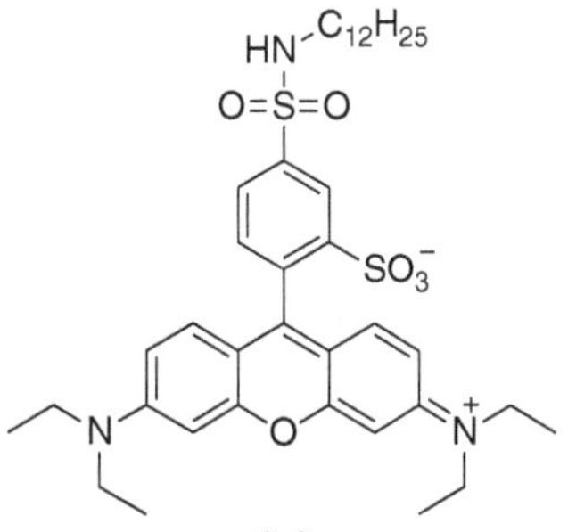

8.1.4 Future Directions

In the past few decades, FLIM has proven to be a powerful tool for bioimaging and has greatly progressed our understanding of health and disease. The sensitivity of fluorescence lifetime sensors towards the environment has

Figure 8.3 Structure of a sensor for lifetime imaging of cellular temperature [20].

deepened our knowledge of the roles that microenvironments play in cellular function. Although intensity-based imaging is likely to predominate bioimaging, the fluorescence lifetime can provide additional information that may be complementary to the information obtained from intensity-based imaging. It may also be worthwhile to revisit old sensors in the literature to investigate whether they have fluorescent lifetime properties useful for FLIM. Indeed, fluorescence quenching and the alleviation of fluorescence quenching, the most widely used methods for achieving fluorescent sensing, commonly result in changes in fluorescence lifetime. This means that it is likely that many reported intensity-based fluorescent sensors do in fact undergo concomitant changes in fluorescence lifetime, although this parameter is not routinely assessed.

A disadvantage of FLIM is its much longer required acquisition times, making it difficult to visualise rapid processes such as cell signalling events. New techniques and imaging systems with faster data acquisition times and high temporal resolution are therefore needed to progress the field of FLIM. Another complication with FLIM is that endogenous fluorophores or drugs with fluorescent properties can interfere with the imaging of exogenous fluorophores. Therefore, in the design of fluorescent sensors for FLIM, sensors with lifetimes longer than 8 ns would be optimal to minimise any interferences. Furthermore, the high sensitivity of fluorescence lifetimes to environmental factors can affect the read out and complicate data interpretation and needs to also be considered for future sensor design. With these design features in mind, the development of new sensors for FLIM will enable users to harness the potential that FLIM provides in the study of biological systems.

While we have limited discussion to organic FLIM-based sensors, this is an area that is well-served by luminescent metal complexes, which typically exhibit nanosecond lifetimes. For example, iridium complexes have been reported for lifetime-based pH sensing [22] and viscosity imaging [23]. Furthermore, polymeric systems have also proved useful for FLIM-based sensing, as large structural changes can have a dramatic effect on fluorescence lifetimes, as seen for a polypropylacrylamide-based temperature sensor that exhibits an almost twofold fluorescence lifetime change across 10 °C [24].

8.2 Near-infrared Sensors

The majority of reported fluorescent sensors absorb and emit in the UV-visible region (170–780 nm). However, fluorescent sensors that emit in the visible region have a number of drawbacks. Firstly, many biomolecules, such as FAD and NADH, also fluoresce in this wavelength region, which can cause high background noise and thus decreases signal-to-noise ratio [25, 26]. Other biomolecules, such as haemoglobin, fats, and water, can also easily absorb visible light, which results in light scattering, thus decreasing laser penetration depth in tissue [27]. Near infrared (NIR) sensors emit light between 650 and 1700 nm are therefore able to circumvent these complications, allowing for deeper tissue penetration (>500 μm) with less damage to cells [27]. NIR sensors therefore lend themselves to new applications in bioimaging, such as *in vivo* imaging and image-guided surgery. This section will focus on commonly used near-IR fluorophore scaffolds and notable examples of sensors in the literature.

NIR sensors fall into two categories: sensors that emit in the first near-infrared window, NIR I (650–950 nm) [28], and sensors which emit in the second near-infrared window, NIR II (1000–1700 nm) [29]. While NIR I sensors are able to achieve a much deeper tissue penetration compared to UV-visible sensors [30], autofluorescence is still observed at these wavelengths, as well as photon scattering that limits tissue penetration to only 1–2 cm [31, 32]. In contrast, there is up to a 1000-fold reduction in light scattering in the NIR II region, which allows for sub 10 μm image resolution [33]. NIR II light also exhibits greater tissue penetration than NIR I light [34].

NIR sensors have been developed based on a range of materials, including quantum dots [35], upconverting nanoparticles [36], single-walled carbon nanotubes [37], semi-conducting polymers [38] and small organic molecules [39]. Amongst small-molecule fluorophores, there are only a limited number of scaffolds reported to date, with many scaffolds exhibiting limited water solubility or poor photophysical properties.

8.2.1 Strategies to Make NIR Sensors

Although many commonly used fluorophores do not emit in the NIR region, one can make synthetic modifications to these fluorophore scaffolds in order to red-shift their fluorescence to the NIR. A number of strategies can be employed to achieve such a bathochromic shift. These include:

1) *Enhancement of intramolecular charge transfer (ICT)* through the introduction of electron donating "push" and electron withdrawing "pull" substituents on the fluorophore core. Push substituents cause destabilisation of the HOMO and pull substituents result in stabilisation of the LUMO, resulting in a decrease in the HOMO-LUMO gap and red-shifted fluorescence.
2) *Expansion of π-electron system* decreases the energy gap between the HOMO and LUMO, and therefore a bathochromic shift can be observed. This can be seen for example with cyanine dyes, where extension of the conjugated polymethine chain results in cyanine dyes with fluorescence emissions at longer wavelengths, or in BODIPY dyes, where substitution on the pyrrole groups with aromatic groups or fusion with aromatic groups leads to a red shift in fluorescence emission.
3) *Incorporation of group 14 metalloles*, such as silicon or germanium, which have low-lying LUMO levels that result in a bathochromic shift in fluorescence [40]. For example, replacement of the xanthene oxygen of rhodamine with silicon red-shifts the excitation wavelength to >600 nm [41].

These strategies are commonly applied to fluorophores, that already have long emission wavelengths to further red shift them to the NIR I or NIR II region. Below we present the most commonly used fluorophore scaffolds for the development of NIR sensors, and selected examples of sensors developed based on each class.

8.2.2 NIR Fluorophore Scaffolds

8.2.2.1 Cyanine Dyes and Their Derivatives

Cyanine dyes (Figure 8.4a) predominate NIR sensor development due to their long absorption and emission wavelengths [42]. Typical cyanine dyes used for NIR sensors contain two heterocyclic systems connected *via* a polymethine chain (closed chain cyanine, Figure 8.4b). Cyanines with longer polymethine chains tend to exhibit longer emission wavelengths.

Indocyanine green (**ICG**) is a heptamethine cyanine featuring negatively charged sulfates for improved water solubility and is currently one of only two FDA-approved NIR dyes

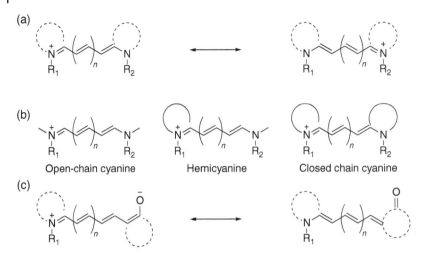

Figure 8.4 General structures of cyanine derivatives. (a) Cyanines have delocalised charge, which applies to (b) open chain cyanines, hemicyanines, and closed chain cyanines. (c) General structure and delocalised charge in merocyanines. Dotted curve line represents any alkyl derivative, while solid curved line represents heterocyclic alkyl derivatives.

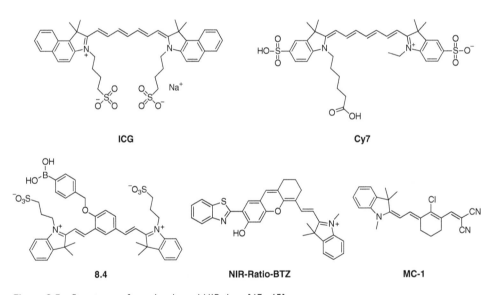

Figure 8.5 Structures of cyanine-based NIR dyes [43–45].

(Figure 8.5). **ICG** has an emission wavelength of 822 nm and has been used in a number of imaging applications including cardiovascular and lymphatic angiography [46, 47], tumour imaging [48], imaging of atheroscleric plaques [49], and for image-guided surgery [50]. However, since **ICG** has a net negative charge, it strongly binds to many proteins, resulting in a high signal to noise ratio [51]. To avoid this non-specific protein binding, much research has been dedicated to the improvement of **ICG** properties [52].

Heptamethine cyanine (**Cy7**) dyes are a sub-class of cyanine dyes that, like ICG, contain a polymethine chain that is seven carbons long (Figure 8.5). They exhibit NIR excitation and

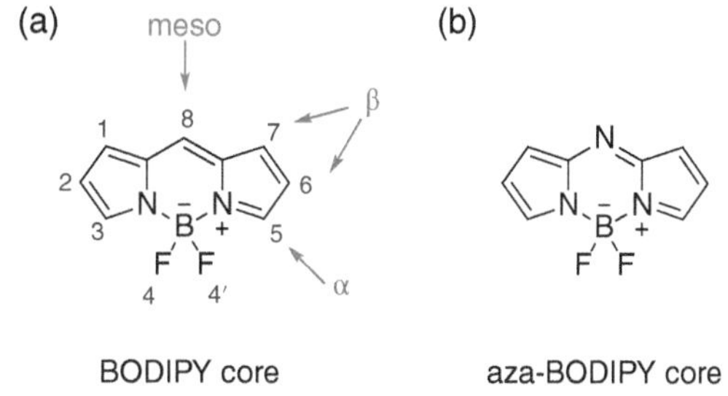

Figure 8.6 (a) General scaffold of BODIPY fluorophores and (b) structure of an aza-BODIPY core.

BODIPY core

aza-BODIPY core

emission wavelengths and are most commonly used for *in vivo* imaging. The introduction of a rigid cyclohexene substituent into the polymethine chain increases the photostability and fluorescence quantum yields in comparison to non-substituted Cy7 dyes [53]. Many commercially available dyes are based on this rigid heptamethine structure, including IR-780, IR-783, and IR-808, with a number of studies reported using them for imaging purposes [54–56]. Due to these improved photophysical properties, this scaffold has frequently been used in sensor development.

The Shabat group reported sensor **8.4**, a turn-on NIR I sensor based on a heptamethine cyanine dye with a cleavable boronate group for the detection of hydrogen peroxide (Figure 8.6) [43]. Reaction with H_2O_2 results in the restoration of NIR fluorescence ($\lambda_{em} = 720$ nm). The incorporation of sulfonated alkyl groups provides increased water solubility and prevents sensor aggregation. The sensor was shown to be sensitive to increased H_2O_2 production in mice treated with lipopolysaccharide (LPS), which induces an inflammatory response.

Hemicyanines are asymmetrical cyanines (Figure 8.4b), which have the additional advantage that functional groups can be readily integrated into the scaffold to tune NIR absorption and emission properties, and to develop targeted or analyte responsive sensors [33, 52]. In comparison to cyanine dyes, hemicyanines display better photostability and higher quantum yields [33].

Yang and co-workers reported a NIR hemicyanine sensor for ratiometric, *in vivo* pH monitoring [44]. The sensor, **NIR-Ratio-BTZ**, features a hydroxyl group on the xanthene ring which acts as the pH sensing unit and a benzothiazole group, which red-shifts the fluorescence emission due to its electron withdrawing nature, enhancing ICT (Figure 8.6). Under acidic conditions, **NIR-Ratio-BTZ** has a fluorescence emission at 672 nm, and under basic conditions, the fluorescence of **NIR-Ratio-BTZ** is red-shifted to 742 nm. The large bathochromic shift between the protonated and unprotonated species is ideal for imaging, as cross-talk is limited, increasing imaging resolution. Cell studies with **NIR-Ratio-BTZ** showed that the sensor could be used to successfully quantify pH levels in HeLa cells incubated at specific pH values. The authors next demonstrated the ability of sensor **NIR-Ratio-BTZ** to monitor pH in nude mice treated with LPS.

Merocyanines are electronically neutral cyanine-type dyes, which therefore exhibit different biological behaviour to their charged analogues (Figure 8.4c). Yan *et al.* utilised a merocyanine scaffold for their sensor **MC-1**, developed for the imaging of amyloid-β (Aβ) plaques, a biomarker for Alzheimer's disease (Figure 8.6) [45]. By replacing the cationic moiety of a cyanine with a dicyanomethylene group, the ability of the sensor to cross the blood–brain barrier was increased. **MC-1** is a turn-on fluorescent sensor, exhibiting increased fluorescence emission at 685 nm upon binding to Aβ plaques. *In vivo* studies were performed in a transgenic mouse model of Alzheimer's disease, with imaging showing that **MC-1** was able to penetrate the blood–brain barrier and a fluorescence signal was observed in transgenic mice, while no signal was observed in wild-type mice.

8.2.2.2 BODIPY Dyes

BODIPY dyes typically emit green-yellow fluorescence, but substitutions on the BODIPY core can result in red-shifting of the fluorescence emission to the NIR region [57]. For example, substitutions with aromatic units at the α, β, or meso positions can extend the π conjugation to give a red shifted fluorescence, or replacement of the meso carbon with an imine nitrogen can yield an aza-BODIPY, which also results in a large red-shift in fluorescence (Figure 8.6) [57]. The easily tuneable nature of the BODIPY scaffold makes them ideal fluorophores for the construction of NIR sensors.

Li *et al.* reported sensor **NIR-BODIPY-Ac**, a BODIPY-based sensor for the detection of cysteine (Figure 8.7) [58]. NIR I emission was achieved by conjugation of an indolinium to the BODIPY core, extending the π-conjugation. The fluorescence quenching acrylate group was appended to the meso position. Michael addition of cysteine to the acrylate group is followed by cyclisation and cleavage, giving a highly fluorescent hydroxy BODIPY product (λ_{em} = 708 nm). **NIR-BODIPY-Ac** was applied to mice, with weaker NIR emission observed in mice treated with *N*-ethyl maleimide, a thiol blocking agent, than in control mice.

Bai *et al.* reported a series of NIR II fluorophores based on aza-BODIPY for tumour imaging [59]. Amongst this series, **NJ1060** had the longest wavelength emission at 1060 nm (Figure 8.7). Due to the low water solubility of the dye and its tendency to aggregate within aqueous solutions, the authors encapsulated the sensor within a Pluronic F-127 matrix, and the resulting nanoparticle able to be homogenously distributed within aqueous solutions without aggregation. Mice bearing a 4T1 tumour were treated with **NJ1060**, and a fluorescence enhancement could be observed in the tumour region in the NIR II window. This scaffold could therefore be used as the basis for design of responsive sensors with NIR II emission.

8.2.2.3 Squaraine Dyes

Squaraines are polymethine dyes featuring a central four-membered ring and a resonance-stabilised zwitterionic structure. Squaraine dyes have narrow absorption and emission

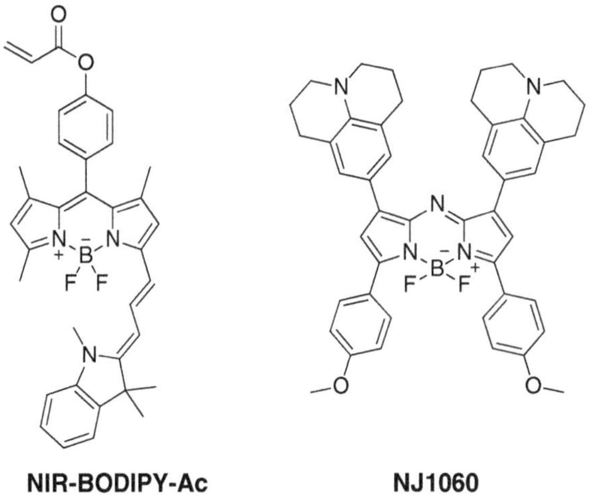

Figure 8.7 Structures of BODIPY-based NIR dyes [58, 59].

NIR-BODIPY-Ac   **NJ1060**

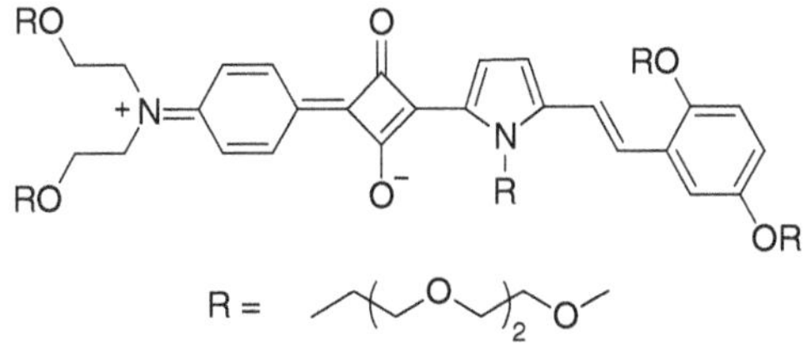

Figure 8.8 Structure of a squaraine-based NIR dye [60].

USq

profiles in the NIR region and are highly fluorescent, making them excellent fluorophores for NIR sensor development [61].

Anees *et al.* reported squaraine-based sensor **USq** for the detection of aminothiols (Figure 8.8) [60]. Biological thiols react with the electron-deficient cyclobutene ring of the squaraine in an addition reaction, leading to a change in maximum emission from 700 to 520 nm. Triethlyene glycol units were included for increased water solubility. **USq** was subjected to *in vivo* studies with both fluorescent and photoacoustic imaging modes. Excitation at 675 nm showed even distribution of sensor **USq** within mice models, with intensity decreasing after feeding due to the metabolism-generated thiol species.

8.2.2.4 Other Dye Scaffolds

Rhodamine dyes typically emit orange fluorescence, but modifications to their scaffold can lead to emissions in the NIR. For example, Koide *et al.* reported **SIR700**, a silicon rhodamine dye for the *in vivo* imaging of malignant gliomas (Figure 8.9) [62]. In addition to substituting the oxygen of the xanthene group with a silicon, the authors also extended the π-conjugation through expansion of the xanthene. For labelling of antibodies and proteins, the authors functionalised the pendant ring with a succinimidyl ester as a reactive site. Through these structural modifications, the authors were able to achieve NIR emission (712 nm). The

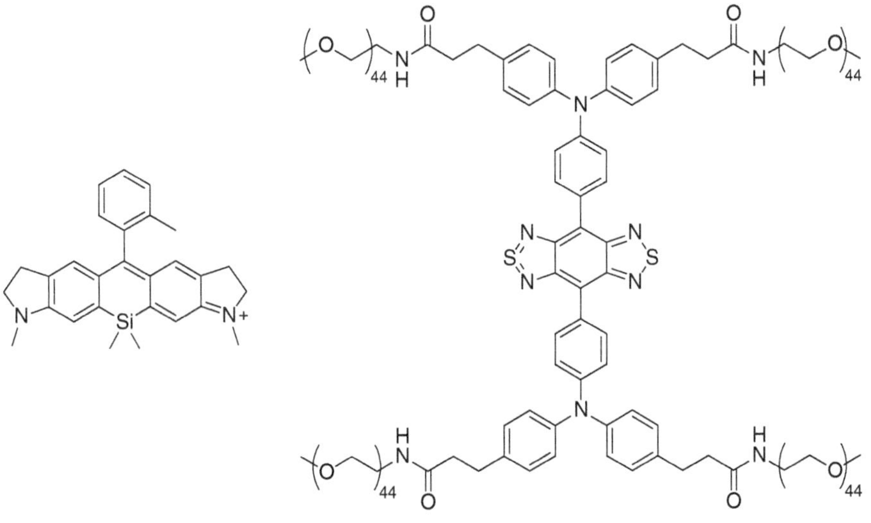

SIR700

CH1055

Figure 8.9 Structures of rhodamine-based [62] and benzobisdiathiazole [63] NIR dyes.

authors next prepared the tumour-targeting analogue of **SIR700** by conjugating it to an anti-tanascin-C antibody, as tanascin-C is a glycoprotein known to be highly expressed in malignant gliomas. The antibody-labelled sensor was applied to mice bearing human malignant meningioma HKBMM xenografts, and a fluorescence signal was observed in the tumours after 24 hours. The signal was still observable after 10 days, demonstrating the utility of this sensor for long-term imaging of tumours with low laser power.

Donor-acceptor-donor (D-A-D) dyes are promising candidates in the field of NIR sensor research because their energy gaps are even lower than classic donor–acceptor fluorophores, resulting in emission in the NIR I/II region [64]. One of the first D-A-D dyes was reported by Antaris *et al.*, who reported NIR II sensor **CH1055**, based on benzobisthiadiazole (Figure 8.9) [63]. The sensor features four peg chains, improving the water solubility and allowing the sensor to be used without the need for encapsulation within a nanoparticle. The sensor **CH1055** has suitable fluorescent properties for NIR II imaging, with emission at 1055 nm, a relatively high quantum yield ($\phi = 0.3$) and good photostability. **CH1055** has good excretion kinetics, with 90% of sensor removed through the renal system within 24 hours of injection, and is non-toxic to cells. In comparison to ICG, **CH1055** showed improved signal-to-noise ratio with increased resolution in the imaging of sentinel lymph nodes and orthotopic glioblastoma brain tumours in mice models. **CH1055** also showed superior resolution compared to NIR I sensors and was also used in a proof-of-concept NIR II imaging-guided tumour removal surgery.

8.2.3 Future Directions

Fluorescent sensors that emit in the NIR I and NIR II regions offer the opportunity to move beyond the culture dish to *in vivo* systems. Furthermore, NIR sensors are also finding use as theranostic agents, such as in photodynamic therapy, and in multimodal imaging, in combination with photoacoustic imaging. However, many NIR I and NIR II sensors suffer from poor chemical and photochemical properties, such as low fluorescent quantum yields, poor photostability, poor water solubility, small Stokes shifts, formation of aggregates, and limited potential for functionalisation with sensing or targeting groups. There are only two FDA approved dyes – ICG and methylene blue – that are currently used as dyes for medical diagnostics, highlighting the gap in the field of NIR sensor development for end users. Therefore, future endeavours should be focussed on designing sensors, which can overcome these disadvantages.

8.3 Dual-analyte Sensing

8.3.1 Introduction

The majority of fluorescent sensors reported to date and discussed in this book have been for the study of a single analyte. In theory, it is possible to simultaneously utilise two selective sensors to separately measure two different analytes. However, any subtle differences in the behaviour of the sensors, for example their cellular uptake, cellular retention, or metabolism and localisation within cells, can impact their fluorescent output and therefore complicate data interpretation. In order to most accurately study the interaction of chemical species within cells, it is therefore advantageous to use a single sensor with two (or more) sensing moieties.

In the design of fluorescent sensors for multiple analyte sensing, there are some important criteria that must be met regarding the mode of response of the sensor. While there are many sensors that are claimed in the literature to be dual-responsive, many fail to meet these key criteria, and it is therefore essential to assess the reported results before choosing to apply such a sensor to biological studies. Dual-responsive sensors need to be able to respond to two different analytes and have one or more fluorescence responses in the presence of either and both analytes.

At the very least, a dual-analyte sensor should provide a clear read-out for the presence of the two analytes (Figure 8.10i). In order for a dual-analyte sensor to be practically useful, however, it is necessary to be able to differentiate between the presence of both analyte A and B, the absence of both analyte A and B, and the presence of either analyte A or B (Figure 8.10ii). Furthermore, as sensor concentration within cells can vary with time and with cellular localisation, it is insufficient to rely only on changes in fluorescence intensity to distinguish between the presence of either or both analytes. Therefore, an ideal fluorescent dual-responsive sensor is one that exhibits a change in emission wavelength rather than change in intensity. This gives rise to a ratiometric response, which bypasses any issues relating to varying sensor concentration [2]. Even more ideal would be a sensor that can give different fluorescent responses for the presence of analyte A only, analyte B only, and both analytes A and B, allowing the user to also determine whether only analyte A or only analyte B is present (Figure 8.10iii).

The design of fluorescent dual-analyte sensors typically consists of one or two fluorophores, and two distinct analyte sensing groups. These sensing groups can operate by a number of different chemical processes.

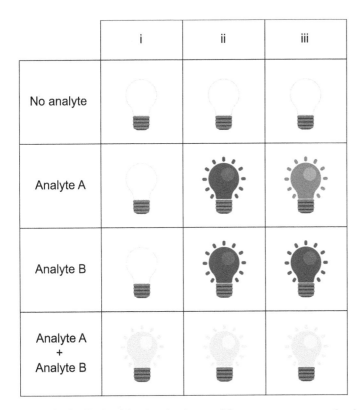

Figure 8.10   Truth table of optimal type of fluorescence response for dual-analyte sensing.

For the purpose of this chapter, the types of dual-analyte sensing sensors will be divided according to the nature of the two sensing processes, and key examples of sensors in the literature, which fall under each of these categories will be discussed below.

8.3.2 Recognition-Based Dual-analyte Sensors

Recognition-based dual-analyte sensors are those in which both analytes are able to interact with the sensor in a reversible manner. As with fluorescent sensors for the detection of a single analyte, a large number of recognition-based dual-analyte sensors are designed for metal ion sensing and for studying pH. In the design of recognition-based dual-analyte sensors, the sensing units for the analytes in question must be distinct from one another.

The first fluorescent molecular logic gate, **8.5**, was reported by de Silva *et al.* in 1993 [65]. **8.5** (Figure 8.11) is an AND fluorescent logic gate which reports on the simultaneous presence of Na^+ and H^+. The sensor contains an anthracene derivate as a fluorophore and a tertiary amine and benzocrown ether as the H^+ and Na^+ sensing group, respectively. In the presence of H^+ or Na^+ alone, **8.5** exhibits a slight fluorescence increases of 1.7-fold and 1.1-fold, respectively, compared to a 6-fold fluorescence turn-on in the presence of both Na^+ and H^+.

The first dual-analyte sensing sensor that was applied in a biological system was reported by Komatsu *et al.* in 2005 (Figure 8.11) [66]. **KCM-1** is a coumarin-based sensor for the simultaneous detection of Ca^{2+} and Mg^{2+}. The Ca^{2+} sensing unit is the chelator BAPTA (O,O'-bis(2-aminophenyl)ethyleneglycol-N,N,N',N'-tetraacetic acid) at the C3 position of the coumarin, while the Mg^{2+} sensing unit is the charged β-diketone at the C7 position. The benefit of having the sensing units at the electron donor and electron acceptor of the coumarin is that shifts in the emission are observed upon metal-ion binding. Binding of Ca^{2+} leads to a 45-nm blue shift in absorption and 5-nm blue shift in emission, and binding to Mg^{2+} leads to a 21-nm red shift in absorption and a 5-nm red shift in emission. In both solution and in cells, excitation of the sensor at 358, 403, and 424 nm enabled quantification of Ca^{2+} and Mg^{2+} concentrations. For biological studies, the acetoxymethyl ester form of the sensor was synthesised to promote cell penetration and retention. **KCM-1** was also used to visualise changes in Ca^{2+} and Mg^{2+} concentrations upon mitochondrial uncoupling.

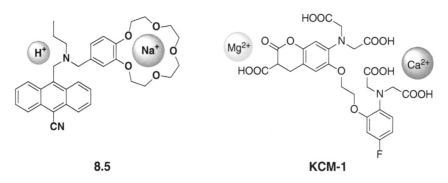

8.5   **KCM-1**

Figure 8.11 Structures of recognition-based dual-analyte sensors [65, 66].

8.3.3 Activity-based Dual-analyte Sensors

Activity-based sensors involve the chemoselective reaction of an analyte with the sensor, forming a new product, which exhibits a change in fluorescence. There are two sub-classes of Activity-based dual-analyte sensors: sensors that have two separate reactive sites and sensors that are sequence-specific, where an analyte reacts with the sensor to reveal a second analyte recognition site. Sensors that contain two reactive sites are favourable because the reaction order is insignificant. Sensors that feature two distinct reactive sites will be discussed in this section.

FP-H$_2$O$_2$-NO is a Förster resonance energy transfer (FRET)-based sensor for the detection of H$_2$O$_2$ and NO (Figure 8.12) [67]. The sensor contains a non-fluorescent phenylboronate-masked coumarin, which upon reaction with H$_2$O$_2$ generates a fluorescent 7-hydroxy coumarin. The NO reactive site is the closed spirolactam of the rhodamine that opens and becomes fluorescent in the presence of NO. The 7-hydroxy coumarin and rhodamine are a FRET pair, so in the presence of both analytes, the emission from both fluorophores can be observed upon excitation of the coumarin. The sensor was applied to HeLa cells and could respond to exogenous and endogenously generated H$_2$O$_2$, NO, and a combination of both analytes.

Another design strategy for activity-based dual-analyte sensors is to use a single fluorophore with two orthogonal masking groups. Finkler *et al.* used this strategy in the design of sensor **8.6** (Figure 8.12) for the simultaneous detection of two different enzymes: porcine liver esterase (PLE) and alkaline phosphatase (AlP) [68]. The sensor is based on a pyrenol, with a carboxylic ethyl ester as the PLE sensing unit and a phosphate group as the AlP sensing unit. Cleavage of the ethyl ester by PLE leads to a blue shift in emission from 472 to 448 nm and cleavage of the phosphate by AlP results in a large red shift in emission to 558 nm. The presence of both analytes leads to a smaller red shift in emission to 536 nm.

8.3.4 Mixed Dual-analyte Sensors

The previous two subsections discussed sensors, that responded to both analytes through a recognition-based manner or a activity-based manner. It is also possible to design a sensor for the simultaneous detection of analytes using a combination of the two above methods, where one analyte is sensed in a recognition-based manner and the other analyte is recognised through a chemical reaction.

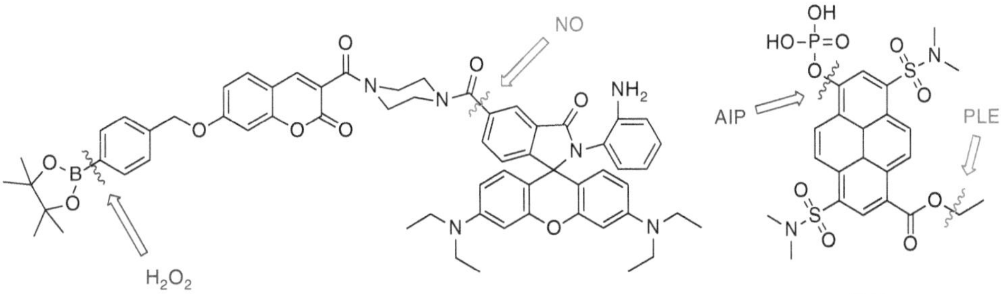

Figure 8.12 Structures of reaction-based dual-analyte sensors [67, 68]. Arrows and wavy lines signify sites of reaction. PLE = porcine liver esterase; AlP = alkaline phosphatase.

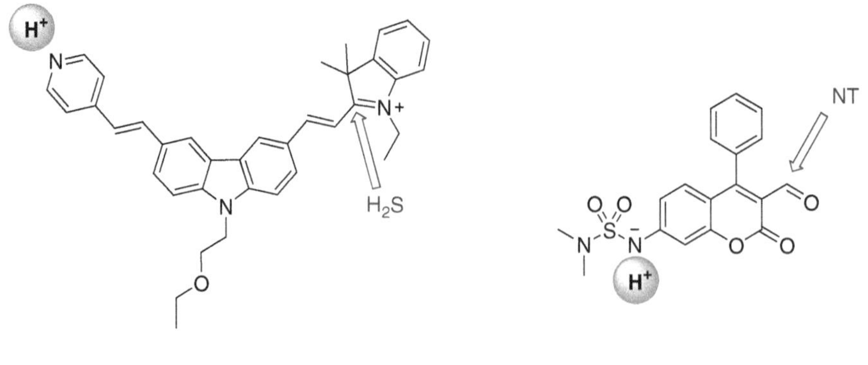

TP-PMVC   **ES517**

Figure 8.13 Structures of mixed dual-analyte sensors [69, 70]. Arrows indicate site of reaction. NT = neurotransmitter.

Sensor **TP-PMVC** is a mixed sensor for the detection of protons and H_2S (Figure 8.13) [69]. **TP-PMVC** is a fluorescent sensor based on carbazole, with a pyridine as the H^+ recognition site and an indolenium unit as the H_2S recognition site. In the absence of both analytes, the sensor is non-fluorescent. At acidic pH, the pyridine unit becomes protonated, resulting in a fluorescence turn on at $\lambda_{em} = 625$ nm due to an increase in ICT efficiency. In the presence of both analytes, the nucleophilic addition of H_2S to the indolenium moiety results in the breaking of the π conjugation, resulting in a blue-shifted emission at $\lambda_{em} = 550$ nm. The pyridine additionally acts as a lysosomal targeting moiety [71]. Cells treated with PMA (paramethoxy-acetophenone), a reagent for lowering endogenous H_2S levels, showed a decrease in fluorescence at the lower wavelength, whereas cells pre-incubated with NaSH showed increased fluorescence. Furthermore, **TP-PMVC** is a two-photon sensor and was successfully applied to mouse liver tissue slices to study its ability to image lysosomal H_2S in deep tissue.

ExoSensor 517 (**ES517**) (Figure 8.13) is a coumarin-based sensor for the simultaneous detection of pH and primary-amine neurotransmitters (NTs) [70]. A sulfonamide acts as the reversible sensing unit for protons and a carbaldehyde group is the reactive site for primary amine NTs. At acidic pH, the sensor is non-fluorescent, and as pH increases and the sulfonamide is deprotonated, an increase in fluorescence at a blue wavelength is observed due to the increase in ICT efficiency. The carbaldehyde unit reacts with the primary amine of NTs to form an iminium ion, resulting in a fluorescence emission at a green wavelength.

8.3.5 Sequence-specific Reactions

An extension of activity-based dual-analyte sensors are sequence-specific reaction-based sensors, where the first analyte interreacts with the sensor in order to reveal a second recognition site for a second analyte. The benefit of sequence specific dual-analyte sensors is their ability to sense an analyte within a specific environment, *e.g.* in an environment that is high in ROS or biothiols. These sensors are also useful for sensing the presence of one specific analyte only when a second-specific analyte is also present. In this case, however, the sequence of analytes reacting with the sensor is important. The sensor is only able to report on analyte A before it is able to report on analyte B and if analyte B is presented to the sensor before analyte A, there will be no signal from the sensor. Alternatively, it can respond to the simultaneous presence of

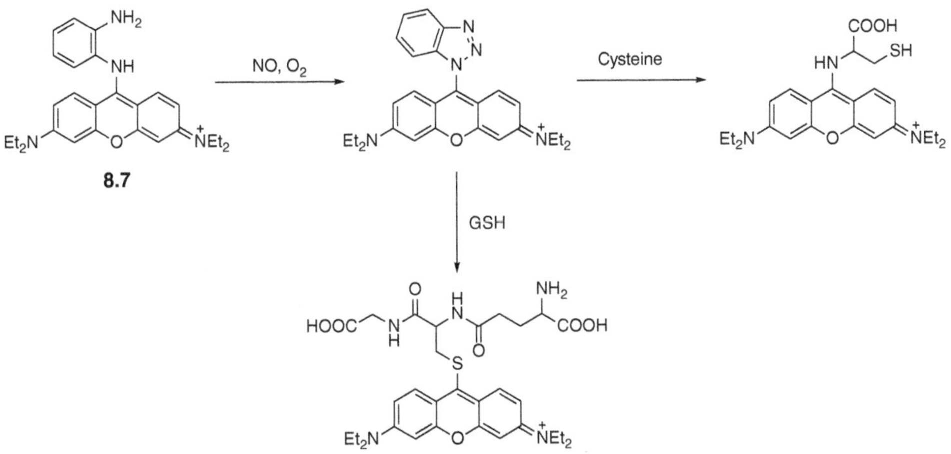

Figure 8.14 Structure of a sequence-specific dual-analyte sensor [72].

A and B. Therefore, it is important to note that sensors that use this sensing mechanism cannot give any definite conclusions regarding biochemical reaction sequences.

Sun *et al.* reported a reaction-based sequence specific sensor **8.7** for the simultaneous detection of NO and biothiols, such as cysteine and glutathione (Figure 8.14) [72]. The sensor is based on a pyronin dye, with *o*-phenylenediamino quenching group directly conjugated to the system as the NO sensing unit. NO generates the formation of a benzotriazole, overcoming the PET-quenching and causing a turn on in fluorescence emission at 616 nm. The benzotriazole can then act as a leaving group in S-acylation reactions, and the sensor is able to sense biothiols through an aromatic substitution reaction. Upon reaction with cysteine, a green fluorescent species is formed ($\lambda_{ex} = 455$ nm, $\lambda_{em} = 536$ nm) and upon reaction with glutathione, a red fluorescent species is formed ($\lambda_{ex} = 570$ nm, $\lambda_{em} = 618$ nm). The sensor was applied to B16 and RAW 264.7 cells and could image exogenously added and endogenously generated NO in cells also treated with cysteine in a dual-channel manner.

8.3.6 Conclusions and Future Directions

The recent influx of fluorescent sensors for dual-analyte sensing in the literature over the past two decades has helped to provide insight into the relationship between analytes and their role in cell signalling and pathology. The most ideal dual-analyte sensing sensors should also be able to discriminate between the absence of any analyte, the presence of either analyte separately and the presence of both analytes. This can be achieved through the design of ratiometric dual-analyte sensors, which will also circumvent problems relating to concentration. Future focus should be on the design of ratiometric sensors that can give unique responses for the presence of either analyte and the presence of the two analytes, in order to maximise data collected and simplify data interpretation.

Dual-analyte sensors are also often discovered serendipitously, and the analytes that are being detected are often not biologically relevant or have no relation to each other in biology. Therefore, future dual-analyte sensors should also be designed with a specific biological question in mind. The two analytes being detected should already be speculated to be related to each other, such as both being involved in a cellular signalling pathway, or both being involved in the development of a certain disease.

8.4 Super-resolution Microscopy

8.4.1 Introduction

Light microscopy techniques have become an essential tool for visualising and characterising cells and tissues. The ability of fluorophores to label specific molecules or cellular substructures in real time and in live samples affords a unique advantage to light microscopy over other imaging techniques. Despite this high degree of specificity, conventional light microscopy techniques are limited by the diffraction limit of light, with spatial resolution of approximately 200 nm in the xy-lateral directions and approximately 500 nm in the z-axial direction [73]. While this is sufficient for resolving larger organelles and cellular substructures on the microscale, it is not sufficient for resolving fundamental cellular activities that occur at the nanoscale, including finer organelle substructure and molecular processes. Small structures at this scale have traditionally been observed with electron microscopy (EM) techniques, but this requires careful sample fixation that may result in artefacts and cannot visualise live dynamics [74, 75]. More recently, advances in optical techniques have allowed the emergence of super-resolution microscopy (SRM), with spatial resolution of up to 1 nm for the highest resolution techniques [76]. The significance of SRM as a technique was recognised with the 2014 Nobel Prize in Chemistry [77].

In recent times, several detailed reviews have covered various aspects of SRM, including principles, applications, and aspects of sensor design [78–80]. The following section is intended to provide a brief overview of SRM techniques, key considerations for the use of SRM, as well as the properties of the fluorescent dyes required for SRM.

8.4.2 Super-resolution Microscopy Techniques

It is a common misconception that SRM refers to a singular technique. Rather, it is a term that describes a group of techniques that can overcome the diffraction limit of light, including both near-field and far-field techniques; Table 8.1 provides a summary of both conventional and SRM microscopy techniques. Far-field microscopy includes conventional widefield and confocal laser scanning techniques; all far-field SRM techniques build on widefield or laser scanning illumination.

In widefield techniques, the entire specimen is exposed to excitation light at the same time, and the resultant emission light is captured to create the image. As the entire sample is illuminated, there can be issues with background fluorescence from out of the plane, though this can be addressed with deconvolution [81]. With confocal laser scanning, the excitation light is focused on a small region of the sample, and the image is scanned pixel-by-pixel. Similarly, the image is built from the emitted photons, but out of plane light is removed *via* a pinhole, reducing background fluorescence [82]. In both techniques, resolution is constrained by the numerical aperture of the objective and wavelength of emitted light, with shorter wavelengths providing higher resolutions. While these techniques are capable of imaging phenomena on the microscale, none of these techniques can bypass the diffraction limit of light and image phenomena on the nanoscale (<100 nm).

Some of the first SRM techniques, such as incoherent interference illumination image interference microscopy (I^5M), incorporate a second opposing objective lens to illuminate the sample simultaneously from both sides. This improves the z-resolution by 5- to 7-fold compared to diffraction-limited techniques; however, the xy-resolution remains unchanged [83]. As with conventional microscopy techniques, dual-lens techniques can be

Table 8.1 Comparison of fluorescence imaging techniques.

General technique	Abbreviations of techniques	Description	Compatible fluorophores	Spatial resolution (nm)	Acquisition time	Dual lens	Other remarks
Widefield	WF	Diffraction limited technique. Entire specimen is exposed to light source	Any	xy ~ 200–300 z ~ 500–700	ms-s	Yes, I^{5}M	Issues with background fluorescence
Confocal laser scanning microscopy	CLSM, LSCM	Diffraction limited technique. A focused laser is used for excitation, scanning the sample pixel-by-pixel. Pinhole allows emission from focal plane only	Any	xy ~ 200–300 z ~ 500–700	ms-s	Yes, 4Pi	
Multiphoton excitation	MPE, 2PE, 3PE	Laser-scanning SRM technique. Simultaneous excitation requires two (or more) photons of a longer wavelength providing the equivalent energy for single-photon excitation. Occurs at focal plane only	Any	xy ~ 200–300 z ~ 500–700	ms-s	Yes, 4Pi	Deeper penetration possible (*e.g.* for tissue)
Structured illumination	SIM, PEM, HELM, LMEM	Widefield SRM technique. Sample is excited multiple times with striped illumination patterns, translating and rotating the pattern between excitations. The resultant interference patterns (Moire fringe) can be processed *via* algorithm to produce reconstructed images	Photostable fluorophores	xy ~ 100–120 z ~ 250–350 [109]	s [110]	Yes, I^{5}S	

Far-field

(Continued)

Table 8.1 (Continued)

General technique	Abbreviations of techniques	Description	Compatible fluorophores	Spatial resolution (nm)	Acquisition time	Dual Lens	Other remarks
Non-linear structured illumination	SSIM, SPEM	Widefield SRM technique. The sample is excited several times with an illumination pattern, shifting the pattern between excitations. This occurs at an intensity that allows depletion of the ground state and saturation of the excited state. The resultant Moire fringe can be processed *via* algorithm to produce a reconstructed image	Photostable fluorophores; to achieve nonlinearity, photoswitchable fluorophores have also been used	xy ~ 50 z not reported [111, 112]	s-min [111, 112]		Issues with photobleaching
Stimulated emission depletion	STED	Laser-scanning SRM technique. Two lasers are employed. The excitation laser excites the fluorophore. The STED laser is used in stimulated emission, used to deplete excited-state fluorophores located around the focal point. Emission is only observed in the small region around the focal point	Fluorophore must be highly photostable to withstand depletion wavelength, and must respond to the depletion beam wavelength	xy ~ 20–60 [113–115] z ~ 90–110 [116, 117] ~1 nm combined with SMLM [76]	ms-min [113]	Yes, 4Pi-STED isoSTED	Issues with photobleaching
Single molecule localisation/ Pointillism	SMLM, PALM, FPALM, STORM, dSTORM, PALMIRA	Widefield SRM technique. Fluorophores are excited randomly over time, resulting in several images of spatially separated subsets of individual molecules. These are then used to produce a reconstructed image	Fluorophores must be photoactivatable, photoconvertible, or photoswitchable	xy ~ 10–30 [118] z ~ 20–50 [119, 120] ~1 nm combined with STED [76]	s-min [121]	Yes, iPALM	Potential for mislabelling; may require TIRF for best results

Near-field	Evanescent wave illumination	TIRF	Total internal reflection of the excitation light at the interface between specimen and glass produces evanescent waves that penetrate ~100 nm into the medium; this excites fluorophores in a thin surface region	Any	xy ~ 200–300 [91]	ms-s [91]	Z-sectioning not possible, limited to surface studies
	Near-field scanning optical microscopy	NSOM/SNOM	Excitation light is forced through an aperture smaller than the wavelength of the excitation light, generating evanescent waves that excites fluorophores in a thin surface region. No objective is used	Any	xy ~ 20–100 [91]	s-min [91]	Z-sectioning not possible, limited to surface studies

2PE = two-photon excitation; 3PE = three-photon excitation; CLSM = confocal laser scanning microscopy; dSTORM = direct STORM; FPALM = fluorescence photoactivation localisation microscopy; HELM = harmonic excitation light microscopy; I³M = incoherent interference illumination image interference microscopy; I⁵S = incoherent interference illumination image interference SIM; iPALM = interferometric PALM; isoSTED = isotropic STED; LMEM = lateral modulated excitation microscopy; LSCM = laser scanning confocal microscopy; MPE = multiphoton excitation; NSOM = nearfield scanning optical microscopy; PALM = photo-activated localisation microscopy; PALMIRA = PALM with independently running acquisition; PEM = patterned excitation microscopy; SIM = structured illumination microscopy; SMLM = single-molecule localisation microscopy; SNOM = scanning nearfield optical microscopy; SPEM = saturated pattern excitation microscopy; SSIM = saturated structured illumination microscopy; STORM = stochastic optical reconstruction microscopy; TIRF = total internal reflection fluorescence; WF = widefield.

used with virtually any fluorophore. They can also be further incorporated with the other far-field SRM techniques described below. SRM far-field techniques can be classified under three major categories: structured illuminated microscopy (SIM), deterministic SRM techniques, and stochastic SRM techniques. Figure 8.18 provides an illustration of the principles of the far-field microscopy techniques.

In SIM, the sample is excited multiple times with non-uniform spatially structured patterns of widefield light, translating and rotating the pattern between excitations [84]. The resultant information can be processed *via* reconstruction methods to produce images with a twofold improvement of resolution in both xy and z-directions on diffraction limited techniques. SIM can theoretically be used with any fluorophore, but image resolution is improved with brighter and more photostable fluorophores (see Table 8.1, Figure 8.18).

Both deterministic and stochastic SRM techniques rely on the isolation of fluorophore emission, though *via* different approaches. Deterministic super-resolution techniques rely on hardware to provoke a non-linear response in fluorophores [85]. Most deterministic super-resolution techniques are RESOLFT (REversible Saturable Optical Fluorescence Transitions) microscopy techniques, which include STED (STimulated Emission Depletion), and SSIM (Simulated Structured Illuminated Microscopy). RESOLFT techniques rely on the reversible photoswitching of fluorophores between two states, generally a fluorescent 'on' state and non-fluorescent 'off' state. Due to constant irradiation, all RESOLFT techniques require highly photostable fluorophores. Some RESOLFT techniques have further requirements; for example, STED techniques use an excitation laser and depletion laser to create a smaller focal excitation spot, and for optimal imaging, the fluorophore must respond to both lasers [86].

Stochastic super-resolution techniques, also known as single molecule localisation microscopy (SMLM) or photoactivated localisation microscopy, require photoactivatable or photoswitchable 'blinking' fluorophores [87]. Due to stochastic activation, only a small subset of fluorophores is excited at a time, producing a series of spatially separated fluorophores. The resultant images can be then used to reconstruct the location of the isolated original molecules. The first techniques, stochastic optical reconstruction microscopy (STORM) [88], photo-activated localisation microscopy (PALM) [89], and fluorescence photoactivation localisation microscopy (FPALM) [90], were reported in the same time period and rely on the same principles; of these, STORM uses fluorescent sensors, whilst PALM and FPALM use photoswitchable fluorescent proteins.

On the other hand, near-field microscopy involves the generation of evanescent waves and relies on the properties of these for improved resolution [91]. These techniques can be used with any type of fluorophore, but the nature of evanescent waves means that only the thin surface section can be captured. Furthermore, 3D sectioning is not possible.

8.4.3 Considerations for Use of Super Resolution Microscopy

The chief advantage of SRM is greater spatial resolution compared to conventional light microscopy techniques, making it most suitable to resolve structures and molecular processes at the sub-organelle level. However, SRM is not necessarily suited for all sample types and biological investigations.

The nature of SRM techniques often limits the thickness of the sample or depth of imaging possible, and thus thin sections of tissue or cell monolayers are most commonly used for investigation. For the same reason, 3D reconstruction can be limited. Thick biological samples remain a challenge for imaging, as these naturally have anisotropic optical properties, though in recent times advances in hardware have helped overcome this issue [92].

While live cell imaging is possible with some SRM techniques, cell morphology and biochemical distribution can change over time, resulting in poor labelling over the relatively long acquisition periods necessary for SRM [93]. Furthermore, some of the excitation energies required for SRM techniques are far beyond the intensities safe to use with live samples. Generally, a compromise with spatial resolution is required in order to improve the temporal resolution, offering real-time insight into biochemical dynamics.

8.4.4 Fluorescent Sensors for Super-resolution Microscopy

An understanding of the photochemistry and properties of fluorophores (see Chapter 1) is useful in the design and use of fluorescent sensors in experiments. In general, fluorophores for SRM should be photostable and bright. Properties including membrane permeability and solubility are also desirable for biological applications, as well as targeted localisation in specific sites of the cell.

While some SRM techniques are compatible with any type of fluorophore, other SRM techniques require fluorophores with specific properties, as described in the following.

1) *Photostability* is required for techniques with extended acquisition times to provide a stable fluorescent signal over time without photobleaching. While photobleaching occurs by many pathways, most known mechanisms involve the triplet state, followed by oxidation. This may be overcome by structural modifications that either reduce the reactivity to reactive oxygen species (ROS) or by triplet-state quenching that reduces triplet state lifetime [94].

   A standard method to reduce reactivity to ROS is to incorporate electron-withdrawing groups (EWGs) into the fluorescent molecule. Commonly used EWGs are fluorine (Figure 8.15, **8.8**) [95] and cyano (Figure 8.15, **S-SO-CN**) [96] groups. Other structural modifications may be performed to reduce oxidation. For example, Song *et al.* reported that the substitution of the dimethylamino groups in a rhodamine derivative with 7-azanorbornane resulted in better photostability (Figure 8.15, **221SR**) [97].

   For reducing the triplet state lifetime, triplet-state quenchers (TSQs) may be used, including 4-nitrobenzyl alcohols, cyclooctatetraene (COT), or Trolox [98]. While this can be used as an additive in fixed cell imaging media, high concentrations (mM) of TSQs are required. TSQs may also be incorporated into the structure of the fluorescent molecule to allow for intramolecular quenching. For example, the proximal covalent linkage of COT

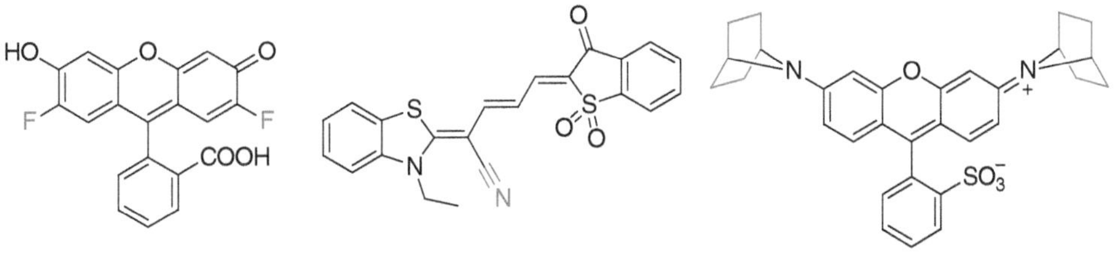

8.8   **S-SO-CN**   **221SR**

Figure 8.15 Structures of fluorescent sensors with improved photostability over parent molecules due to modifications (red) that reduce oxidation [95–97].

has been shown to increase the photostability of several classes of fluorophores, including cyanines, carbopyranines, and rhodamines [99].

2) *Stochastic blinking* properties are required for SMLM techniques, such as STORM. This stochastic activation may include the irreversible photoactivation of a molecule from a non-fluorescent 'dark' state to a fluorescent 'light' state (photoactivable), irreversible photoconversion from a 'light' state to a red-shifted 'light' state (photoconvertible), or reversible switching between a 'dark' and 'light' state (photoswitchable) [100]. The original STORM technique used a photoactivatable dye pair consisting of a reporter dye and a shorter wavelength activator dye [88]; the reporter is typically a cyanine (*e.g.* AlexaFluor 647, Cy5, Cy7) and the activator is generally a blue-shifted dye (*e.g.* AlexaFluor 405, Cy2, Cy3) [101]. On the other hand, the direct STORM (dSTORM) technique requires only a single dye and can be extended to standard organic fluorophores [102]. SMLM techniques generally require specialised buffer solutions, with the formulation dependent on the photochemistry of the chosen dyes, though most include reducing agents (*e.g.* thiols like 2-mercaptoethanol) and oxygen-scavenging systems (*e.g.* GLOX, an enzymatic scavenger) [101]. Powerful laser irradiation conditions are also generally required.

Spontaneous blinking fluorophores is a recent development that enables stochastic activation without specific laser irradiation or buffer conditions, allowing SMLM imaging under live, physiological conditions. **HMSiR** was the first reported spontaneously blinking fluorophore (Figure 8.16), with the mechanism of blinking based on the intramolecular spirocyclisation reaction of the rhodamine derivative [103].

3) *Brightness* is a fluorophore parameter that refers to emissive fluorescent output, which can be quantified using the extinction coefficient and fluorescence quantum yield. Brighter fluorophores allow lower levels of laser irradiation to be used, reducing damage to biological samples. In SMLM techniques, a high contrast ratio between the 'light' and 'dark' states is also desirable, as this allows for more precise localisation and thus higher resolution; a ratio of ~100 : 1 is required, with many popular SMLM fluorophores having a contrast ratio greater than ~1000 : 1 [104].

Several methods for improving the brightness of fluorophores have been explored. The suppression of competing non-radiative processes *via* structural modification is a general method for improving brightness. For example, Grimm *et al.* demonstrated that the replacement of the *N,N*-dialkyl group with azetidines in a rhodamine-based fluorophore resulted in improved brightness and photostability (Figure 8.17, **JF549**); this

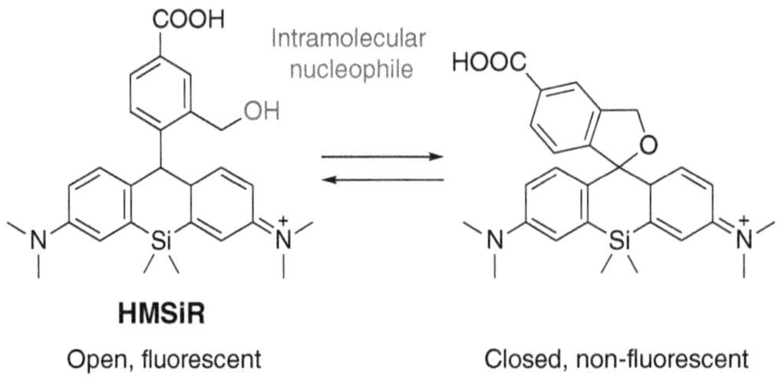

HMSiR
Open, fluorescent   Closed, non-fluorescent

Figure 8.16 Structure and mechanism of **HMSiR**, the first reported spontaneously blinking fluorophore [103].

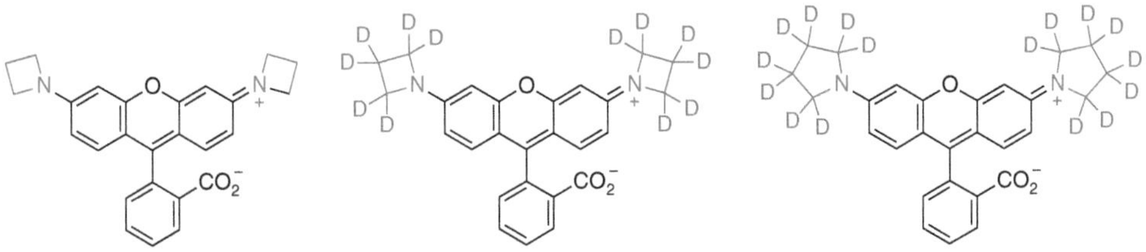

Figure 8.17 Structures of rhodamine derivatives with structural substitutions (red) to enhance brightness [105, 106].

modification could also be applied to a range of fluorophores including fluorescein and coumarins [105]. The same group also found that deuteration of the azetidine gave small improvements in brightness and photostability (Figure 8.17, **8.9**), while the use of deuterated pyrrolidine gave larger improvements in these properties and could also be used in a range of fluorophores (Figure 8.17, **8.10**) [106].

SRM achieves its greatest potential when imaging biomolecules and events that occur at the nanoscale that cannot be readily imaged with other techniques. Thus, the attachment of fluorophores to biomolecules of interest or localisation to a specific subcellular compartment is often desired. Different labelling strategies are required for different cargo, such as fluorescent proteins and synthetic sensors. For proteins, conjugating a protein of interest to a fluorescent protein is a common strategy. Similarly, fusion with a self-labelling protein with subsequent application of an external fluorescent ligand may also be performed (see Chapter 7). For small-molecule fluorophores, a number of organelle-targeting strategies have been developed (see Chapter 3). For fixed samples, immunofluorescence labelling techniques may be employed.

8.4.5 Future Directions

The breaking of the diffraction limit and development of SRM techniques has revolutionised the visualisation of molecules and cellular ultrastructure. Over the span of only a few decades, SRM has become relatively accessible and is increasingly used for cellular imaging. Key challenges include reducing photobleaching and toxicity while increasing temporal resolution, both of which are necessary for improving live cell imaging capabilities. Innovations in sample preparation (*e.g.* available fluorophores, labelling methods), as well as in instrumentation will allow for future improvements in the field.

Diffraction-limited techniques require a large number of photons, and SRM techniques have generally evolved to maximise the number of photons the sample can tolerate. In recent times, an approach called MINFLUX combined the stochastic switching of SMLM and depletion beam of STED, using minimum emission flux rather than maximising photon output; this achieved ~1 nm resolution [76]. Improvements in 3D imaging will allow for a better understanding of biological systems compared to 2D monolayers. For example, lattice light-sheet (LLS) microscopy utilises an ultrathin structured light sheet to scan successive planes of a sample at a rate of hundreds of planes per second, rapidly generating a 3D image [107]. To improve temporal resolution, multiplane imaging with the simultaneous acquisition of multiple focal planes can reduce acquisition time [108] (Figure 8.18).

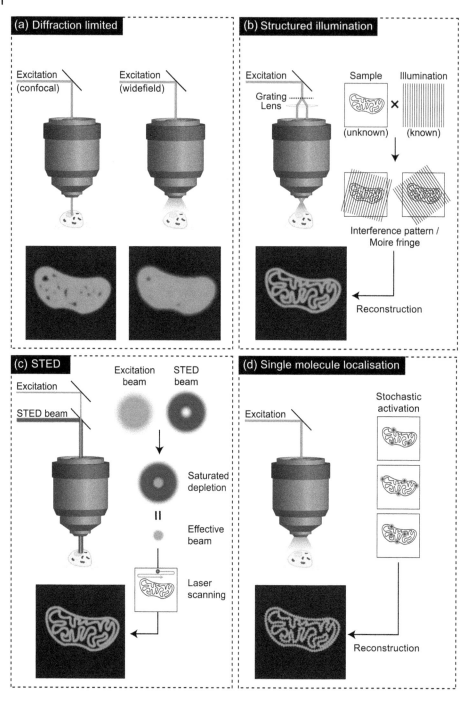

Figure 8.18 Illustrations of the principles of far-field fluorescence microscopy techniques. (a) Describes the diffraction limited techniques, confocal laser scanning microscopy, and widefield microscopy. (b–d) Describes the super resolution microscopy techniques, (b) showing structured illumination microscopy, (c) showing stimulated emission depletion, and (d) showing single-molecule localisation microscopy.

8.5 Multimodality

8.5.1 Introduction

The structure and function of cells can be studied with a number of imaging techniques that rely on different physical phenomena to interact with the sample, resulting in capabilities that differ in sensitivity, spatial and temporal resolution, as well as the structural and functional information obtained. Of these techniques, fluorescence imaging has become an important scientific and diagnostic tool, in large part due to the small-molecule sensors that can be targeted to specific cellular substructures and detect changes in their chemical environment [122]. However, the sensing capacity of fluorescent sensors is limited to one or two analytes of interest. Furthermore, crosstalk, or the 'bleed' of fluorophores across multiple channels in multicolour imaging can complicate the investigation of multiple species. Fluorescent sensors are thus less effective for studying the general chemical composition of cellular environments. The combination of fluorescence imaging with one or more additional imaging modalities can enable collection of complementary and orthogonal information [123, 124]. Importantly, multimodal imaging can also offer validation of results seen in each modality.

As a technique, fluorescence imaging is unique in that it is widely used in both *in vitro* analysis of cells and tissues, as well as *in vivo* imaging of animal models and in clinical treatment. It can thus be combined with a range of *in vivo* imaging techniques (*e.g.* positron emission tomography (PET), single-photon emission computed tomography (SPECT), X-ray computed tomography (CT), magnetic resonance imaging (MRI), photoacoustic imaging (PAI)), as well as *in vitro* methods (*e.g.* vibrational spectroscopy, synchrotron X-ray techniques, mass spectrometry, electron microscopy).

The following sections will detail how fluorescence techniques can be used in conjunction with other techniques, providing a short summary and illustrative examples of current developments in multimodal imaging. Generally, this involves the development of fluorescent sensors that have additional reporters for other modalities. While it is not always necessary to include contrast agents for each modality in the same sensor, this allows for the colocalisation of signal for each modality and bypasses the need for multiple agents with different clearance and retention rates [125].

8.5.2 Radioisotope Techniques

The two most common clinical imaging techniques are positron emission tomography (PET) and single-photon emission computed tomography (SPECT). The vast majority of dual modality fluorescence-PET and fluorescence-SPECT sensors are aimed at *in vivo* models and eventual clinical use, with the PET or SPECT providing initial whole-body imaging and fluorescence imaging providing cellular imaging at the superficial or histologic levels [123]. The construction of such sensors is relatively straightforward, as it involves incorporation of the relevant radioisotope into a fluorescent scaffold. For SPECT, this will be a γ-emitting radioisotopes [126] such as ^{99m}Tc, ^{111}In, and ^{177}Lu, while for PET, positron-emitting radioisotopes are required, such as ^{18}F, ^{64}Cu, and ^{68}Ga [127].

While PET and SPECT can be performed *in vitro*, simple radiolabelling can also be used to provide correlation and quantification of uptake that is not possible with fluorescence alone. For example, Paulus *et al.* first examined uptake of an ^{18}F-labelled BODIPY-C_{16}/triglyceride using fluorescence, and quantified this using radiolabelling, finding uptake was significantly higher in brown adipose tissue cells compared to white adipose tissue [128].

For *in vivo* applications, the presence of both a fluorophore and radioisotope allow for dual labelling. Several bimodal fluorescence-PET or fluorescence-SPECT sensors are designed around a radiometal ion [129, 130], with different radiometal ions having their own preferred chelators and ligands [127]. Figure 8.19 shows some of the common radionuclides used for dual modality sensors, and their preferred chelators. As some chelators can be used for both PET and SPECT, it is theoretically possible to use the same core structure with different radiometal ions.

As fluorescence and PET/SPECT have comparable levels of sensitivity, a one-to-one ratio of fluorophore to radioisotope in the final sensor is suitable. For *in vivo* imaging, NIR fluorophores are often used to allow for deeper penetration. As the attachment of two relatively bulky imaging labels can significantly impact the localisation of small molecules, targeting groups tend to be larger biomolecules. For example, Edwards *et al.* created a fluorescent and targeted chelator, **LS172**, by attaching DOTA and the NIR dye cypate to the cyclic octapeptide Y3-octreotate, which binds to the somatostatin receptor (Figure 8.20) [131]. Subsequent chelation of ^{64}Cu or ^{117}Lu allowed for fluorescence-PET and fluorescence-SPECT, respectively. Alternatively, if antibodies are used as the targeting group, there are multiple possible sites of attachment for the fluorophore and radiolabel, though maintaining a stoichiometric ratio of labels can be challenging [123].

	DOTA	NOTA	DTPA
PET	^{64}Cu, ^{68}Ga	^{64}Cu, ^{68}Ga	
SPECT	^{111}In, ^{177}Lu		^{111}In

Figure 8.19 Selected chelator scaffolds and radionuclides used for dual modality fluorescence-PET and fluorescence-SPECT sensors. DOTA = dodecane tetraacetic acid; DTPA = diethylenetriaminepentaacetic acid; NOTA = nonane triacetic acid.

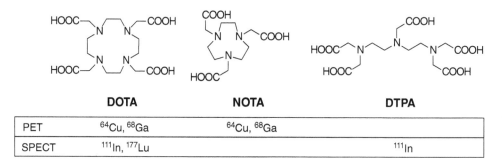

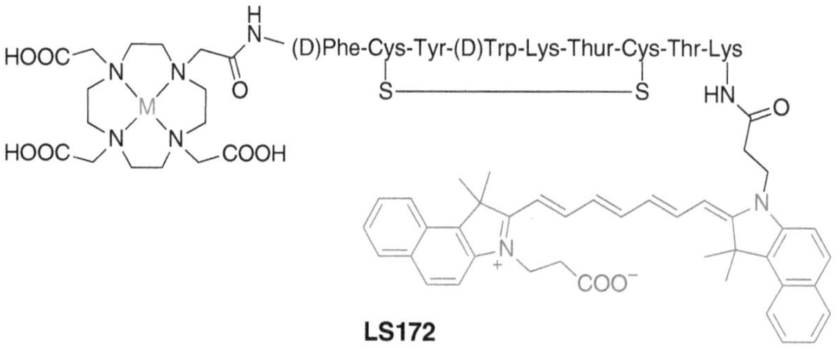

LS172

Figure 8.20 Structure of a dual-labelled peptide for fluorescence-PET/SPECT imaging [131]. M = ^{64}Cu (PET) or ^{117}Lu (SPECT). Position of radioisotope is highlighted in blue and fluorophore highlighted in red.

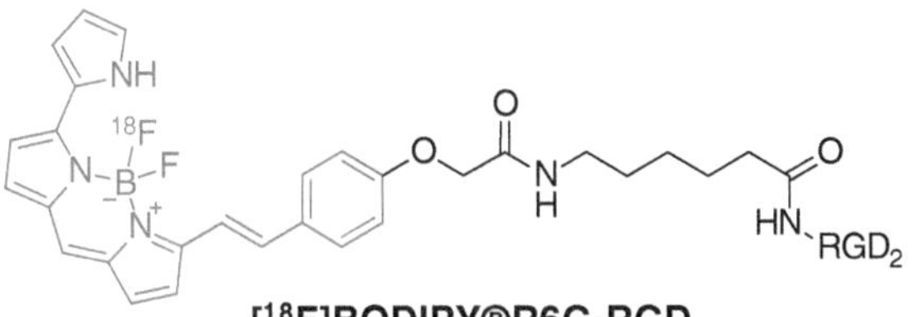

Figure 8.21 Examples of ^{18}F-labelled NIR fluorophores [133, 134]. Radioisotope highlighted in blue and fluorophore highlighted in red.

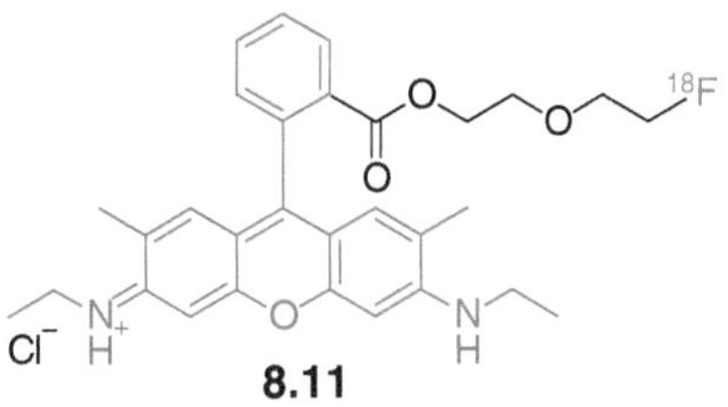

[18]F is the most commonly used radionuclide for PET. The relatively small size of ^{18}F allows it to be readily incorporated to NIR organic fluorophore scaffolds with minimal perturbation, and it may be installed *via* a number of synthetic routes [132]. For example, Liu *et al.* produced **[18F]BODIPY®R6G-RGD2** (Figure 8.21), the first example of a [18F]BODIPY as a dual-modality imaging agent. They radiofluorinated the commercial dye BODIPY RG6 and then conjugated it to a PEGylated dimer of cyclic RGDyK (RGD2) to achieve tumour targeting; tumour uptake was observed in both modalities [133]. Bartholoma *et al.* synthesised a range of ^{18}F-labelled rhodamine derivatives, conjugating the rhodamine lactones with ^{18}F-labelled diethylene glycol esters [134]. A rhodamine 6G derivative, **8.11**, was the most promising myocardial perfusion agent in both modalities (Figure 8.21).

8.5.3 Computed Tomography

Computed tomography (CT) provides three-dimensional anatomical reconstructions scans from a series of X-ray images taken at multiple angles throughout the body [135]. While CT can be a label-free technique, highly X-ray absorbing contrast media has been used to increase sensitivity and selectivity. The most commonly used contrast agents include iodinated molecules [136], though several lanthanide [137] and other metal-based sensors have been developed in recent times, including those in nanoparticulate form [138]. The incorporation of a more sensitive imaging technique, such as fluorescence, can improve sensitivity and selectivity and provide additional information.

Relatively few examples of dual-modality fluorescence-CT sensors exist [123], despite the potential of the technique. Nanoemulsion platforms incorporating iodinated oil have also been used; for example, Huang *et al.* prepared liposomes incorporating iohexol (X-ray contrast agent) and Cy5.5 (fluorescent dye), allowing the visualisation and tracking of therapeutic vehicles at the whole-body, tumour, and cellular scales with high sensitivity. The vast majority of fluorescence-CT sensors are nanomaterials [139]. A possible limiting factor is the toxicity of many CT contrast agents [140], and the use of higher energy X-rays in CT compared to other imaging techniques.

8.5.4 Magnetic Resonance Imaging

Magnetic resonance imaging (MRI) is a form of 3D imaging reliant on nuclear magnetic resonance (NMR). Anatomical features can be distinguished due to the different water content of tissue types, providing contrast, but the technique is relatively insensitive and targeted contrast agents are commonly used [141]. Contrast agents are typically paramagnetic gadolinium(III) [142] or manganese(II) complexes [143], but iron oxide nanoparticles [144] and scaffolds incorporating ^{19}F [145] have recently gained popularity. Incorporation of a fluorescent component can enable not only labelling of structures of interest but also responsive analyte detection in both modalities [146].

For paramagnetic metals, it is possible to append an organic fluorophore to the chelator. For example, You and co-workers re-purposed the existing zinc-responsive fluorescein indicator Zinpyr-1 (ZP1); they chelated it to form the manganese complex **ZP1Mn$_2$** and showed that the displacement of manganese from ZP1 by zinc allowed it to serve as a dual sensor *in vitro* [146] (Figure 8.22a). Notably, no significant modification was required as manganese displacement by zinc served as the trigger for signal output in both modalities. Other ion exchange type sensors are also possible; Zhang *et al.* developed a gadolinium(III) complex **Nap-DO3A** that exhibited a turn-on increase in relaxation rate in MRI and a quenching of fluorescence upon addition of copper(II) [147] (Figure 8.22b).

Rather than imaging water protons by ^{1}H MRI, it is also possible to use other nuclei. ^{19}F MRI is an attractive alternative as there are only trace levels of biologically occurring

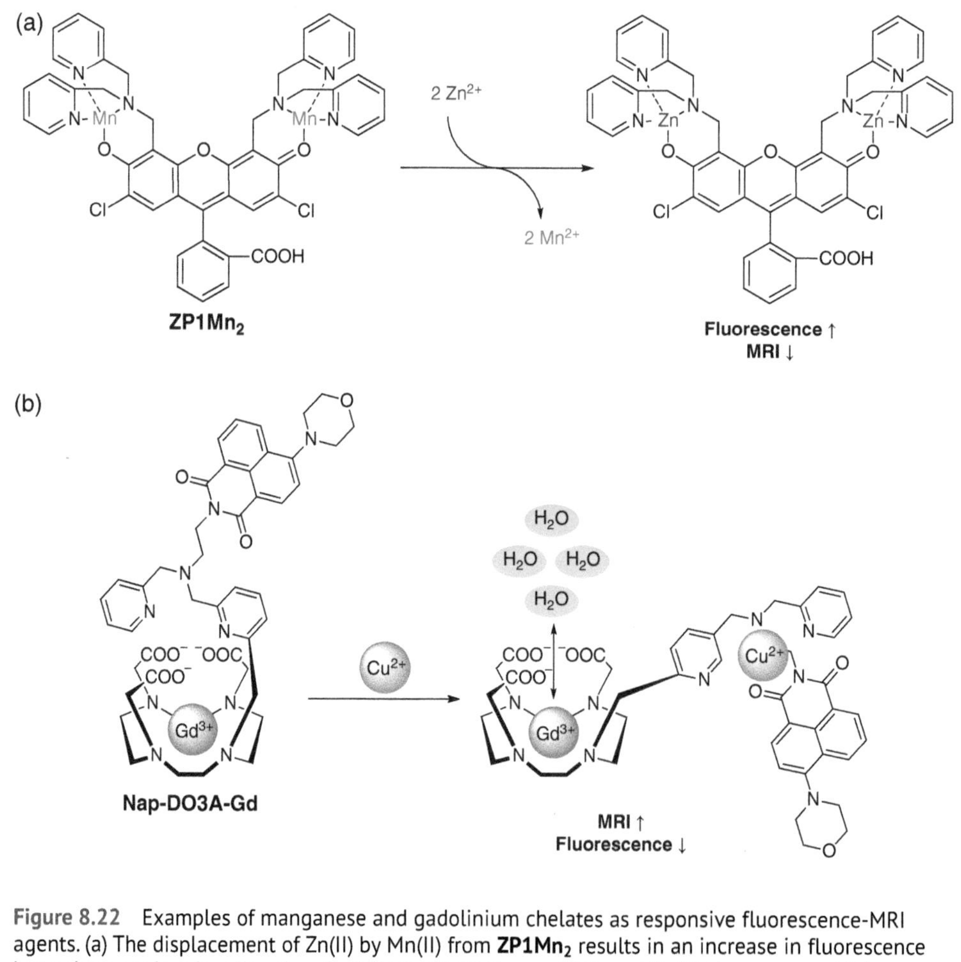

Figure 8.22 Examples of manganese and gadolinium chelates as responsive fluorescence-MRI agents. (a) The displacement of Zn(II) by Mn(II) from **ZP1Mn$_2$** results in an increase in fluorescence but a decrease in relaxation time for MRI [146]. (b) Upon addition of Cu(II), the sensor exhibits a turn-off fluorescence response due to Cu(II) binding the dipicolylamine and increase in relaxivity due to exposure of Gd(III) to bulk water [147].

fluorine, so only the contrast agent is tracked [145]. Nakamura *et al.* prepared a novel bimodal agent, mFLAME, using mesoporous silica nanoparticles as delivery system, encapsulating perfluoro-15-crown-5-ether in the core and with the fluorophore, Cy5, covalently attached to the mesoporous silica shell [148]. A folate-functionalised variant of mFLAME allowed for tumour targeting and loading of doxorubicin allowed potential use as a theranostic.

8.5.5 Photoacoustic Imaging

Photoacoustic imaging (PAI) involves the use of pulsed lasers to induce thermoelastic expansion, which produces ultrasonic waves that can be detected by ultrasound instrumentation [149, 150]. PAI has high spatial resolution and deep tissue penetration due to the combination of ultrasonic detection and laser excitation. While PAI can make use of the contrast differences between tissue types [151], exogenous agents such as nanomaterials and organic fluorophores may also be used [152]. As PAI techniques are unaffected by optical scattering, dual fluorescence-PAI can offset the poor tissue penetration of fluorescence alone. PAI also provides anatomical and functional information, while fluorescence provides better temporal resolution and sensitivity with surface visualisation of tissues and organs, providing complementary information.

A single organic fluorophore can be used for both PAI and fluorescence modalities. These are generally fluorophores in the near-infrared I window (650–950 nm) that match the laser range used by commercial PAI instruments [153]. Imaging in the near-infrared II window is also possible and offers a similar penetration in both modalities [154]. Indocyanine green (ICG) has recently been trialled as a candidate for dual modality imaging and was able to map sentinel lymph nodes and lymphatic vessels in a rat model of breast cancer, helping to evaluate tumour metastasis [155].

A number of responsive dual fluorescence-PAI sensors have been reported, including an asymmetrical squaraine dye **USq**, which could reversibly respond to thiol-containing biomolecules such as glutathione, and monitor thiol levels both *in vitro* and in a mouse model *in vivo* (Figure 8.8) [60]. Chan and co-workers reported **HyP-1**, an aza-BODIPY dye with aniline *N*-oxide, that could be converted irreversibly to the corresponding aniline red-HyP-1 under hypoxic conditions (Figure 8.23) [156]. This was demonstrated in both modalities *in vitro* and an *in vivo* animal model.

An alternate strategy to preparing dual fluorescence-PAI sensors is to incorporate separate labels for each modality, as higher quantum yields are beneficial for fluorescence but lower quantum yields beneficial for PAI. Lozano *et al.* formulated a liposome-gold nanorod hybrid vesicle system; the NIR dye NIR-797 covalently linked to the lipid to label the liposome bilayer allows for fluorescence imaging, while gold nanorods incorporated *via* surface-functionalisation allow for PAI [157].

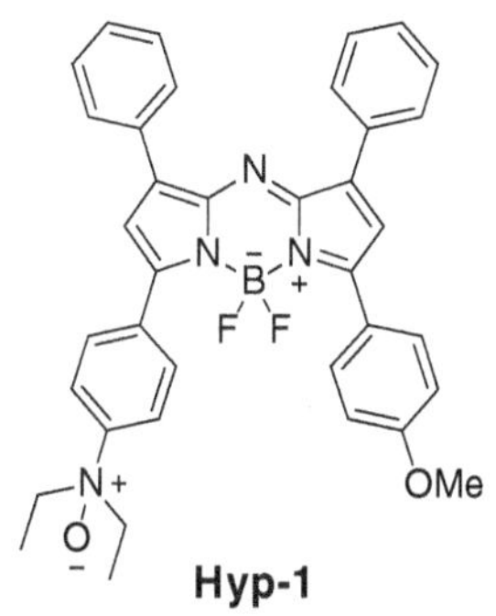

Hyp-1

Figure 8.23 Structures of a NIR fluorophore that can be used for bimodal fluorescence-PAI imaging [156].

8.5.6 Vibrational Spectroscopy

Molecular vibrational states, which are diagnostic of the unique composition of chemical bonds in a sample, can be sensed in a variety of ways, most commonly through infrared (IR) or Raman spectroscopy. For a molecule to be IR active, there must be a change in dipole moment within the molecule (*i.e.* electrons are not shared evenly in the molecule) [158]. On the other hand, for a molecule to be Raman active, there must be a change in polarisability (*i.e.* a change in the electron cloud surrounding the molecule) [159]. Despite an ability to capture information about the major classes of biomolecules, including carbohydrates, proteins, lipids and nucleic acids, vibrational spectroscopy techniques have relatively low spatial and temporal resolutions. Multimodal studies thus combine the specificity offered by fluorescence with the broader characterisation of the cellular environment offered by vibrational spectroscopy.

Combining fluorescence and Raman techniques is not straightforward, as fluorescence emission can overwhelm the Raman signal, requiring advanced instrumentation and imaging protocols [160]. A common approach in examining non-stained biological samples is combining Raman spectroscopy with tissue autofluorescence; this has been used with clinical samples to provide a more accurate diagnostic than in each modality separately [161–163]. Fluorescent sensors have also been used as a confirmatory technique to support the conclusions of separately performed Raman experiments, or as a screening tool to select regions of interest for subsequent Raman analysis [164]. To enhance signal, many fluorescent sensors for Raman studies utilise surface-enhanced Raman scattering (SERS); the fluorophore is conjugated to a nanoparticle scaffold, with distinct signals in both modalities [165, 166].

The biologically silent Raman region, from 1800 to 2800 cm^{-1}, offers the opportunity for exogenous labelling with groups such as alkynes and nitriles, enabling the tracking of small molecule uptake in cells [167, 168]. This approach can also be adopted to develop responsive sensors, although this is a relatively novel field of research, with few sensors reported to date. Li *et al.* reported a mitochondrial targeted sensor, **AIE-SRS-mito** (Figure 8.24), used in aggregation-induced emission (AIE) and stimulated Raman scattering [169]. Lin *et al.* reported **NpCN1** (Figure 8.24), a bimodal naphthalimide-based sensor, that successfully reported on the chemical composition and distribution of lipid droplets in cells [170].

As with Raman spectroscopy, many fluorescence and infrared multimodal studies involve the use of commercial fluorophores as a preliminary technique to narrow down areas of interest in biological samples or as a confirmatory technique to correlate localisation [171]. Bimodal fluorescence-infrared studies have been performed on luminescent metal complexes, such as those based on rhenium(I) carbonyl complexes, which provide a strong peak in the biologically silent region [172]. As the region from 1800 to 2800 cm^{-1} is relatively biologically silent for both Raman and infrared modalities, it is theoretically possible to use the same tag for studies in both modalities, though this has not been widely studied to date.

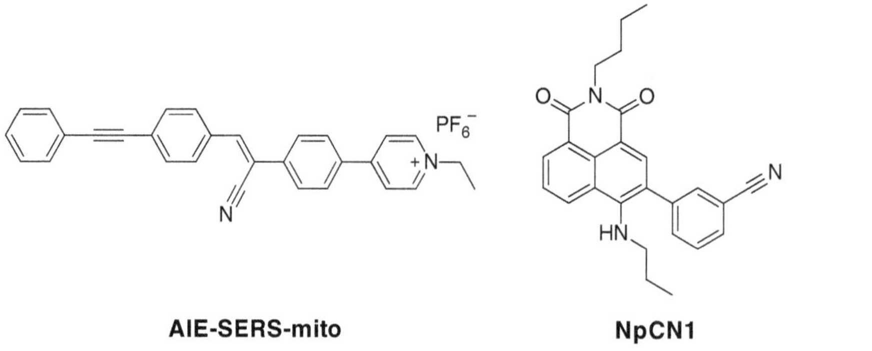

AIE-SERS-mito　　　　　　　**NpCN1**

Figure 8.24　Small-molecule fluorophores with biorthogonal Raman tags [169, 170].

8.5.7 Synchrotron X-ray Techniques

X-ray fluorescence (XRF) techniques allow for the quantification of elemental composition in a sample. This may be applied to X-ray fluorescence microscopy (XFM), allowing the visualisation of elemental distribution in a sample [173]. On the other hand, X-ray absorption spectroscopy (XAS) techniques are element specific, providing information about the chemical state and local atomic structure of the element [174]. X-ray absorption near edge structure (XANES) provides information about the oxidation state and coordination environment of a metal, while extended X-ray absorption fine structure (EXAFS) provides information about the local atomic environment [175].

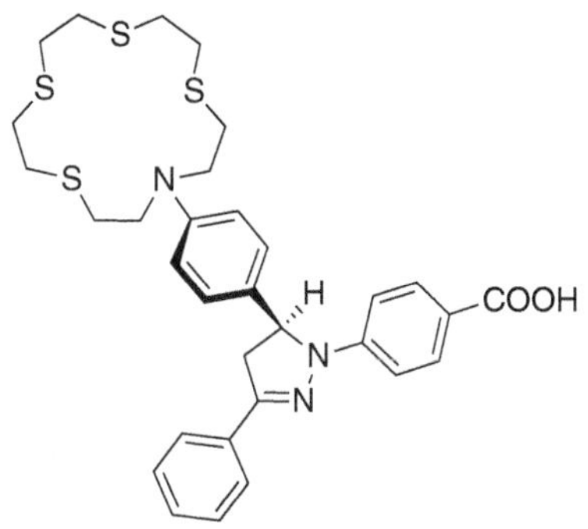

CTAP-1

Figure 8.25 Structure of a probe used for dual optical fluorescence-XFM studies [176].

XFM and XAS techniques are powerful, label-free techniques that can visualise the spatial distribution of elements and their speciation throughout a sample. However, they are generally low throughput and require synchrotron access to perform and are also insensitive to light elements. Furthermore, it is often necessary to perform additional correlative techniques to identify subcellular compartments where the element is associated, such as *via* fluorescence imaging.

Synchrotron X-ray techniques are complementary to the specific analyte information provided by fluorescent sensors, particularly in cases of validating results of metal sensors. For example, Yang *et al.* reported **CTAP-1**, a selective fluorescent sensor for Cu(I) (Figure 8.25), and carried out sequential fluorescence microscopy and XFM studies of the same **CTAP-1**-stained sample [176]. They found a high degree of colocalisation between **CTAP-1** optical fluorescence and the total copper and sulfur pools. XANES studies showed that copper pool was largely low coordinate, monovalent copper inconsistent with coordination to metallothionein or glutathione, suggesting that Cu(I) is coordinated to another sulfur donor ligand.

To definitively track the presence of a fluorescent sensor in synchrotron X-ray techniques requires the incorporation of an element that does not have significant biological concentrations. Such studies often involve metal complexes used as imaging agents or therapeutics, including luminescent metal complexes [177]. Heavy atom tags like iodine also offer a potential way of tagging and following fluorescent molecules in XFM [177].

On the other hand, a fluorescent ligand may be used to track the uptake of a metal complex, such as those used for therapeutic purposes. Hambley and co-workers investigated the uptake of a set of platinum compounds incorporating a fluorescent anthraquinone ligand in spheroid tumour models, **8.12** and **8.13** [178, 179] (Figure 8.26). Optical fluorescence studies showed fluorescence predominantly around the periphery of the spheroids [178], while XFM revealed high platinum levels in the centre of the spheroid [179]. This suggested ligand exchange occurred within the peripheral cell layers.

8.5.8 Mass Spectrometry

Mass spectrometry (MS) is frequently used to characterise biomolecules including proteins, sugars, lipids, and oligonucleotides. This technique can be applied as an imaging technique;

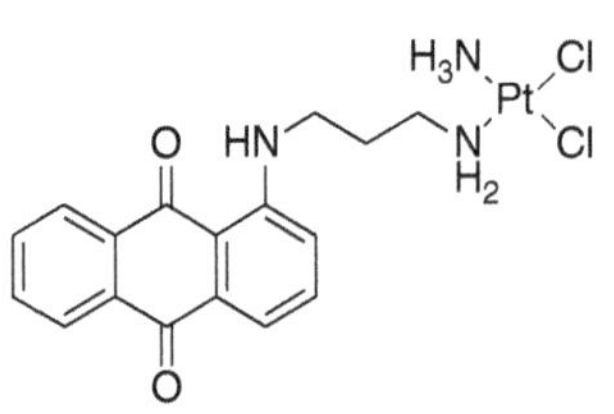

Figure 8.26 Metal complexes studied in bimodal fluorescence-XFM studies [178, 179].

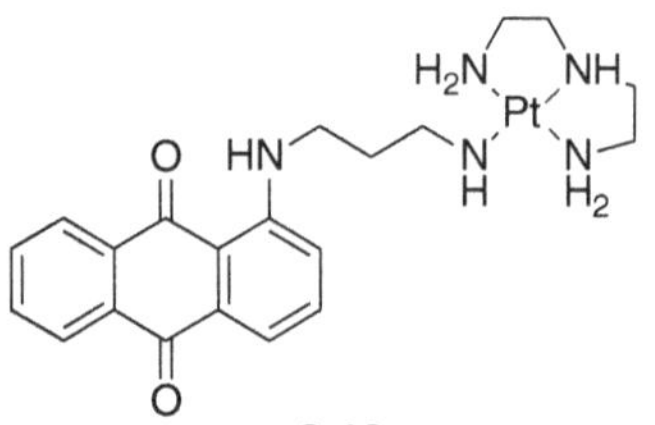

thin samples or sections are carefully prepared, followed by point-by-point collection of mass spectra from a defined region of interest [180]. Mass spectrometry imaging (MSI) can also involve tandem MS (MS/MS), a technique that couples multiple mass spectrometers. After a sample is ionised and the ions are separated by the first spectrometer, ions with a certain m/z value can be selected for further fragmentation and detection by an additional mass spectrometer, with enhanced structure determination capabilities [181]. A variety of ionisation methods are possible, including secondary ion mass spectrometry (SIMS) [182], matrix-assisted laser desorption ionisation (MALDI) [183] and desorption electrospray ionisation (DESI) [184]. MSI has a high degree of chemical specificity but lacks spatial and temporal resolution compared to other imaging modalities. Correlative techniques may also be required to match data from MSI to cellular substructures.

In cell and tissue samples, molecules of interest can be labelled *via* immunofluorescence (IF) techniques [185]. More recently, fluorescent sensors have been used to correlate localisation of analytes. Chandra *et al.* investigated the subcellular localisation of a fluorinated boron neutron capture agent, F-BPA, for anticancer treatment, showing *via* SIMS imaging that intracellular boron and fluorine were co-compartmentalised [186]. Using rhodamine 123 to correlate mitochondrial localisation, they found that cells treated with F-BPA exhibited low boron and fluorine signals in the mitochondria-rich perinuclear region compared to the nucleus and remaining cytoplasmic region. Pour *et al.* investigated hippocampal zinc concentration in a rat animal model after traumatic brain injury, finding that time of flight SIMS imaging was sensitive to bound zinc, while fluorescence imaging with the sensor Fluozin-3 was sensitive to free zinc [187]. This experiment provided complementary information on localisation.

8.5.9 Electron Microscopy

Electron microscopy (EM) provides the highest spatial resolution of all imaging techniques, below 1 nm [188]. EM can therefore provide the finest detail of cellular ultrastructure and even resolve biomolecules. However, these high levels of magnification results in a small field of view. There are several types of microscopes for EM, including the scanning electron microscope (SEM) [189] and transmission electron microscope (TEM) [188].

Live imaging is not possible with EM [75], and complex sample preparation must be performed prior to imaging [190]. This must be considered carefully in including stains and contrast agents for multimodal studies. There is a general requirement for fixation, dehydration, and mounting of the sample for SEM techniques or embedding and sectioning the sample for TEM techniques (see Figure 8.27). It is common to add a general histological stain to allow for recognition of general structures, cell types, or macromolecules, allowing for the

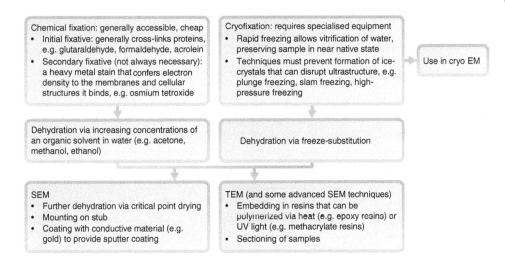

Figure 8.27 General procedure for sample preparation for electron microscopy.

selection of regions of interest for electron microscopy [188]. It is also possible to label specific molecules *via* antibodies. Immunogold labelling is the classical technique, where colloidal gold nanoparticles are attached to secondary antibodies [191]. During imaging, the high electron density of gold provides higher electron scatter and thus high contrast 'dark spots'. This labelling may be performed prior to fixation or at various stages during the sample preparation process, with various advantages and disadvantages to each stage [188].

In terms of multimodality, correlative light and electron microscopy (CLEM) has been an established technique for several decades [192]. Traditionally, CLEM has been carried out sequentially on separate instruments, though in recent times, instruments capable of performing both techniques have been constructed [193]. Fluorescence imaging may be performed on samples prior to fixation and preparation for EM, giving contextual information to inform high-resolution EM imaging [194]. Alternatively, samples prepared for EM with fluorescent labels may be imaged with both techniques. Here, fluorescence imaging is commonly used to identify regions of interest for further EM on the same sample, allowing the identification of dynamic processes at known processes, rare events, or structures [194]. These identification processes can thus increase EM sample size and throughput.

A key concern is establishing sample preparation protocols and labelling methods compatible with both techniques. Combinatorial sensors allow for recognition in both EM and fluorescence modalities, generally *via* immunostaining. FluoroNanogold™ is a nanosensor containing an affinity label (*e.g.* an antibody for the protein of interest) bound to both a fluorophore (*e.g.* FITC) and a dense colloidal gold nanoparticle as used in immunogold labelling [195]. Quantum dots may also be attached to an affinity label; their natural fluorescence properties and electron dense crystalline structures allow for recognition in both modalities [196].

A variety of systems sensitive to photooxidation have also been developed for labelling structures in CLEM, generally involving diaminobenzidine (DAB) as a reagent in the sample preparation process [197]. A fluorophore may act as a photooxidiser, producing highly reactive singlet oxygen that reacts with DAB [198]. DAB polymerises after oxidation and forms a highly insoluble, electron-dense structure after treatment with osmium tetroxide.

Genetically encoded self-labelling proteins such as SNAP-tag and HaloTag can also be used for molecular labelling. Perkovic *et al.* conjugated synthetic fluorophores to SLPs in live cells, demonstrating that fluorescence was conserved post-embedding and preparation for EM, even in the presence of the heavy metal stain, uranyl acetate [199].

8.5.10 Three or More Modalities

Thus far, the sensors and techniques discussed in this section have involved fluorescence and one other modality. It is also possible to perform imaging with three or more modalities. The development of contrast or imaging agents for multiple techniques becomes increasingly difficult as more modalities are incorporated, especially as some modalities rely on competing physical processes and have different levels of sensitivity. For small-molecule sensors, additional functionality is most easily included in the form of small tags (*e.g.* biorthogonal Raman tags, ^{19}F for MRI) or based on NIR dyes with pre-existing multifunctionality (*e.g.* dual fluorescence-PAI sensors). Nanomaterial-based systems can be easily functionalised [200] and present an alternate method of incorporating multiple contrast agents. Both small-molecule and nanomaterial-based sensors have been demonstrated for three or more modalities including fluorescence.

Utilising the fact that many NIR fluorescent scaffolds provide photoacoustic signal, Liu *et al.* synthesised an ^{18}F-labelled NIR croconium dye with intrinsic sensitivity for amyloid-β plaques, [^{18}F]CDA-3, with the ^{18}F acting as a PET contrast agent (Figure 8.28) [201]. Trimodal fluorescence-PAI-PET imaging was achieved in an *in vivo* mouse model, providing selective imaging of brain plaques.

BDPF is a fluorinated aza-BODIPY used in fluorescence-PAI-MRI imaging (Figure 8.28) [202]. The aza-BODIPY core is both an NIR fluorophore and a photoacoustic

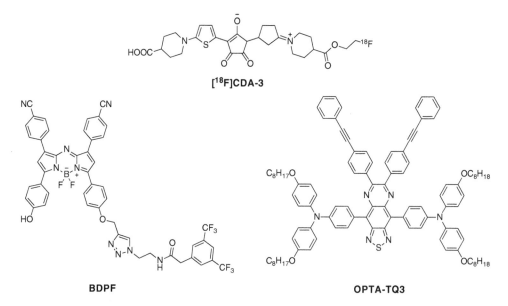

[^{18}F]CDA-3

BDPF

OPTA-TQ3

Figure 8.28 Structures of small-molecule sensors used for trimodality studies [201–203].

contrast agent, while trifluoromethyl groups provide signal for ^{19}F MRI, allowing for *in vitro* imaging of cancer cells and *in vivo* imaging of tumours in all three modalities.

Qi *et al.* synthesised **OTPA-TQ3**, containing intramolecular rotors and functionalities for fluorescence-PAI-Raman (Figure 8.28) [203]. Intramolecular rotation boosted the AIE effect, resulting in bright NIR fluorescence in aqueous media, and excited-state intramolecular motions and high absorption coefficient provided strong PA signal generation capability. Raman signal was provided by the diphenylacetylene group, and intramolecular motions in high-energy state also enhanced this. To improve biocompatibility, **OPTA-TQ3** was encapsulated into nanoparticles, allowing pre-operative fluorescence-PAI to provide tumour information and intraoperative fluorescence-Raman imaging to delineate tiny residual tumours.

Some *in vivo* imaging techniques have established multimodal applications, *e.g.* PET-MRI, PET-CT, SPECT-CT, MRI-CT [204, 205]. Adding fluorescence as a modality to these techniques can add an additional degree of complementary information. For trimodal imaging, most reported sensors are based on nanomaterial containing a separate contrast agent for each modality. For example, Yang *et al.* synthesised tungsten sulfide nanosheets (a CT contrast agent), onto which SPIO nanoparticles (a MRI contrast agent) were adsorbed; this was then coated with a mesoporous silica shell, then functionalised with a PEG and the fluorophore Cy5.5, creating trimodal CT-MRI-fluorescence sensor WS$_2$-IO-@MS-PEG [206]. Sensors were also loaded with doxorubicin. This allowed for trimodal imaging of tumour cells and theranostic drug delivery. Trimodal sensors have been demonstrated for PET-MRI-fluorescence [207], SPECT-MRI-fluorescence [207], PET-CT-fluorescence [208], and MRI-CT-fluorescence [209].

8.5.11 Future Directions

It is clear that fluorescence is a complementary modality to other imaging techniques. Ideally, the combined methods should provide information that cannot be achieved with either modality alone and at a high resolution, with high sensitivity and specificity to targets of interest. Several strategies have been formulated to design sensors suitable for use in multiple modalities, including both small-molecule organic sensors and nanomaterial-based systems. Nonetheless, this field is still nascent. Many sensors identify novel synthetic methods or are proof-of-concept type designs; however, these sometimes fall short of providing applications that necessitate or benefit from the multimodal sensor design. Nonetheless, this can still inform future sensor and biological experiment design. Due to difficulties in attaching multiple functionalities, many multimodal sensors are untargeted or rely on intrinsic properties to achieve localisation; the most notable class is probably the affinity of many nanoparticles for cancer tumours. Improvements in areas of chemical synthesis will drive the development of targeted multimodal sensors. The further development of imaging instruments, especially those that allow imaging of two or more modalities at the same time, will also extend the applications of multimodal sensors. While there are applications that necessitate separate instruments (*e.g. in vivo* imaging of a tumour compared to *in vitro* analysis of tissue samples from the tumour), instruments that can perform simultaneous data acquisition can provide co-registered imaging with different structural and/or functional information from each modality. This could provide new avenues in biomedical and biochemical imaging, from the preclinical stage up to clinical diagnosis.

References

1 Sarder, P., Maji, D., and Achilefu, S. (2015). *Bioconjugate Chemistry* 26 (6): 963–974.
2 Lee, M.H., Kim, J.S., and Sessler, J.L. (2015). *Chemical Society Reviews* 44 (13): 4185–4191.
3 Lakowicz, J.R. (1999). *Principles of Fluorescence Spectroscopy*, 2nde. New York: Kluwer Academic/Plenum.
4 Berezin, M.Y. and Achilefu, S. (2010). *Chemical Reviews* 110 (5): 2641–2684.
5 Yasuda, R. In: Alfano, R. R., Shi, L., editors. *Neurophotonics and Medical Spectroscopy*, Elsevier, Amsterdam: 2019. p. 53–64.
6 Becker, W. (2012). *Journal of Microscopy* 247 (2): 119–136.
7 Glasgow, B.J. (2016). *Experimental Eye Research* 147: 12–19.
8 Datta, R., Heaster, T.M., Sharick, J.T. et al. (2020). *Journal of Biomedical Optics* 25 (7): 1–43.
9 Lakowicz, J.R., Szmacinski, H., Nowaczyk, K., and Johnson, M.L. (1992). *Proceedings of the National Academy of Sciences* 89 (4): 1271–1275.
10 Blacker, T.S., Mann, Z.F., Gale, J.E. et al. (2014). *Nature Communications* 5: 3936.
11 Skala, M.C., Riching, K.M., Gendron-Fitzpatrick, A. et al. (2007). *Proceedings of the National Academy of Sciences* 104 (49): 19494–19499.
12 Drezek, R., Brookner, C., Pavlova, I. et al. (2001). *Photochemistry and Photobiology* 73 (6): 636–641.
13 Sun, Y., Hatami, N., Yee, M. et al. (2010). *Journal of Biomedical Optics* 15 (5): 056022.
14 Hille, C., Berg, M., Bressel, L. et al. (2008). *Analytical and Bioanalytical Chemistry* 391 (5): 1871–1879.
15 Almutairi, A., Guillaudeu, S.J., Berezin, M.Y. et al. (2008). *Journal of the American Chemical Society* 130 (2): 444–445.
16 Haidekker, M.A. and Theodorakis, E.A. (2007). *Organic & Biomolecular Chemistry* 5 (11): 1669–1678.
17 Abdel-Mottaleb, M.S.A. (1989). *Journal of Photochemistry and Photobiology A: Chemistry* 48 (1): 87–93.
18 Kuimova, M.K., Yahioglu, G., Levitt, J.A., and Suhling, K. (2008). *Journal of the American Chemical Society* 130 (21): 6672–6673.
19 Ogle, M.M., Smith McWilliams, A.D., Jiang, B., and Martí, A.A. (2020). *ChemPhotoChem* 4 (4): 255–270.
20 Jenkins, J., Borisov, S.M., Papkovsky, D.B., and Dmitriev, R.I. (2016). *Analytical Chemistry* 88 (21): 10566–10572.
21 Karstens, T. and Kobs, K. (1980). *The Journal of Physical Chemistry* 84 (14): 1871–1872.
22 Ma, Y., Liang, H., Zeng, Y. et al. (2016). *Chemical Science* 7 (5): 3338–3346.
23 Hao, L., Li, Z.W., Zhang, D.Y. et al. (2019). *Chemical Science* 10 (5): 1285–1293.
24 Okabe, K., Inada, N., Gota, C. et al. (2012). *Nature Communications* 3: 705.
25 Frangioni, J.V. (2003). *Current Opinion in Chemical Biology* 7 (5): 626–634.
26 Schaefer, P.M., Kalinina, S., Rueck, A. et al. (2019). *Cytometry. Part A* 95 (1): 34–46.
27 Weissleder, R. (2001). *Nature Biotechnology* 19: 316–317.
28 Sordillo, L., Pu, Y., Pratavieira, S. et al. (2014). *Journal of Biomedical Optics* 19 (5): 056004.
29 Kenry, D.Y. and Liu, B. (2018). *Advanced Materials* 30 (47): 1802394.
30 Smith, A.M., Mancini, M.C., and Nie, S. (2009). *Nature Nanotechnology* 4 (11): 710–711.
31 Cheong, W.-F., Prahl, S.A., and Welch, A.J. (1990). *IEEE Journal of Quantum Electronics* 26 (12): 2166–2185.
32 Gao, X., Cui, Y., Levenson, R.M. et al. (2004). *Nature Biotechnology* 22 (8): 969–976.

33 Li, J.B., Liu, H.W., Fu, T. et al. (2019). *Trends in Chemistry* 1 (2): 224–234.

34 Zhao, J., Zhong, D., and Zhou, S. (2018). *Journal of Materials Chemistry B* 6 (3): 349–365.

35 Hong, G., Robinson, J.T., Zhang, Y. et al. (2012). *Angewandte Chemie* 51 (39): 9818–9821.

36 Dou, Q.Q., Guo, H.C., and Ye, E. (2014). *Materials Science & Engineering. C, Materials for Biological Applications* 45: 635–643.

37 Welsher, K., Liu, Z., Sherlock, S.P. et al. (2009). *Nature Nanotechnology* 4 (11): 773–780.

38 Hendriks, K.H., Li, W., Wienk, M.M., and Janssen, R.A. (2014). *Journal of the American Chemical Society* 136 (34): 12130–12136.

39 Escobedo, J.O., Rusin, O., Lim, S., and Strongin, R.M. (2010). *Current Opinion in Chemical Biology* 14 (1): 64–70.

40 Koide, Y., Urano, Y., Hanaoka, K. et al. (2011). *ACS Chemical Biology* 6 (6): 600–608.

41 Fu, M., Xiao, Y., Qian, X. et al. (2008). *Chemical Communications* 15: 1780–1782.

42 Sun, C., Du, W., Wang, B. et al. (2020). *BMC Chemistry* 14 (1): 21.

43 Karton-Lifshin, N., Segal, E., Omer, L. et al. (2011). *Journal of the American Chemical Society* 133: 10960–10965.

44 Li, Y., Wang, Y., Yang, S. et al. (2015). *Analytical Chemistry* 87 (4): 2495–2503.

45 Yan, J.W., Zhu, J.Y., Zhou, K.X. et al. (2017). *Chemical Communications* 53 (71): 9910–9913.

46 Mieog, J.S., Troyan, S.L., Hutteman, M. et al. (2011). *Annals of Surgical Oncology* 18 (9): 2483–2491.

47 Yamamoto, M., Orihashi, K., Nishimori, H. et al. (2012). *European Journal of Vascular and Endovascular Surgery* 43 (4): 426–432.

48 Tummers, Q.R., Verbeek, F.P., Prevoo, H.A. et al. (2015). *Surgical Innovation* 22 (1): 20–25.

49 Vinegoni, C., Botnaru, I., Aikawa, E. et al. (2011). *Atherosclerosis* 3 (84): 84ra45.

50 Gotoh, K., Yamada, T., Ishikawa, O. et al. (2009). *Journal of Surgical Oncology* 100 (1): 75–79.

51 Hong, G., Antaris, A.L., and Dai, H. (2017). *Nature Biomedical Engineering* 1 (1).

52 Chen, H., Dong, B., Tang, Y., and Lin, W. (2017). *Accounts of Chemical Research* 50 (6): 1410–1422.

53 Feng, L., Chen, W., Ma, X. et al. (2020). *Organic & Biomolecular Chemistry* 18 (46): 9385–9397.

54 Yi, X., Yan, F., Wang, F. et al. (2015). *Medical Science Monitor* 21: 511–517.

55 Yang, X., Shao, C., Wang, R. et al. (2013). *The Journal of Urology* 189 (2): 702–710.

56 Tan, X., Luo, S., Wang, D. et al. (2012). *Biomaterials* 33 (7): 2230–2239.

57 Ni, Y. and Wu, J. (2014). *Organic & Biomolecular Chemistry* 12 (23): 3774–3791.

58 Li, S.J., Fu, Y.J., Li, C.Y. et al. (2017). *Analytica Chimica Acta* 994: 73–81.

59 Bai, L., Sun, P., Liu, Y. et al. (2019). *Chemical Communications* 55 (73): 10920–10923.

60 Anees, P., Joseph, J., Sreejith, S. et al. (2016). *Chemical Science* 7 (7): 4110–4116.

61 Zhang, X., Bloch, S., Akers, W., and Achilefu, S. (2012). *Current Protocols in Cytometry*: Chapter 12:Unit12.27.

62 Koide, Y., Urano, Y., Hanaoka, K. et al. (2012). *Journal of the American Chemical Society* 134 (11): 5029–5031.

63 Antaris, A.L., Chen, H., Cheng, K. et al. (2016). *Nature Materials* 15 (2): 235–242.

64 Zhu, S., Tian, R., Antaris, A.L. et al. (2019). *Advanced Materials* 31 (24): e1900321.

65 de Silva, P.A., Gunaratne, N.H.Q., and McCoy, C.P. (1993). *Nature* 364: 42–44.

66 Komatsu, H., Miki, T., Citterio, D. et al. (2005). *Journal of the American Chemical Society* 127 (31): 10798–10799.

67 Yuan, L., Lin, W., Xie, Y. et al. (2012). *Journal of the American Chemical Society* 134 (2): 1305–1315.

68 Finkler, B., Riemann, I., Vester, M. et al. (2016). *Photochemical & Photobiological Sciences* 15 (12): 1544–1557.

69 Liu, Y., Meng, F., He, L. et al. (2016). *Chemical Communications* 52 (43): 7016–7019.

70 Klockow, J.L., Hettie, K.S., Secor, K.E. et al. (2015). *Chemistry—a European Journal* 21 (32): 11446–11451.

71 Miao, F., Song, G., Sun, Y. et al. (2013). *Biosensors & Bioelectronics* 50 (15): 42–49.

72 Sun, Y.Q., Liu, J., Zhang, H. et al. (2014). *Journal of the American Chemical Society* 136 (36): 12520–12523.

73 Huang, B., Bates, M., and Zhuang, X. (2009). *Annual Review of Biochemistry* 78: 993–1016.

74 McIntosh, J.R. (2001). *Journal of Cell Biology* 153 (6): F25–F32.

75 de Jonge, N. and Peckys, D.B. (2016). *ACS Nano* 10: 9061–9063.

76 Balzarotti, F., Eilers, Y., Gwosch, K.C. et al. (2017). *Science* 355 (6325): 606–612.

77 Möckl, L., Lamb, D.C., and Bräuchle, C. (2014). *Angewandte Chemie International Edition* 53 (51): 13972–13977.

78 Schermelleh, L., Ferrand, A., Huser, T. et al. (2019). *Nature Cell Biology* 21 (1): 72–84.

79 Wang, L., Frei, M.S., Salim, A., and Johnsson, K. (2019). *Journal of the American Chemical Society* 141 (7): 2770–2781.

80 Fernández-Suárez, M. and Ting, A.Y. (2008). *Nature Reviews Molecular Cell Biology* 9: 929–943.

81 Verdaasdonk, J.S., Stephens, A.D., Haase, J., and Bloom, K. (2014). *Journal of Cellular Physiology* 229 (2): 132–138.

82 Croix, C.M.S., Shand, S.H., and Watkins, S.C. (2005). *BioTechniques* 39 (6S): S2–S5.

83 Gustafsson, M.G., Agard, D.A., and Sedat, J.W. (1999). *Journal of Microscopy* 195 (1): 10–16.

84 Heintzmann, R. and Huser, T. (2017). *Chemical Reviews* 117 (23): 13890–13908.

85 Sharma, R., Singh, M., and Sharma, R. (2020). *Spectrochimica Acta Part A: Molecular and Biomolecular Spectroscopy* 231: 117715.

86 Blom, H. and Widengren, J. (2017). *Chemical Reviews* 117 (11): 7377–7427.

87 Möckl, L. and Moerner, W.E. (2020). *Journal of the American Chemical Society* 142 (42): 17828–17844.

88 Rust, M.J., Bates, M., and Zhuang, X. (2006). *Nature Methods* 3 (10): 793–796.

89 Betzig, E., Patterson, G.H., Sougrat, R. et al. (2006). *Science* 313 (5793): 1642–1645.

90 Hess, S.T., Girirajan, T.P.K., and Mason, M.D. (2006). *Biophysical Journal* 91 (11): 4258–4272.

91 Schermelleh, L., Heintzmann, R., and Leonhardt, H. (2010). *Journal of Cell Biology* 190 (2): 165–175.

92 Juette, M.F., Gould, T.J., Lessard, M.D. et al. (2008). *Nature Methods* 5 (6): 527–529.

93 Godin Antoine, G., Lounis, B., and Cognet, L. (2014). *Biophysical Journal* 107 (8): 1777–1784.

94 Wysocki, L.M. and Lavis, L.D. (2011). *Current Opinion in Chemical Biology* 15: 752–759.

95 Sun, W.-C., Gee, K.R., Klaubert, D.H., and Haugland, R.P. (1997). *The Journal of Organic Chemistry* 62 (19): 6469–6475.

96 Toutchkine, A., Nguyen, D.-V., and Hahn, K.M. (2007). *Organic Letters* 9 (15): 2775–2777.

97 Song, X., Johnson, A., and Foley, J. (2008). *Journal of the American Chemical Society* 130 (52): 17652–17653.

98 Dave, R., Terry, D.S., Munro, J.B., and Blanchard, S.C. (2009). *Biophysical Journal* 96 (6): 2371–2381.

99 Zheng, Q., Jockusch, S., Rodríguez-Calero, G.G. et al. (2016). *Photochemical & Photobiological Sciences* 15 (2): 196–203.

100 van de Linde, S. and Sauer, M. (2014). *Chemical Society Reviews* 43 (4): 1076–1087.

101 Xu, J., Ma, H., and Liu, Y. (2017). *Current Protocols in Cytometry* 81: 12.46.1–12.46.27.

102 Heilemann, M., van de Linde, S., Mukherjee, A., and Sauer, M. (2009). *Angewandte Chemie International Edition* 48 (37): 6903–6908.

103 Uno, S.-n., Kamiya, M., Yoshihara, T. et al. (2014). *Nature Chemistry* 6 (8): 681–689.

104 Li, H. and Vaughan, J.C. (2018). *Chemical Reviews* 118 (18): 9412–9454.

105 Grimm, J.B., English, B.P., Chen, J. et al. (2015). *Nature Methods* 12 (3): 244–250.

106 Grimm, J.B., Xie, L., Casler, J.C. et al. (2021). *JACS Au* 1 (5): 690–696.

107 Chen, B.-C., Legant, W.R., Wang, K. et al. (2014). *Science* 346 (6208): 1257998.

108 Geissbuehler, S., Sharipov, A., Godinat, A. et al. (2014). *Nature Communications* 5 (1): 5830.

109 Gustafsson, M.G.L., Shao, L., Carlton, P.M. et al. (2008). *Biophysical Journal* 94 (12): 4957–4970.

110 Fiolka, R., Shao, L., Rego, E.H. et al. (2012). *Proceedings of the National Academy of Sciences* 109 (14): 5311–5315.

111 Rego, E.H., Shao, L., Macklin, J.J. et al. (2012). *Proceedings of the National Academy of Sciences* 109 (3): E135–E143.

112 Li, D., Shao, L., Chen, B.-C. et al. (2015). *Science* 349 (6251): aab3500.

113 Bottanelli, F., Kromann, E.B., Allgeyer, E.S. et al. (2016). *Nature Communications* 7 (1): 10778.

114 Göttfert, F., Wurm Christian, A., Mueller, V. et al. (2013). *Biophysical Journal* 105 (1): L01–L03.

115 Jacquemet, G., Carisey, A.F., Hamidi, H. et al. (2020). *Journal of Cell Science* 133 (11).

116 Klar, T.A., Jakobs, S., Dyba, M. et al. (2000). *Proceedings of the National Academy of Sciences* 97 (15): 8206–8210.

117 Osseforth, C., Moffitt, J.R., Schermelleh, L., and Michaelis, J. (2014). *Optics Express* 22 (6): 7028–7039.

118 Khater, I.M., Nabi, I.R., and Hamarneh, G. (2020). *Patterns* 1 (3): 100038.

119 Huang, B., Wang, W., Bates, M., and Zhuang, X. (2008). *Science* 319 (5864): 810–813.

120 von Diezmann, L., Shechtman, Y., and Moerner, W.E. (2017). *Chemical Reviews* 117 (11): 7244–7275.

121 Lin, Y., Long, J.J., Huang, F. et al. (2015). *PLoS One* 10 (5): e0128135.

122 Wu, D., Sedgwick, A.C., Gunnlaugsson, T. et al. (2017). *Chemical Society Reviews* 46 (23): 7105–7123.

123 Zhao, J., Chen, J., Ma, S. et al. (2018). *Acta Pharmaceutica Sinica B* 8 (3): 320–338.

124 Hackett, M.J., Aitken, J.B., El-Assaad, F. et al. (2015). *Scientific Advances* 1 (11): e1500911.

125 Louie, A. (2010). *Chemical Reviews* 110 (5): 3146–3195.

126 Wadas, T.J., Wong, E.H., Weisman, G.R., and Anderson, C.J. (2010). *Chemical Reviews* 110 (5): 2858–2902.

127 Price, E.W. and Orvig, C. (2014). *Chemical Society Reviews* 43 (1): 260–290.

128 Paulus, A., Maenen, M., Drude, N. et al. (2017). *PLoS One* 12 (8): e0182297-e.

129 An, F.-F., Chan, M., Kommidi, H., and Ting, R. (2016). *American Journal of Roentgenology* 207 (2): 266–273.

130 Rijpkema, M., Oyen, W.J., Bos, D. et al. (2014). *Journal of Nuclear Medicine* 55 (9): 1519–1524.

131 Edwards, W.B., Xu, B., Akers, W. et al. (2008). *Bioconjugate Chemistry* 19 (1): 192–200.

132 Preshlock, S., Tredwell, M., and Gouverneur, V. (2016). *Chemical Reviews* 116 (2): 719–766.

133 Liu, S., Li, D., Zhang, Z. et al. (2014). *Chemical Communications* 50 (55): 7371–7373.

134 Bartholomä, M.D., Zhang, S., Akurathi, V. et al. (2015). *Nuclear Medicine and Biology* 42 (10): 796–803.

135 Pelc, N.J. (2014). *Annals of Biomedical Engineering* 42 (2): 260–268.

136 Lee, S.Y., Rhee, C.M., Leung, A.M. et al. (2015). *The Journal of Clinical Endocrinology and Metabolism* 100 (2): 376–383.

137 Heffern, M.C., Matosziuk, L.M., and Meade, T.J. (2014). *Chemical Reviews* 114 (8): 4496–4539.

138 Liu, Y., Ai, K., and Lu, L. (2012). *Accounts of Chemical Research* 45 (10): 1817–1827.

139 Huang, H., Dunne, M., Lo, J. et al. (2013). *Molecular Imaging* 12 (3): 148–160.

140 Hasebroock, K.M. and Serkova, N.J. (2009). *Expert Opinion on Drug Metabolism and Toxicology* 5 (4): 403–416.

141 Shen, C. and New, E.J. (2013). *Current Opinion in Chemical Biology* 17 (2): 158–166.

142 Zhaoda, Z., Shrikumar, A.N., and Thomas, J.M. (2005). *Current Medicinal Chemistry* 12 (7): 751–778.

143 Pan, D., Schmieder, A.H., Wickline, S.A., and Lanza, G.M. (2011). *Tetrahedron* 67 (44): 8431–8444.

144 Korchinski, D.J., Taha, M., Yang, R. et al. (2015). *Magnetic Resonance Insights* 8 (Suppl 1): 15–29.

145 Tirotta, I., Dichiarante, V., Pigliacelli, C. et al. (2015). *Chemical Reviews* 115 (2): 1106–1129.

146 You, Y., Tomat, E., Hwang, K. et al. (2010). *Chemical Communications* 46 (23): 4139–4141.

147 Zhang, X., Jing, X., Liu, T. et al. (2012). *Inorganic Chemistry* 51 (4): 2325–2331.

148 Nakamura, T., Sugihara, F., Matsushita, H. et al. (2015). *Chemical Science* 6 (3): 1986–1990.

149 Steinberg, I., Huland, D.M., Vermesh, O. et al. (2019). *Photoacoustics* 14: 77–98.

150 Wang, L.V. and Hu, S. (2012). *Science* 335 (6075): 1458–1462.

151 Weber, J., Beard, P.C., and Bohndiek, S.E. (2016). *Nature Methods* 13 (8): 639–650.

152 Wu, D., Huang, L., Jiang, M.S., and Jiang, H. (2014). *International Journal of Molecular Sciences* 15 (12): 23616–23639.

153 Knox, H.J. and Chan, J. (2018). *Accounts of Chemical Research* 51 (11): 2897–2905.

154 Zhang, R., Wang, Z., Xu, L. et al. (2019). *Analytical Chemistry* 91 (19): 12476–12483.

155 Kim, C., Song, K.H., Gao, F., and Wang, L.V. (2010). *Radiology* 255 (2): 442–450.

156 Knox, H.J., Hedhli, J., Kim, T.W. et al. (2017). *Nature Communications* 8 (1): 1794.

157 Lozano, N., Al-Jamal, W.T., Taruttis, A. et al. (2012). *Journal of the American Chemical Society* 134 (32): 13256–13258.

158 Sulé-Suso, J., Skingsley, D., Sockalingum, G.D. et al. (2005). *Vibrational Spectroscopy* 38: 179–184.

159 Matthäus, C., Bird, B., Miljković, M. et al. (2008). Infrared and Raman Microscopy in cell biology. In: *Methods in Cell Biology* (ed. J.J. Correia and H.W. Detrich), 275–308. Academic Press.

160 Das, N.K., Dai, Y., Liu, P. et al. (2017). *Sensors* 17 (7): 1592.

161 Huang, Z., Lui, H., McLean, D.I. et al. (2005). *Photochemistry and Photobiology* 81 (5): 1219–1226.

162 Wang, H., Fu, Y., Zickmund, P. et al. (2005). *Biophysical Journal* 89 (1): 581–591.

163 Bocklitz, T.W., Salah, F.S., Vogler, N. et al. (2016). *BMC Cancer* 16: 534.

164 van Manen, H.J., Kraan, Y.M., Roos, D., and Otto, C. (2005). *Proceedings of the National Academy of Sciences of the United States of America* 102 (29): 10159–10164.

165 Navas-Moreno, M., Mehrpouyan, M., Chernenko, T. et al. (2017). *Scientific Reports* 7 (1): 4471.

166 Jeong, S., Kim, Y.-i., Kang, H. et al. (2015). *Scientific Reports* 5: 9455.

167 Zhao, Z., Shen, Y., Hu, F., and Min, W. (2017). *Analyst* 142 (21): 4018–4029.

168 Yamakoshi, H., Dodo, K., Okada, M. et al. (2011). *Journal of the American Chemical Society* 133 (16): 6102–6105.

169 Li, X., Jiang, M., Lam, J.W.Y. et al. (2017). *Journal of the American Chemical Society* 139 (47): 17022–17030.

170 Lin, J., Graziotto, M.E., Lay, P.A., and New, E.J. (2021). *Cells* 10 (7): 1699.

171 Miller, L.M., Dumas, P., Jamin, N. et al. (2002). *Review of Scientific Instruments* 73 (3): 1357–1360.

172 Hostachy, S., Policar, C., and Delsuc, N. (2017). *Coordination Chemistry Reviews* 351: 172–188.

173 Kopittke, P.M., Punshon, T., Paterson, D.J. et al. (2018). *Plant Physiology* 178 (2): 507–523.

174 Newville, M. (2014). *Reviews in Mineralogy and Geochemistry* 78 (1): 33–74.

175 Yano, J. and Yachandra, V.K. (2009). *Photosynthesis Research* 102 (2-3): 241–254.

176 Yang, L., McRae, R., Henary, M.M. et al. (2005). *Proceedings of the National Academy of Sciences of the United States of America* 102 (32): 11179–11184.

177 Wedding, J.L., Harris, H.H., Bader, C.A. et al. (2017). *Metallomics* 9 (4): 382–390.

178 Bryce, N.S., Zhang, J.Z., Whan, R.M. et al. (2009). *Chemical Communications* 19: 2673–2675.

179 Zhang, J.Z., Bryce, N.S., Lanzirotti, A. et al. (2012). *Metallomics* 4 (11): 1209–1217.

180 Buchberger, A.R., DeLaney, K., Johnson, J., and Li, L. (2018). *Analytical Chemistry* 90 (1): 240–265.

181 Glish, G.L. and Vachet, R.W. (2003). *Nature Reviews Drug Discovery* 2 (2): 140–150.

182 Benninghoven, A. and Sichtermann, W.K. (1978). *Analytical Chemistry* 50 (8): 1180–1184.

183 Caprioli, R.M., Farmer, T.B., and Gile, J. (1997). *Analytical Chemistry* 69 (23): 4751–4760.

184 Takáts, Z., Wiseman, J.M., Gologan, B., and Cooks, R.G. (2004). *Science* 306 (5695): 471–473.

185 Kang, S., Shim, H.S., Lee, J.S. et al. (2010). *Journal of Proteome Research* 9 (2): 1157–1164.

186 Chandra, S., Kabalka, G.W., Lorey, D.R. et al. (2002). *Clinical Cancer Research* 8 (8): 2675–2683.

187 Dowlatshahi Pour, M., Ren, L., Jennische, E. et al. (2019). *Journal of Analytical Atomic Spectrometry* 34 (8): 1581–1587.

188 Winey, M., Meehl, J.B., O'Toole, E.T., and Giddings, T.H. Jr. (2014). *Molecular Biology of the Cell* 25 (3): 319–323.

189 Golding, C.G., Lamboo, L.L., Beniac, D.R., and Booth, T.F. (2016). *Scientific Reports* 6 (1): 26516.

190 Weston, A.E., Armer, H.E.J., and Collinson, L.M. (2009). *Journal of Chemical Biology* 3 (3): 101–112.

191 Stirling, J.W. (1990). *The Journal of Histochemistry and Cytochemistry* 38 (2): 145–157.

192 de Boer, P., Hoogenboom, J.P., and Giepmans, B.N.G. (2015). *Nature Methods* 12 (6): 503–513.

193 Liv, N., Zonnevylle, A.C., Narvaez, A.C. et al. (2013). *PLoS One* 8 (2): e55707.

194 KL, M.D. (2009). *Journal of Microscopy* 235 (3): 273–281.

195 Robinson, J.M., Takizawa, T., and Vandré, D.D. (2000). *The Journal of Histochemistry and Cytochemistry* 48 (4): 487–492.

196 Alivisatos, A.P., Gu, W., and Larabell, C. (2005). *Annual Review of Biomedical Engineering* 7 (1): 55–76.

197 Ellisman, M.H., Deerinck, T.J., Shu, X., and Sosinsky, G.E. (2012). *Methods in Cell Biology* 111: 139–155.

198 Deerinck, T.J., Martone, M.E., Lev-Ram, V. et al. (1994). *Journal of Cell Biology* 126 (4): 901–910.

199 Perkovic, M., Kunz, M., Endesfelder, U. et al. (2014). *Journal of Structural Biology* 186 (2): 205–213.

200 Subbiah, R., Veerapandian, M., and Yun, K.S. (2010). *Current Medicinal Chemistry* 17 (36): 4559–4577.

201 Liu, Y., Yang, Y., Sun, M. et al. (2017). *Chemical Science* 8 (4): 2710–2716.

202 Liu, L., Yuan, Y., Yang, Y. et al. (2019). *Chemical Communications* 55 (42): 5851–5854.

203 Qi, J., Li, J., Liu, R. et al. (2019). *Chem* 5 (10): 2657–2677.

204 Wang, G., Kalra, M., Murugan, V. et al. (2015). *Medical Physics* 42 (10): 5879–5889.

205 Martí-Bonmatí, L., Sopena, R., Bartumeus, P., and Sopena, P. (2010). *Contrast Media & Molecular Imaging* 5 (4): 180–189.

206 Yang, G., Gong, H., Liu, T. et al. (2015). *Biomaterials* 60: 62–71.

207 Yang, C.-T., Ghosh, K.K., Padmanabhan, P. et al. (2018). *Theranostics* 8 (22): 6210–6232.

208 Xing, H., Bu, W., Zhang, S. et al. (2012). *Biomaterials* 33 (4): 1079–1089.

209 van Schooneveld, M.M., Cormode, D.P., Koole, R. et al. (2010). *Contrast Media & Molecular Imaging* 5 (4): 231–236.

Index